CHEMISTRY OF THE ENVIRONMENT

CHEMISTRY OF THE ENVIRONMENT

Editor

Raphaela Schnur

scitus
academics

Chemistry of the Environment
Edited by **Raphaela Schnur**

ISBN: 9781681178905
Library of Congress Control Number: 2017948918

SCITUS Academics LLC,
616, Corporate Way, Suite 2, 4766,
Valley Cottage, NY 10989

www.scitusacademics.com

Printed in United States of America.

Preface

The field of environmental chemistry is both very broad and highly interdisciplinary Environmental issues such as climate change, water pollution and renewable energy make the news headlines and has become increasingly important in everyday life. Environmental chemistry is the study of chemical processes occurring in the environment which are impacted by humankind's activities. These impacts may be felt on a local scale, through the presence of urban air pollutants or toxic substances arising from a chemical waste site, or on a global scale, through depletion of stratospheric ozone or global warming. Many people perceive chemistry and the chemical industry as harmful to the environment. However, many new advances and scientific researches in the field of chemistry are helping us to develop more environment friendly materials and applications, while preserving the quality and the lifestyle we expect.

Over the years, the industry and wider public have become aware of the damaging effects of some past practices and the need to protect the environment. In the past, few were aware of the potentially negative effects our modern lifestyle might have on the environment, and rather saw only the positive potential for creating new, useful materials and products. These impacts may be felt on a local scale, through the presence of urban air pollutants or toxic substances arising from a chemical waste site, or on a global scale, through depletion of stratospheric ozone or global warming.

'Chemistry of the Environment' covers current research and advances in the field of environmental chemistry, including atmospheric chemistry, the chemistry of water pollution, and green chemistry. It presents the coverage of the atmosphere as a distinct sphere of the environment and the practice of industrial ecology as it applies to chemical science. It includes widespread coverage of both natural environmental sources and anthropogenic sources, their impacts on health, and their role in energy production in a balanced and scientifically rational way. It focuses on developing a fundamental understanding of the nature of these chemical processes, so that humankind's activities can be accurately evaluated. This book is intended to upper undergraduate and graduate level students studying environmental chemistry, chemical engineering, industrial ecology, and general biology. This book is of great benefit to environmental chemists and researchers, biochemists, and elementary organic chemists.

Table of Contents

CHAPTER 1

Monitoring of environmental exposure to polycyclic aromatic hydrocarbons: A review

K. Srogi

Institute for Chemical Processing of Coal, Zamkowa 1, 41-803 Zabrze, Poland

ABSTRACT

Polycyclic aromatic hydrocarbons (PAHs) are a large group of organic compounds with two or more fused aromatic rings. They have a relatively low solubility in water, but are highly lipophilic. Most of the PAHs with low vapour pressure in the air are adsorbed on particles. When dissolved in water or adsorbed on particulate matter, PAHs can undergo photodecomposition when exposed to ultraviolet light from solar radiation. In the atmosphere, PAHs can react with pollutants such as ozone, nitrogen oxides and sulfur dioxide, yielding diones, nitro- and dinitro-PAHs, and sulfonic acids, respectively. PAHs may also be degraded by some microorganisms in the soil. PAHs are widespread environmental contaminants resulting from incomplete combustion of organic materials. The occurrence is largely a result of anthropogenic emissions such as fossil fuel-burning, motor vehicle, waste incinerator, oil refining, coke and asphalt production, and aluminum production, etc. PAHs have received increased attention in recent years in air pollution studies because some of these compounds are highly carcinogenic or mutagenic. Eight PAHs (Car-PAHs) typically considered as possible carcinogens are: benzo(a)anthracene, chrysene, benzo(b)fluoranthene, benzo(k)fluoranthene, benzo(a)pyrene

(B(a)P), dibenzo(a,h)anthracene, indeno(1,2,3-cd)pyrene and benzo(g,h,i) perylene. In particular, benzo(a)pyrene has been identified as being highly carcinogenic. The US Environmental Protection Agency (EPA) has promulgated 16 unsubstituted PAHs (EPA-PAH) as priority pollutants. Thus, exposure assessments of PAHs in the developing world are important. The scope of this review will be to give an overview of PAH concentrations in various environmental samples and to discuss the advantages and limitations of applying these parameters in the assessment of environmental risks in ecosystems and human health. As it well known, there is an increasing trend to use the behavior of pollutants (i.e. bioaccumulation) as well as pollution-induced biological and biochemical effects on human organisms to evaluate or predict the impact of chemicals on ecosystems. Emphasis in this review will, therefore, be placed on the use of bioaccumulation and biomarker responses in air, soil, water and food, as monitoring tools for the assessment of the risks and hazards of PAH concentrations for the ecosystem, as well as on its limitations.

Keywords: Polycyclic, aromatic, hydrocarbons Plant, Soil, Sediment, Water, Food, Air, PAH biomonitoring, Seasonal trend, PM2.5, PM2.5–10, Pollution control

1. INTRODUCTION

Polycyclic aromatic hydrocarbons (PAHs) from incomplete combustion or pyrolysis organic material sources are ubiquitous in the global environment and are typically more concentrated near urban centers (Hyötyläinen and Oikari 2004). Therefore vehicular emissions may not be the only source of PAHs in these samples. Other possible sources are tire wear debris, asphalt particles (Binet et al. 2002) and stationary combustion sources or crematoria (Santasiero et al. 2005). Sources of PAHs in urban atmosphere include automobiles, re-suspended soils, refineries and power plants (Omar et al. 2002; Yang et al. 2002; Dyke et al. 2003). PAHs are also occurring in large amounts in sedimentary rocks and petroleum (Lichtfouse et al. 1997, 1999; Henner et al. 1999). Additional contributions to ambient air levels arise from tobacco smoking (Sakai et al. 2002), while the use of heating sources can increase PAH concentrations in indoor air (WHO 1987). Natural sources such as forest fires and volcanic eruptions are less important. Their fate is determined by their physico-chemical properties, especially nonpolarity and hydrophobicity responsible for their persistence in the environment. PAHs from fuel oils tend to be quickly adsorbed onto particles and to accumulate in sediments (Tolosa et al. 2004). PAH refers to a group of a few hundreds of compounds with two or more fused benzene rings.

PAHs have received increased attention in recent years in air pollution studies because some of these compounds are highly carcinogenic or mutagenic (IARC 1983). Eight PAHs (Car-PAHs) typically considered as possible carcinogens are: benzo(a)anthracene, chrysene, benzo(b) fluoranthene, benzo(k)fluoranthene, benzo(a)pyrene (B(a)P), dibenzo(a,h)anthracene, indeno(1,2,3-cd)pyrene and benzo(g,h,i)perylene (Menzie et al. 1992). In particular, benzo(a)pyrene has been identified as being highly carcinogenic (Kuo et al. 1998; Wang et al. 2002). The US Environmental Protection Agency (EPA) has promulgated 16 unsubstituted PAHs (EPA-PAH) as priority pollutants.

To understand the extent of human exposure to B(a)P and other PAHs, reliable sampling and analytical methods are necessary (US EPA 1999). These indicate that particle-bound PAHs are considered to be a significant hazardous substance to human health through breathing. In view of this health concern, monitoring the level of particle-bound PAHs in urban areas has become more important (Chetwittayachan et al. 2002). B(a)P is often used as a marker for total PAHs exposure in industry and in the environment (Lin et al. 2002). Because of its lowest boiling point and highest volatility, naphthalene occurs almost completely in the vapor-phase. Therefore, its surface deposition will be relatively small in comparison with other PAHs. The observed high levels of naphthalene in domestic environments in Taipei might be related to the popular use of camphor balls in the wardrobe (Li and Ro 2000).

The highest concentrations of atmospheric PAHs can be found in the urban environment, due to the increasing vehicular traffic and the small dispersion of the atmospheric pollutants. The risk associated with human exposure to atmospheric PAHs is highest in the cities, considering the population density (Caricchia et al. 1999). Despite the drastic reduction of urban particulate pollution in cities resulting from the improvement of coal usage and the shift toward other fossil fuels (oil or natural gas) for domestic heating, the densification of the urban net combined with population growth and the increasing importance of traffic have contributed to reinforce urban particulate pollution. Furthermore, particles produced by cars are much smaller than coal particles and found in the breathable size fraction (Manoli et al. 2002). There is much information on the multi-ringed heavier PAHs but have left the lighter vapor-phase PAH components rather neglected. Although these lighter compounds have weaker carcinogenic/mutagenic properties, they are the most abundant in the urban atmosphere and react with other pollutants to form more toxic derivatives (Park et al. 2002). Thus, the implication of human exposure to mixtures of PAHs, rather than to individual substances, is important. The levels of individual PAHs vary over several orders of magnitude and are generally in the range between <0.1 and 100 ng/m^3. PAHs are mainly adsorbed to airborne particulate matter (WHO 1998).

There are various industrial workplaces for which a significant increase of certain cancer diseases has been found that may be attributed to an unusually high exposure to PAH. For instance, PAH exposure is high in coke plants, coal tar and pitch producing and manufacturing industries, aluminium plants, iron and steel foundries, creosote-, rubber-, mineral oil-, soot- and carbon black-producing or manufacturing companies. As highly exposed occupational groups, chimney sweeps, roadmen (pavement-tarring) and roofers (roof-tarring) are also under increased risk (Jacob and Seidel 2002). Automotive byproducts are a common denominator for many observations for both PAHs and metals. The motor vehicle is increasingly recognized as a contributor to air pollution. Motorized vehicle emissions are strongly associated with health problems as shown by research on how proximity to roads affects mortality (Hoek et al. 2002). Brunekreef and Holgate (2002) reviewed the health consequences of air pollution and demonstrate a strong association between air pollution, asthma and chronic obstructive pulmonary disease. The PAHs exhausted as fine particles by motor vehicles are deposited in and accumulate in soil. Comparing busy streets with residential streets in the inner-city and in the suburbs provides evidence about the importance of motor traffic as a source of soil PAHs.

Generally, environmental exposure of organisms is assessed by monitoring their environment (sediment/soil, water, and air). Biomonitoring, however, can provide an assessment of the integrated uptake through all exposure routes. This is important since only the bioavailable chemical is likely to be assimilated/concentrated in tissues, body fluids, and excreta. Owing to the complexity of biological samples, however, intricate analytical protocols are traditionally used for analyses. Conventional biomonitoring methods are often time-consuming, labor intensive, and expensive. Currently, urinary 1-OHP is tested as biomarker for the assessment of low level PAHs exposure of general population. Reports by Vyskocil et al. (1997), Zhao et al. (1992) Kanoh et al. (1993) and Jongeneelen (1994) suggested that levels of 1-OHP reflect the levels of environmental PAH exposure from different sources such as air, food and the environment of the home life. However, in study (Vyskocil et al. 1997), it was shown that other factors than air pollution contribute more substantially to overall exposure to PAH and it was speculated that diet could be one of the most important factors. For example, the highest peak of 1-OH pyrene and benzo(a)pyrene-type metabolite concentrations, were found in significantly elevated levels in the bile of cat-shark collected in Lenga in San Vicente Bay. These probably reflect boat traffic and combustion-based (steel, petrochemical) industries present in these areas. However, the naphthalene-type metabolites were found in all sampled cat-shark bile in the assessed areas, reflecting that petrogenic PAH are ubiquitous in the whole embayment system of Chile (Fuentes-Rios et al. 2005). Cormier et al. (2000) reported a similar situation in white sucker from Ohio state (USA), where naphthalene-type metabolites were detected in all sampled fish bile including fish from the reference areas.

The scope of this review will be to give an overview of PAH concentrations in various environmental samples and to discuss the advantages and limitations of applying these parameters in the assessment of environmental risks in ecosystems and human health. As it well known, there is an increasing trend to use the behavior of pollutants (i.e. bioaccumulation) as well as pollution-induced biological and biochemical effects on human organisms to evaluate or predict the impact of chemicals on ecosystems. Emphasis in this review will, therefore, be placed on the use of bioaccumulation and biomarker responses in air, soil, water and food, as monitoring tools for the assessment of the risks and hazards of PAH concentrations for the ecosystem, as well as on its limitations.

2. EXPOSURE TO PAH

The composition of PAHs in association with airborne particles changes significantly according to their emission sources, e.g. traffic or other processes of combustion. PAH concentration ratios can be used to identify possible emission sources (Li and Kamens 1993; Venkataraman et al. 1994). Exposure to indoor particulate matter (PM) has been recognized as a significant health problem since most people spend approximately 90% of their time indoors. Particles of concern to human health are those known as inhalable particles ($PM_{2.5}$, PM with aerodynamic diameter less than 2.5 μm). Most particulate-phase PAH are adsorbed onto fine particles (<2.5 μm), which are deposited slowly and, depending on atmospheric conditions and chemical reactivity, may be transported over long distances and pollute even remote areas. PAHs on particles less than 1 μm were result from combustion and other high temperature sources (Sheu et al. 1997; Kiss et al. 1998).

As it well-know low molecular weight PAHs (two- and three-rings) occur in the atmosphere in the vapour phase whereas multi-ringed PAHs (five-rings) are bound to particles. Intermediate molecular weight PAHs (four-rings) are partitioned between the vapour and particulate phases, depending on atmospheric temperature (Howsam et al. 2000, 2001).

In the following sections, main sources of PAHs in ecosystem include plants, air, water, food and soils will be described.

3. PLANTS

Many plants have a relatively large surface area covered with waxes that facilitates the accumulation of hydrophobic chemicals. The use of plants as passive samplers of organic compounds in the atmospheric has been suggested by many authors (Niu et al. 2003; De Nicola et al. 2005; Lodovici et al. 1998; Müller et al. 2001). Indeed, PAHs are characterised by a low to moderate volatility according to their molecular weight (Sverdrup et al. 2003). Moreover, Schreiber and Schönherr (1992) reported that the main factor affecting the absorption of volatilized PAHs through the cuticle is the extent of leaf area exposed to the atmosphere.

PAHs from a polluted atmosphere are generally transferred to plants by particle-phase deposition on the waxy leaf cuticle or by uptake in the gas phase through stomata (Kipopoulou et al. 1999; Lehndorff and Schwark 2004). Leaf features (surface, cuticular waxes, hairs, number of stomata) play an important role in PAH uptake and accumulation (Jouraeva et al. 2002). Howsam et al. (2000) found higher PAH concentrations in hairy leaves compared to hairless leaves collected in the same area. Hairs increase the leaf surface that is able to capture particulate from the air; in addition hairs favour a boundary layer of stagnant air on the leaf surface, improving particle retention (Rauret et al. 1994). Moreover, some studies highlight the role of other leaf components, such as lipids and epicuticular waxes, in leaf accumulation of PAHs that are lipophilic compounds (Howsam et al. 2000). Besides emission sources, PAH air concentrations depend also on atmospheric conditions (Menichini 1992; Caricchia et al. 1999). Indeed, the highest PAH air concentrations are measured in winter due to the larger number of emission sources (De Nicola et al. 2005), the stability of atmospheric conditions and the presence of fewer degradation phenomena (Menichini 1992; Park et al. 2002). For example, total PAH concentrations in *Quercus ilex* leaves collected in Naples showed a clear temporal trend, with the highest values in winter, comparable to the trend found in PAH airborne particulate. At the control site leaf PAH concentrations [from 300 (September 2001 and May 2002) to 500 ng/g dw—May 2001 and January 2002] showed lower values and smaller temporal variations than the urban sites (the concentrations of leaf total PAHs ranged from 1,100 to 2,076 ng/g dw in May 2001, from 1,349 to 1,930 ng/g dw in September 2001 and from 1,038 to 1,962 ng/g dw in May 2002). This suggests the ability of *Quercus ilex* leaves to act as monitors of PAH gradients in time despite the variation in leaf morphological features due to plant phenology. The high leaf concentrations of total PAHs in January are mainly attributable to the contribution of medium molecular weight PAHs to the total (De Nicola et al. 2005).

The temporal trend agreed with the seasonal variations observed in PAH leaf concentrations by other authors (Nakajima et al. 1995; Lodovici et al. 1998). Seasonal variations in concentrations of PAHs with higher values in winter than in summer are found in Azalea leaves (Nakajima et al. 1995) and in leaves of *Laurus nobilis* collected in different sites of Tuscany (Lodovici et al. 1998). PAH leaf concentrations were on average twofold higher in leaves collected in the winter samplings. Although it was not possible to establish a linear quantitative relationships between leaf and air PAH concentrations, the higher values detected in winter might be attributable to a greater deposition of these contaminants on leaf surface, due to a large number of emission sources (Menichini 1992), increased atmospheric stability (Caricchia et al. 1999), greater condensation (Simonich and Hites 1995) and lesser photolysis processes (Papageorgopoulou et al. 1999). Niu et al. (2003) highlighted for spruce needles, that show leaf surface rich in wax components, an enhanced photolysis of PAHs under sunlight irradiation.

PAHs can also enter plant tissues by partitioning from contaminated soil to the roots and translocation into the shoot. However, the extent of PAH uptake by roots is still debated in the literature: because of their high lipophility and low solubility, PAHs would be adsorbed through the epidermis of roots in contact with soil particles, but not drawn along the inner root (Kipopoulou et al. 1999). By contrast, Fismes et al. (2002) reported a direct relationship between PAH concentrations in soil and plants or tree *Laurus nobolis* (Lodovici et al. 1998) and suggested a soil-to-root transfer predominant upon the atmosphere-to-plant pathway. Heavy pollutions of halophytic plants, especially *Salicornia* species, have been reported further to the significant impacts of sediments by fuel oil (Meudec et al. 2006). The edible *Salicornia fragilis* being common along the Atlantic coast of Europe is potentially exposed to accidental marine pollution because of its low localisation on intertidal area.

Meudec et al., (2006) reported the bioaccumulation of PAHs from oil-polluted sediments in the shoots of *Salicornia fragilis*. As high concentrations of PAH were detected in plants, and a soil–root–shoot transfer was suggested. Total PAHs concentrations in *Salicornia fragilis* tissues were from 3.5 to 40.9 mg/kg dw. As expected, the highest concentration was recorded in plants grown on the most polluted sediment. Total PAHs in roots of plants exposed to 0.2% were found to be 33.5 mg/kg dw, about ten times the level in shoots (Meudec et al. 2006). PAH compounds at concentrations of up to 43 μg/g (naphthalene) were detected in roots and up to 0.2 μg/g (naphthalene) in leaves of plants (*Spartina alterniflora*) grown in pots of contaminated soil. The concentrations in the roots are one to two orders of magnitude lower than the concentrations in the soil and are linearly related to the soil concentrations. The concentrations in the leaves are very low-three to four orders of magnitude lower than the soil, and do not have a direct linear relationship to the soil concentrations No difference was found in leaf PAH concentrations between plants grown in contaminated water and control plants (Watts et al. 2006).

According to Kipopoulou et al. (1999), the strong adsorption of high-weight PAHs upon the root epidermis would prevent them from being drawn into the inner root conversely to low- and medium-weight PAHs. This theory is contradicted by the study of Fismes et al. (2002) who found high abundance in foliar extracts of high-weight benzo(a)pyrene or benzo(ghi)perylene, after translocation from roots to aerial parts (Meudec et al. 2006).

Henner et al. (1999) identified plants growing on soils from three ancient gasworks, from an ancient coking factory and from a prepared decontamination bed filled with gas-work soil. All plants were typical pioneer weeds that were commonly found in heterogeneous, low fertility soils. They were well adapted to drought and all showed a dense, deep root network, even through "aged" tars. Observation of tar shows of weathered soils in industrial fields revealed that the location of the contamination can be very heterogeneous, some areas being highly polluted while others much less. Nonetheless, all locations showed a well-flourishing vegetation, some what enhanced by fertilisers at the decontamination bed. Noteworthy, at some locations, plants were even able to grow roots through "aged", non-smelling solid tar layers located at about 3 ± 10 cm depth. There were only two exceptions showing the absence of plants: first, a small soil area, e.g. 20 m^2, located at an ancient gaswork field, where high amounts of foul smelling liquid tar occur between 0 and 20 cm depth, and second, a hillock of several tons of "benzene smelling" gas-work soils which had been excavated 6 months ago. They concluded that the inhibition of plant growth is due to volatile, water-soluble compounds.

Grasses and annual herbs have been the primary focus of experiments evaluating the potential of plants to remediate PAH contaminated soils (Davis et al. 2002). Trees have received very little attention with regard to PAHs, although their perennial life-history and extensive root systems suggest they may be desirable for use in phytoremediation. It has also been speculated that trees may have greater rates of rhizodeposition (Grayston et al. 1996) including fine root turnover (Gill and Jackson 2000) and exudation. For example, the total PAH concentrations in *Melaleuca* leaves from the urban background and in grass were lower than PAH levels determined in sugar maple leaves and white pine needles collected from suburban and urban areas of the USA (Simonich and Hites 1995), while the leaves in the *Melaleuca* leaves collected close to the major intersection were higher (Müller et al. 2001).

According to Domeňo et al. (2006) lichens could be used as good bioindicators of air pollution. Twelve out of the 16 PAHs studied were found in lichens *Xanthoria parietina* samples. All of them in a concentration range from 25 to 40 ng/g. The highest concentrations in lichens *Xanthoria parietina* were found for dibenzo(a,h)anthracene and benzo(k)fluoranthene, followed by benzo(a)anthracene, chrysene and fluorene. The reason of non-detection in lichens of other PAHs (five or more rings in their structure) present in the atmosphere in high concentrations may be of being almost exclusively adsorbed on suspended particulate matter. Concerning the origin of the PAHs found in the lichen, benzo(a)pyrene is usually emitted from catalyst and noncatalyst automobiles. Benzo(a)anthracene and chrysene are often resulted from the combustion of both diesel and natural gas. In both cases the origin suggests the traffic road as a major source of these compounds, which fits to other studies in which benzo(a)pyrene and dibenzo(a,h)anthracene indicate traffic emission and identify traffic as the main source of urban PAH emission. Also, Capuano et al. (2005) the highest PAH value (101.3 ng/g) in the pine needles was observed, coming from a pine tree situated close to a heavily congested road. The concentration of PAHs was highest in the needles collected during the soot episodes when 140 ng/g dw was found, which was significantly higher than the 32 ng/g dw measured during the same period in 1992. In addition, a greater proportion of non-volatile PAHs, which in the atmosphere are mainly associated on particles, were found in the needles collected in connection with the soot episodes. Consequently, PAHs in the particle phase appeared to be an important parameter governing the uptake of PAH in the needles.

The concentrations of PAHs found in the different compartments of environmnet (Brorström-Lundén and Löfgren 1998) are shown in Table 1. The authors were found that the content of PAHs in the spruce needles was highest in the samples collected in January–February, when compared with the needles collected in April–May.

The total concentrations in plant samples from other urban and industrial areas are shown in Table 2 (Bakker et al. 2000). The variation in plant concentrations is large, arising from differences between species, differences between sampled plant organs (needles, leaves and fruit) and differences in sample treatment. For example, washing of the plant surfaces has a large influence on concentrations of particle-bound PAHs.

Table 1. The concentrations of PAHs found in the different compartments of environment (Brorström-Lundén and Löfgren 1998)

Period (1995) Ambient (C) Precipitation (mm)		Site 1	Site 2
Air	ng m^{-3}	4.7	2.2
Spruce nedles			
Forest edge	ng g^{-1} dw^{-1}	41	28
Inside the forest	ng g^{-1} dw^{-1}	38	24
Deposition			
Open fidel	ng m^{-2} day^{-1}	370	360
Thoughfall	ng m^{-2} day^{-1}	500	700
Litterfall	ng g^{-1} dw^{-1}	200	200
Run-off	ng L^{-1}	4.3	3.9
Sil (humus)	ng g^{-1} dw^{-1}	ND	2,500

4. FOOD

Food appears to be the main source of PAH intake for humans not occupationally exposed to PAH. Consumers might be exposed to PAHs by eating grilled or charred meats, contaminated cereals, flour, bread and vegetables (Grova et al. 2006). In vivo studies suggest a transfer in intestinal epithelium by diffusion, which appears extensively governed by the physicochemical properties of PAHs, particularly lipophilicity. However, other mechanisms, such as metabolism, are considered to intervene (Cavret and Feidt 2005; Phillips 1999).

It should be noted that, food–animal transfer pathways of PAHs are so far poorly known due to the absence, of investigations involving tracers (Laurent et al. 2001). For example, Laurent et al. (2002) reported a study of portal absorption of PAHs using two ^{14}C-tagged compounds: ^{14}C-phenanthrene and ^{14}C-benzo(a)pyrene in the growing pig. These two compounds differ either by the lipophilicity, either by the water solubility, either by fused benzene rings number. The analysis of portal and arterial blood radioactivity showed that both PAHs were absorbed with a maximum concentration at 4–6 h, or 5–6 h after milk ingestion. Then, the blood radioactivity decreased to reach background levels 24 h after milk ingestion. Furthermore, the portal and arterial blood radioactivities were higher for phenanthrene (even if the injected load was the lowest) than these of benzo(a)pyrene, in agreement with their lipophilicity and water solubility difference (Laurent et al. 2001). Main ^{14}C absorption occurred during the 1–3 h time period after ingestion for ^{14}C-phenanthrene and during the 3–6 h time period for ^{14}C-benzo(a)pyrene. ^{14}C portal absorption rate was high for ^{14}C-phenanthrene (95%), it was close to 33% for ^{14}C-benzo(a)pyrene (9%). These results indicate that the two studied molecules have a quite different behaviour during digestion and absorption. Phenanthrene is greatly absorbed and its absorption occurs via the blood system, whereas benzo(a)pyrene was partly and weakly absorbed respectively. However, these two molecules are mainly absorbed via the portal vein.

Concentrations of benzo(a)pyrene in meat and meat products have been restricted to 1 μg/kg in Germany and for food and beverages to 30 ng/kg in Italy. Most of the foodstuffs are less contaminated so that 1 μg/kg rather appears to be a pessimistic value, although occasionally far higher concentrations have been found, especially in marine food (oysters, mussels, smoked fish) and vegetables (green kale). The maximum daily benzo(a)pyrene intake from food can be extrapolated to be 1 μg (Jacob and Seidel 2002).

Kishikawa et al. (2003) reported that the average concentrations of total PAHs in commercial milk, infant formula and human milk were 0.99, 2.01 and 0.75 μg/kg, respectively. Several studies achieved in environmental conditions show that PAHs can be excreted in the milk of ruminants (Grova et al. 2000, 2006). Grova et al. (2006) reported the impact of chronic exposure to PAHs on milk contamination was evaluated by oral administration of a mixture of fluorene, phenanthrene, anthracene, fluoranthene, pyrene, chrysene, benzo(k)fluorene, benzo(a)pyrene and benzo(g,h,i)perylene at 0.02 mg/kg to lactating goats for 28 days. The results evidence several major points: (1) benzo(k)fluorene, benzo(a)pyrene and benzo(g,h,i)perylene were not detected in the milk; (2) unexpectedly, the concentration of fluorene, phenanthrene, anthracene, fluoranthene, pyrene and chrysene did not change with time; (3) monohydroxylated PAH metabolites (-OH), namely 2-OH-fluorene, 3-OH-phenanthrene and 1-OH-pyrene were detected shortly after administration. The concentrations of 2-OH-fluorene and 3-OH-phenanthrene reached, respectively, maxima of 0.41 and 0.22 ng/mL during the first exposure week, whereas the concentration of 1-OH pyrene increased to reach a maximum of 0.97 ng/mL on day 14, then slightly decreased during the last two exposure weeks. Those findings suggest a lack of activation of a metabolism that could lead to an excretion of PAHs into milk under native forms. However, a slight increase in concentration could induce the metabolism, which should lead to an increase in the excretion of metabolites into the milk. In spite of the absence of a significant transfer of parent PAHs to milk, the appearance of metabolites in milk raises questions of their impact on human health.

Table 2. Total concentrations of PAHs plants (ng/g dry wt.) in (sub)urban areas and in the vicinity of human (industrial) activities (after Bakker et al. 2000)

Location	Number of PAH's	Plant	∑PAH concentration in plant (ng g dry wt.)	Reference
Highway (S)	16	Lettuce	17–90	Larssen and Sahlberg (1982)
Highway (CAN)	17	Onions[a], beet[a], tomatoes[a]	10–1,900[b]	Wang and Meresz (1982)
Al smelter (S)	16	Lettuce	320–920	Larsen and Sahlberg (1982)
Highway (S)	16	Kale	500 (at 50 m)[b] 5,000(at 10 m)[b]	Brorström-Lunden and Skärby (1984)
Busy city street (S)	16	Kale	5,000(at 50 m)[b] 14,000(at 10 m)[b]	Brorström-Lunden and Skärby (1984)
Urban (USA)	10	Pine	800–1,600	Simonich and Hites (1994)
Urban (USA)	10	sugar maple	500–1,100	Simonich and Hites (1994)
Urban (I)	10	Bay tree[a]	73–880	Lodovici et al. (1994)
Suburban (UK)	16	Pine	20–3,100[c]	Tremolada et al. (1996)
Suburban (USA)	18	Maple	510 ± 100	Wagrowski and Hites (1997)
Urban (USA)	18	Maple	1600 ± 210	Wagrowski and Hites (1997)
Urban (RBD)	13	Kale	1,000–5,000	Franzaring (1997)
Urban (UK)	16	Grass	153 ± 8	Meharg et a.l (1998)
Polyprop.fire (UK)	16	Grass	2,400[d]	Meharg et al. (1998)
Industrial (GR)	16	Various vegetables[a]	25–239	Kipopoulou et al. (1999)

[a] Washed leaves

[b] Original concentrations expressed in μg g fresh wt. converted to μg/g dry wt. assuming a fresh wt. dry wt. of 10

[c] Concentration correlated to number of inhabitants

[d] Highest concentration

Also, cereal products may contain high levels of PAH because of drying methods. Cereals and vegetables are the major dietary sources of PAHs—except for high consumption of meat cooked over an open flame (Phillips 1999). Fiedler et al. (2002) examined green tea leaves, and they found that the total PAH concentrations of samples ranged from 497 to 517 μg/kg, compared with brick tea leaves ranged from 1,048 to 1,162 μg/kg. These concentrations are comparable to those in vegetation grown along the motorway (772 μg/kg) and an urban area (1,901 μg/kg) at North West Italy (Ardito et al. 1999).

The bioavailability of organic trace pollutants adsorbed to sediments is rather low, polluted sediments still represent an important source of contamination for freshwater animals. Bioaccumulation of compounds in fish (Reynolds et al. 2003), therefore, is a result of the different uptake and elimination processes of the compounds combined with metabolic clearance (biotransformation) (Verweij et al. 2004). Recently, Kong et al. (2005) examined tilapia (*Oreochromis mossambicus*), a freshwater fish purchased from the markets in Hong Kong (with fish supplied from the mainland) showed higher concentrations of PAHs (76.5 ng/g wet wt) than those collected from Hong Kong fish ponds (60.1 ng/g wet wt). This suggested that the fishponds in the mainland may be more polluted due to the rapid socio-economic development in the region during the past 20 years. In another works catfish (*Clarias fuscus*) exhibited the highest concentrations of total PAHs in ventral muscle among ten fish species. Grass carp, mud crap and bighead carp had low concentrations of PAHs. Among marine fish species, no significant difference was observed among different species and between ventral and axial muscles in tissue concentrations of total PAHs. The concentrations of total PAHs in the muscles ranged from 15.5 to 118 ng/g wet wt. It also shows that the percentage of each PAH congener in different species of freshwater and marine fish. Nearly all the 16 PAHs except benzo(a)pyrene, indeno(1,2,3-cd)pyrene, dibenzo(a,h)anthracene and benzo(g,h,i)perylene were detected in the fish samples. Higher percentages of two- and three-rings PAHs were found in all species. For example, naphthalene had the highest proportion among all compounds in both ventral and axial muscles of freshwater and marine fish, ranging from 14 to 72% of total PAHs. However, no significant difference in the distribution of PAHs between ventral and axial muscles was observed in marine fish. Among different congeners of PAHs, two-ring (naphtalene) and three-ring PAH (phenanthrene and fluorene) dominated the distribution in fish muscles (Cheung et al. 2007). Naphthalene was the most prevalent parent compound (22–72% in marine fish and 14–52% in freshwater fish) because it has a higher solubility than other congeners with a larger molecular weight of PAHs.

Table 3. Biota-sediment accumulation factors (BSAFs) of polycyclic aromatic hydrocarbons (PAHs) in fish (after Van der Oost et al. 2003)

Species	Mean fish–sediment concentrations ratio (range) of compounds										Dimensions	Reference
	Fluorene	Phenanthrene	Fluoranthene	Chrysene	Sum PAH 2[a]	Sum PAH3[a]	Sum PAH4[a]	Sum PAH5[a]	Sum PAH6[a]	Sum PAH		
Antarctic fish *Notothenia gibberifrons*										0.24–1.25	DW:DW	McDonald et al. (1995)
Brown bullhead *Ictalurus nebulosus*	2–10	2–6	2–5	1–2						1–6	FW:DW	Baumann and Harshbarger (1995)
										0.01–0.1	LW:OM	Van der Oost et al. (1991)
Eel *Anguilla anguilla*					0.24–2.9	0.09–1.6	0.01–0.4	0.003–0.06	0.02–0.20	0.04–0.56	LW:OM	Van der Oost et al. (1994)
Killifish *Fundulus heteroclitus*	0.05–0.36	0.004–0.05	0.0005–0.001	0.0001–0.001						0.001–0.012	DW:DW	Elskus and Stegeman (1989)
Lake trout *Salvelinus namaycush*	0.00011	0.00016	0.00710	0.00033						–	LW:OM	Burkhard and Lukasewycz (2000)
Pike *Esox lucius*										0.02–0.09	LW:OM	Van der Oost et al. (1991)
Roach *Rutilus rutilus*										0.02–0.13	LW:OM	Van der Oost et al. (1991)
					0.5–2.3	0.1–0.6	0.02–0.14	0.01–0.06	0.01–0.13	0.02–0.13	LW:OM	Van der Oost et al. (1994)
Sunfish *Lepomis macrochirus*										0.00001–0.8	LW:OM	Thomann and Komlos (1999)

Symbols and abbreviations: *DW* dry weight, *FW* fresh weight, *LW* lipid weight, *OM* organic matter or organic carbon

[a] number of aromatic rings

[a]number of aromatic rings

Polycyclic aromatic hydrocarbons were also analyzed in fish: *Mullus barbatus, Scomber scombrus, Micromesistius poutassou* and *Merluccius merluccius*) in several pools coming from the Central Adriatic Sea. Atlantic mackerel, European hake and blue whiting showed the highest PAH concentrations, ranging from 44.1 to 63.3 ng/g wt, the group of invertebrate organisms showed a level of contamination about three times lower than those of the vertebrate groups (Perugini et al. 2007).

It should be noted that the bioaccumulation of PAHs by various marine organisms has been extensively reviewed by Meador et al. (1995). BSAFs of different PAH compounds in fish are listed in Table 3 (Van der Oost et al. 2003). BSAFs for the PAHs were calculated as the lipid-normalised PAH concentration in the bivalves divided by the TOC-normalised PAH concentration in the sediments BSAFkg OC/kg lipid = $(C_o/f_1)/(C_s/f_{OC})$—where C_o is the concentration of contaminant in the organism (both wet and dry weight commonly used), C_s is the contaminant concentration in the sediment (generally dry weight), f_l is the lipid fraction of the tissue, and f_{OC} is the organic carbon fraction of the sediment (Simpson et al. 2006).

Vyskocil et al. (2000) demonstrated that food consumption represents the main source of PAH exposure for the children even in the big cities that are not heavily polluted by PAH from industrial sources. Some published reports become available of other studies of multipathway PAH exposure. Butler et al. (1993) measured personal exposure to benzo(a)pyrene in 15 adult individuals living in a town (population of 17,000) in New Jersey (USA). Compared to dietary exposures, the proportional contribution of inhalation exposures during late summer was about 5%, which corresponds well to results of study performed by Vyskocil et al. (2000). Also Van Rooij et al. (1994) demonstrated in the study on Netherlands adult volunteers that the inhalation of ambient air is relatively unimportant for total pyrene intake accounting for less than 1%. Lodovici et al. (1995) compared total dietary and inhalation PAH intake for adult population living in a town of medium size in central Italy (the average of the sum of 15 PAH was 27 ng/m^3). The intake by inhalation was 11% of the total intake. All these results confirm that food is the major source of exposure of humans to PAH in areas which are not highly polluted by PAH.

In summary, Phillips (1999) in your paper considered the current state of knowledge regarding the relative importance of different sources of PAHs to diet and the use of biomarkers to detect human exposure to PAHs.

5. WATER

PAHs enter surface waters mainly via atmospheric fallout, urban run-off, municipal effluents, industrial effluents and oil spillage or leakage. Atmospheric fallout includes wet and dry deposition of particles and vapors. PAHs, as semivolatile organic compounds, exist in both the gaseous and the particulate phase in air, and are subject to both vapor and particle washout from the atmosphere during precipitation. Atmospheric deposition is considered to be an important input of PAHs to surface waters. It has been estimated that 10–80% of PAH inputs to the world's oceans is from atmospheric sources. As a consequence, urban run-off contains PAHs deposited on surfaces, as well as mobile-related PAHs from gasoline and oil drips or spills, exhaust products, tyre particles, and bitumen from road surfaces. Higher concentrations of PAHs in urban runoff were found during autumn and winter, due to the high incidence of vehicles in the streets, coupled with the use of heating systems (Manoli and Samara 1999).

In another study concerning the distribution and the budget of PAHs in western Mediterranean seawater, it has been found that PAH content in the dissolved phase was maximal off-shore of Barcelona (1,800 pg/L) and in the Ebro river plume (2,170 pg/L), while concentrations in remote stations were lower (400–890 pg/L). PAHs associated with suspended particulate matter were evenly distributed in subsurface waters, and their concentrations ranged from 200 to 750 pg/L, maximizing at the Gibraltar and Sicilian Straits and at a frontal zone located midway between the Iberian Peninsula and the Balearic Islands. Slightly higher concentrations were found at the continental shelf (457–772 pg/L) than in the open sea (212–605 pg/L) (Dachs et al. 1997). Higher concentrations for dissolved-phase PAHs have also been observed in the Rhone delta, NW Mediterranean. PAHs in the dissolved phase were found up to 30 and 50 ng/L in summer and winter, respectively, whereas the corresponding particulate concentrations were much lower (up to 14 and 19 ng/L) (Bouloubassi and Saliot 1991).

According to the World Health Organization study in 1997, the concentration of individual PAHs in surface and coastal waters is generally in the neighborhood of 0.05 μg/L and concentration above this point indicates some contamination; also, a study carried out by the World Health Organization in 1993 revealed that benzo(a)pyrene concentration of 0.7 μg/L corresponds to an excess lifetime cancer risk of 10–5. According to studies done in the USA, in four major cities, the total PAHs in drinking water ranged between 4.7 and 600 μg/L (ASTDR 1995) as against obtained by Anyakora et al. (2005) (see Table 4). As shown in Table 4. the high-molecular-mass PAHs such as benzo(ghi)perylene, dibenzo(a,h)anthracene and indeno(1,2,3-cd)pyrene were not detected in the water sample. This can be attributed to their lower water solubility.

Table 4. PAHs concentration in water (Anyakora et al. 2005)

Compound	Water (μg/ml)
Naphthalene	0.55
Acenapthylene	0.34
Acenapthene	0.4
Flourene	0.33
Phenanthrene	1.46
Anthracene	0.35
Flouranthene	0.54
Pyrene	0.67
Benz[a]anthracene	0.56
Chrysene	1.32
Benzo[b]flouranthene	2.38
Benzo[k]flouranthene	1.82
Benzo[a]pyrene	1.72
Dibenz[a,h]anthracene	0
Benzo[ghi]perylene	0
Indeno[1,2,3-cd]pyrene	0

Groundwater is naturally filtered as it flows through various soil matrices, and PAHs adsorb well to organic soil. Groundwater concentrations of carcinogenic PAHs reported for US groundwaters ranged from 0.2 to 6.9 ng/L, while the corresponding concentrations in surface waters were between 0.1 and 800 ng/L and most frequently between 2 and 50 ng/L (Menzie et al. 1992). Regarding the chlorination of drinking water, it has been found that this disinfection technique may lead to formation of oxygenated and chlorinated PAHs, i.e. compounds that are more toxic than the parent PAHs (Manoli and Samara 1999).

The presence of PAHs in drinking water may be due to the surface or groundwater used as raw water sources, or to the use of coal tar-coated pipes in public water supply systems, European Community directive 80/778/EEC (1980) states a maximum level for PAHs in drinking water of

0.2 μg/L with fluoranthene, benzo(k)pyrene, benzo(b)fluoranthene, benzo(k)fluoranthene, benzo(ghi)perylene and indeno(1,2,3-cd)pyrene as reference compounds. Maximum levels for the sum of these PAHs in surface waters can reach 1 μg/L, depending on the surface water treatment process.

Besides emission sources, PAH water concentrations depend also on the depth in the sediment core (see Table 5). It can be seen that a range of PAHs at varying concentrations were found in the porewater samples. Generally, it was the four-ring compounds that were found in the highest concentrations, the largest contribution to this arising from the compounds fluoranthene and pyrene. Naphthalene also dominated in many samples. Total PAH concentration in porewater varied widely with depth in the sediment core and several regions of high concentration can be recognised, the highest (742 μg/L) occurred at between 50 and 52.5 cm (King et al. 2004).

Table 5. PAH concentration (ng/L) in porewater samples from a sediment core from the Mersey Estuary (King et al. 2004)

Depth (cm)	Naphthalene	Acenaphthene	Flourene	Penanthrene	Anthracene	Flouranthene	Pyrene	∑PAHs
0–2.5	67	75	26	58	<20	77	106	409
2.5–5	275	43	29	44	42	62	109	604
5–7.5	<3	46	17	17	<20	28	58	166
7.5–10	<3	66	105	126	<20	100	125	522
10–12.5	27	21	<2	<17	<20	14	32	95
12.5–15	<3	16	28	45	<20	85	86	260
15–17.5	62	16	33	45	<20	75	92	323
17.5–20	33	<6	10	12	57	15	32	159
20–22.5	19	20	6	7	<20	13	32	96
22.5–25	65	23	<2	12	50	15	33	198
25–27.5	29	19	32	<17	<20	111	275	467
27.5–30	69	71	52	94	85	162	189	721
30–32.5	41	20	7	9	<20	18	34	129
32.5–35	35	16	6	10	<20	17	31	116
35–37.5	30	<6	7	15	<20	19	36	107
37.5–40	54	20	<2	6	<20	15	30	126
40–42.5	107	49	35	76	71	181	220	739
42.5–45	<3	<6	<2	<17	<20	73	83	156
45–47.5	159	<6	<2	<17	<20	37	38	234
47.5–50	65	<6	<2	7	<20	26	31	129
50–52.5	268	78	29	74	<20	130	162	742
52.5–55	103	47	<2	<17	<20	100	162	353

6. PAH IN AMBIENT AND INDOOR AIR

The risk associated with human exposure to atmospheric PAHs is highest in cities, considering the density of population, increasing vehicular traffic, and scarce dispersion of the atmospheric pollutants (Rockens et al. 2000). Presently, limit values of 1 or 10 ng benzo(a)pyrene per m^3 air are recommended or mandatory in various countries (e.g. Italy or Germany). Actually, these concentrations are seldom found or exceeded nowadays in ambient air of rural or even of urban living areas (Jacob and Seidel 2002).

Based on calculated PAH diagnostic ratios, the major source of PAHs exposure was from vehicle exhaust (Liu et al. 2007; Zhu et al. 1997). In Taiwan region, several significant contributor to PAHs sources had been sampled such as stationary industrial combustion: steel and iron industries (Yang et al. 2002) with a mean total PAHs concentrations measured to be 1,020 $\mu g/m^3$, traffic vehicles exhaust: motorcycle (Yang et al. 2005) and highway toll station (Tsai et al. 2004) with a mean total-PAHs concentrations ranged from 8,280 to 12,300 ng/m^3. Fang et al. (2004a, b, c) indicated that mean of total PAHs levels at industrial, urban, and rural areas in central Taiwan region ranged from 1,232 to 1,650, 700 to 1,740, and 610 to 831 ng/m^3, respectively. However, the $PM_{2.5}$ (fine particulate) and $PM_{2.5-10}$ (coarse particulate) total PAHs concentrations at the Tunghai University Pastureland (Taiwan) sampling site were found to be 180.62 and 164.98 ng/m^3, respectively. In general, the concentrations of polycyclic aromatic hydrocarbons were higher in spring and winter than those of summer and autumn for either $PM_{2.5}$ or $PM_{2.5-10}$ in Taiwan (Fang et al. 2005). Table 6 showed the results of comparison of atmospheric PAHs concentrations in PM_{10} particulates with other studies around the world. In general, the results indicate that special sampling sites such as Taiwan urban, Taiwan traffic, Taiwan Petrochemical industrial, Flanders urban, Taichung pastureland and Taichung temple possess extremely high total PAHs average concentrations. When compared with the traffic, urban, petrochemical industrial, pastureland and temple sampling sites, Asian countries (Taiwan) have higher total PAHs concentrations than Europe (Italy and Belgium, except Urban Flanders, Belgium), South-East Asia (India and Malaysia) and America (Mexico) (Fang et al. 2006).

Table 6. Comparison of atmospheric PAHs concentrations (ng/m^3) in PM_{10} particulates at different sites around the world (after Fang et al. 2006)

Location	Survey year	Total PAH's concentrations	BaP concentrations	Particle size	Literature
Traffic, Tainan, Taiwan	1994–1995	912	37.01	PM_{10} ($\sum$21 PAH)	Sheu and Lee (1997)
Urban, Tainan, Taiwan	1994–1995	203.9	10.5	PM_{10} ($\sum$21 PAH)	Sheu and Lee (1997)
Petrochemical industrial, Tainan, Taiwan	1994–1995	116.8	2.39	PM_{10} ($\sum$21 PAH)	Sheu and Lee (1997)
Residential, Naples, italy	1996–1997	22.7	0.9	PM_{10} ($\sum$15 PAH)	Caricchia et al. (1999)
Traffic, Naples, italy	1996–1997	54.8	2.97	PM_{10} ($\sum$15 PAH)	Caricchia et al. (1999)
Industrial, Naples, italy	1996–1997	39.5	2.75	PM_{10} ($\sum$15 PAH)	Caricchia et al. (1999)
Urban, Mumbai, India	1995	24.5	1.8	PM_{10} ($\sum$18 PAH)	Kulkarni and Venkatraman (2000)
Industrial, Mumbai, India	1995	38.8	2.1	PM_{10} ($\sum$18 PAH)	Kulkarni and Venkatraman (2000)
Urban, Kuala Lumpur, Malaysia	1998–1999	6.3 ± 4.4	–	PM_{10} ($\sum$17 PAH)	Omar et al. (2002)
Rural, Kuala Lumpur, Malaysia	2000	0.3 ± 0.2	–	PM_{10} ($\sum$17 PAH)	Omar et al. (2002)
Traffic, Hong Kong	2000–2001	44.54	2.13	PM_{10} ($\sum$16 PAH)	Guo et al. (2003)
Industrial, Hong Kong	2000–2001	23.86	1.3	PM_{10} ($\sum$16 PAH)	Guo et al. (2003)
Urban, Flanders, Belgium	2000–2001	93.025	0.82	PM_{10} ($\sum$16 PAH)	Rockens et al. (2000)
Industrial, Flanders, Belgium	2000–2001	55.125	0.76	PM_{10} ($\sum$16 PAH)	Rockens et al. (2000)

In general, summertime PAH levels tend a minimum, reflecting changes in prevailing meteorology, i.e. lower inversions during winter, and lack of a major PAH source, residential fuel combustion for heating. It has been suggested that in urban areas only the particle associated with PAH were highest in winter, while the vapor phase compounds were highest in summer (Baek et al. 1991). However, in Brisbane the phenanthrene, fluorene and pyrene levels in both vapor and particle phase were highest in sample collected at lowest mean sampling temperature (Müller et al. 1998). Bodzek et. al. (1993) observed higher winter/summer ratios for particle-associated PAHs with a lower condensation degree (from fluorene to chrysene) and lower concentration ratios for the higher condensed PAHs BFs (B(b)Fl + B(k)Fl), B(e)Py and B(ghi)Pe. The authors attributed this difference to the temperature dependent distribution of the more volatile PAHs between the particle and the gaseous phase (Broman et al. 1991). Papageorgopoulou et al. (1999) PAHs were determined in the ambient air of six towns in Greece. Correlations of total PAH with other atmospheric pollutants were largely different among towns indicated that the relative contribution of emission sources is different in each location. In the warm months PAHs were significantly correlated with vehicular pollutants thus suggested traffic as major PAH emitting source. The same was also deduced from the comparison of the ambient PAH profiles to the profiles of particular sources. The contribution of residential heating was significant in most towns during winter. Principal component analysis of the data did not result in a clear distinction between towns suggested that all were influenced by similar source types.

Exposure of on-duty traffic police in Beijing to PAHs was investigated by Liu et al. (2007). The authors used a personal sampling technique in measuring both particulate and gaseous phase PAHs. The results were then compared with those from two control sites away from the street. Exposure levels to gaseous and particulate PAHs for the traffic police were found to be 1,525 ± 759 and 148 ± 118 ng/m^3, respectively, representing 2–2.5 times higher levels than those at the control sites. However, Park et al. (2002) found negative correlation between vapor phase PAHs and temperature when investigating atmospheric PAHs in Seoul, Korea. It is believed that changes in the atmospheric mixing layer at higher temperatures leads to stronger convection and dilution, and Panshin and Hites (1994), Wania et al. (1998) even developed experimental equations to quantitatively describe the relationship between vapor PAHs and temperature.

The total concentration of 17 PAHs ranged between 0.84 and 152 ng/m^3, with an average of 116 ng/m^3, in urban area were 1.1–6.6 times higher than those measured in suburban area of Beijing (China). It suggested a serious pollution level of PAHs in Beijing. PAHs concentrations increased with decreasing the ambient temperature. Approximately 68.4–84.7% of PAHs were adsorbed on particles having aerodynamic diameter 62.0 μm. Coal combustion for domestic heating was probably major contributor to the higher PAHs loading in winter, whereas PAHs in other seasons displayed characteristic of mixed source of gasoline and diesel vehicle exhaust. Biomass burning and road dust are minor contributors to the PAHs composition of these aerosol particles. Except for source emission, other factors, such as meteorological condition, photochemical decay, and transportation from source to the receptor site, should to be involved in the generation of the observed patterns (Zhou et al. 2005).

Rehwagen et al. (2005) concluded that the results showed a PAH seasonal behaviour in both regions of La Plata (Italy), with lower contents in summer and higher ones in winter. Highest concentrations of total PAHs were found in the industrial area in La Plata. The size distribution of particles demonstrates the greater relevance of smaller particles. More than 50% of PAHs were associated with particles of less than 0.49 μm. Moreover, this particle size fraction was associated with traffic, whereas other sources of combustion were related also to particles between 0.49 and 0.95 μm. Considering the ratio of benzo(ghi)perylene /benzo(a)pyrene as an indicator for traffic influence, it was observed that La Plata City was more affected than Leipzig by the same proportion in summer and in winter. The benzo(g)perylene/indeno(123-cd)pyrene ratio was lower in winter than in summer in both places and indicates the presence of domestic combustion sources. It is important to point out the significance of using fingerprint compound ratios to identify possible sources of pollution with PAHs bound to particles (Rehwagen et al. 2005).

In a similar study on exposure to airborne particles and PAHs close to a busy street in Gliwice (Poland). Bylina et al. (2005) found that the average concentration of PAHs near the street was more than 1.5 times greater than that 100 m away from the street. Although the total exposure was much lower than that observed in Beijing (only 192 ng m^{-3} in the spring), the results highlight the significant role exhaust gas emissions play in exposure to PAHs on busy streets (Liu et al. 2007).

For example, air concentrations of pyrene and of the sum of 12 PAH compounds measured indoors and outdoors in two kindergartens are presented in Table 7. The outdoor total PAH concentrations were 12 times higher in "polluted" kindergarten than in kindergarten situated in a "non polluted" area and 3 times lower than mean concentration measured near the highway in Montreal. On the contrary, no difference was found between the total PAH concentrations measured indoors in both kindergartens. Similar tendencies were observed for pyrene concentrations (Vyskocil et al. 2000).

Table 7. PAH concentrations in area (Vyskocil et al. 2000)[a]

Place	Total PAH[b] (ng/m^3)		Pyrene (ng/m^3)	
	Outdoor	Indoor	Outdoor	Indoor
Kindergarten in a polluted area	36.1; 17.9; 14.8	2.8; 2.6; 3.6	5.1; 1.6; 1.3	0.4; 0.3; 0.5
Kindergarten in a nonpolluted area	1.6; 2.2; 1.9	1.0; 2.8; 2.5	0.3; 0.4; 0.4	0.2; 0.5; 0.5
Montreal highway Décairie (1989–1991)[c]	62.7			
London (1991)[d]	166			

[a] Individual values

[b] Total of 12 PAH: phenanthrene, anthracene, fluoranthene, pyrene, benzo(a)anthracene, chrysene, benzo(b)fluoranthene, benzo(k)fluoranthene, benzo(a)pyrene, dibenzo(a,h)anthracene, benzo(ghi)perylene, indeno(1,2,3-c,d) pyrene

[c] Sum of 12 PAH calculated by the authors of this paper from the results presented by Ringuette et al. (1993)

[d] Sum of 15 PAH (Halsall et al. 1994)

Sharma et al. (2007) reported that concentrations of total PAHs in air were 667.73 ± 399.38 and 672.16 ± 387.69 ng/m^3 (Delhi, India) in the years 2002 and 2003, respectively. Data also reveal that the concentrations of PAHs are higher in the winter months than in the monsoon and summer months. In earlier studies also the concentrations of particle associated PAHs in urban areas have been found to peak in winter. The elevated concentrations of PAH in winter can be attributed to increased emissions from higher consumption of fossil fuels for heating purposes, more traffic volumes and less dispersion due to meteorological conditions prevailing in winter (Sharma et al. 2007).

Indoor air quality is of particular interest in the world. With the different functions and ventilation conditions, the concentrations of PAHs in indoor air were: bedroom > kitchen > living room > balcony. Indoor air of kitchen is polluted by PAHs that come from not only outdoor air but also indoor emission sources such as cooking, gas home appliances (Liu et al. 2001). Except for cooking practice, cooking oil-fumes should make a significant contribution to PAHs concentrations of indoor air including two parts: (1) once heated PAHs evaporate from the polluted oil into air; (2) at high temperature organic compounds are partially cracked to smaller unstable fragments (pyrolysis), mostly radicals recombine to give relatively stable PAHs (pyrosynthesis) (Moret and Conte 2000).

According to Zhu and Wang (2003) the average concentration of 12 PAHs in air of hotel and commercial kitchens of Hangzhou was 17 μg/m^3, which comprised mainly of three- and four-ring PAHs. The corresponding average in domestic kitchens was 7.6 μg/m^3 and consisted mainly of two- and three-ring PAHs. Anthracene was more abundant in domestic kitchen air than in commercial kitchen air. Food boiling produced the least amount of PAHs. The high-fat food produced more PAHs when broiled, so did low fat-level food when fried except pyrene and anthracene. In commercial kitchens, PAHs came from two sources, cooking practice and oil-fumes, however the cooking practice had a more predominant contribution to PAHs in commercial kitchen air. In domestic kitchens, except for cooking practice and oil-fumes, there were other PAHs sources, such as smoking and other human activities in the domestic houses, where three- to four-ring PAHs mainly came from cooking practice. Two-ring PAHs, naphthalene, was the most predominant kind, mostly resulting from the evaporation of mothball containing a large quantity of naphthalene, used to prevent clothes against moth. The level of air PAHs in commercial kitchens was very high. The exposure to BaP in domestic kitchens was 0.0044–0.016 μg/day, while it increased to 0.32–0.94 μg/day in commercial kitchens. Meanwhile, the production of PAHs from three common cooking oils was compared, in which lard produced more PAHs than rape-seed oil and soybean oil. An increase in cooking temperature generally increased the production of most PAHs.

Table 8. Range and geometric mean concentrations (ng/m^3) of PAH compounds in a kitchen using a kerosene stove (Pandit et al. 2001)

Compound	Indoor (I)	Outdoor (O)	Mean I/O[a]
Naphthalene	ND–183.5 (67.2)	ND–16.6 (6.4)	10.5
Acenapthylene	2.6–45.1 (28.8)	0.6–16.5 (6.5)	4.3
Acenepthene	4.6–22.6 (14.1)	4.6–18.2 (8.9)	1.6
Fluorene	ND–13.6 (4.6)	0.7–2.2 (1.2)	3.8
Phenanthrene	ND–45.2 (10.3)	1.0–2.2 (1.4)	7.3
Anthracene	ND–9.8 (2.3)	2.3–3.0 (2.8)	0.8
Fluoranthene	0.5–48.3 (16.1)	1.4–6.1 (3.6)	4.5
Pyrene	1.4–45.5 (12.8)	2.0–7.8 (4.5)	2.8
Chrysene	0.9–6.8 (2.0)	0.3–1.1 (0.5)	4.0
Benzo(a)anthracene	1.1–9.7 (2.9)	0.4–1.6 (0.8)	3.6
Benzo(k)fluorenthene	0.6–49.5 (16.9)	0.5–5.4 (3.8)	4.4
Benzo(a)Pyrene	ND–17.6 (6.9)	ND–2.0 (0.9)	7.6
Total PAH	25.3–373.9 (164.2)	23.0–45.6 (36.7)	4.5

Mean concentrations are given in parentheses

ND not detected

[a] The I/O ratio has been calculated from mean indoor concentration to mean outdoor concentration

The concentration ranges of individual PAH compounds along with their mean observed for indoor and outdoor locations are given in Table 8 (Pandit et al. 2001). The high concentrations observed for the lower molecular weight PAHs such as naphthalene and acenaphthylene in the indoor sample accounted for 58% of the total PAH while the same accounted for only 35% of the total PAH in outdoor samples. The outdoor samples were collected in an open area 5–6 m away from the respective houses. Indoor concentration levels of some of the compounds like nepthalene, phenanthrene and benzo(a)pyrene were found to be a magnitude higher than the respective outdoor levels. Except for anthracene (I/O = 0.8), the I/O ratio of individual PAH compounds were always higher than one, varying from 1.6 to 10.5. The concentration of the carcinogenic benzo(a)pyrene in the cooking environment using kerosene as fuel ranged between 0.2 and 17.6 ng/m^3 and are found to be lower than 33–186 ng/m^3 of benzo(a)pyrene from other cooking fuels like coal, wood and cattle dung as reported by Raiyani et al. (1993). Even then, it exceeds the air quality standard of 1 ng/m^3 specified by a few countries (though not India) most of the time. The indoor concentration levels suggest that exposure to individual PAH compound during cooking period is 2–10 times higher than ambient exposure. However, based on 6-h cooking and 18-h non-cooking periods, the average daily exposure was only two times higher than ambient air (Pandit et al. 2001).

Chiang et al. (1999) successfully identified four PAHs in fumes from these cooking oils (Chiang et al. 1999). B(a)P, DB(ah)A, and B(a)A are considered probable human carcinogens (group 2A); B(b)FA is considered a possible human carcinogens (group 2B) by IARC (1992). The carcinogenic contents (μg/m^3) of fumes from various oils is shown in Table 9. Exposure to PAHs is considered an etiologic factor of lung cancer (Boffetta et al. 1997). Safflower oil, vegetable oil, and corn oil are frequently used by cooks in Taiwan and during the past 10 years. These results indicated that exposure to cooking oil fumes could possibly increase exposure to PAHs, which may be linked to the increased risk of lung cancer reported in Chinese Women. The concentration of B(a)P in cooking oil fumes exceeds that reported in cigarette smoke by a factor of 6.5 times (Guerrin 1993).

In general, a ratio of outdoor to indoor air PAH concentrations are about 2:1 in a room which is not polluted by tobacco smoking (Van Rooij et al. 1994). It must be, however, noted that the other prominent indoor combustion sources of PAH, such as heating and cooking were absent in kindergartens. Sakai et al. (2002) clearly demonstrated that cigarette smoking and meat cooking are the major emission sources of PAH in indoor air of the Tokyo area. It was also shown that PAH air pollution in the Tokyo area is serious both indoors and outdoors. Most large factories have already relocated to the outskirts of the Tokyo metropolitan area, since the Metropolitan government pollution control policies began in 1969 (TMG 1969). Thus, at present, motor vehicle exhaust is considered the largest emission source of PAH in the Tokyo metropolitan area. Average concentrations of PM_{10} and PAH in the air along the highway between Zurich and Basal on a fine summer afternoon were 55.0 μg/m^3 and 127.5 ng/m^3, respectively (Siegmann et al. 1999). Although this PM_{10} average is close to that in the suburban highway in the study, the PAH average is one half that of the suburban highway around Tokyo. Also, as the ventilation system was operating in Tokyo highway tunnels, it is speculated that the suburban and metropolitan highways in the Tokyo area are much more polluted by PAH. On this account, it was disclosed that a recent increase in the number of motor vehicles in Tokyo (TMG 2000; JEA 2000) has resulted in a serious increase in the outdoor concentrations of PAH there.

Table 9. The Carcinogenic Contents (μg/m^3) of Fumes from various oils (Chiang et al. 1999)

Carcinogens	Cooking oil		
	Safflower	Vegetable	Corn
BaP	22.7 ± 1.5	21.6 ± 1.3	18.7 ± 0.9
DBahA	2.8 ± 0.2	3.2 ± 0.1	2.4 ± 0.2
BbFA	1.8 ± 0.3	2.6 ± 0.2	2.0 ± 0.1
BaA	2.5 ± 0.1	2.1 ± 0.4	1.9 ± 0.1

B(a)P Benzo(a)pyrene, *B(a)A* Benzo(a)fluoranthene, *B(b)FA* Benzo(b)fluoranthene, *B(ghi)P* Dibenzo(ghi)perylene

Gas and particle-phase pollutant emission was measured in the Caldecott Tunnel in the San Francisco Bay Area in the summer of 1996. Diesel trucks were the major source of lighter PAH, whereas light-duty gasoline vehicles were the dominant source of higher molecular weight PAH. Size-resolved measurements of particulate PAH showed significant fractions of diesel derived PAH to be present in both the ultra-fine size mode (<0.12 μm) and the accumulation mode (0.12–2 μm). In contrast, the gasoline engine-derived PAH emissions were found almost entirely in the ultra-fine mode (Miguel et al. 1998).

7. SOIL

Soil system seems to be the important long-term repository for PAHs and is considered to be a steady indicator of the environmental pollution state. Accumulation of PAHs in soils may lead to further potential contamination of vegetables and food chains (Kipopoulou et al. 1999; Mueller and Shann 2006), and then cause direct or indirect exposure to human. Moreover, leaching, evaporation and migration are possible PAHs sources of atmospheric or groundwater contamination. Therefore, knowledge of soil contamination with PAHs is needed to avoid any food production risk and to restrict the deleterious effect of these contaminants. Sorption and degradation are key processes that affect the fate and transport of PAHs in the environment (Magee et al. 1991). Sorption also plays a key role in controlling transport of PAHs during leaching, as well as in transformation and bio-accumulation processes (Reeves et al. 2004). For example, Amellal et al. (2006a) studied the sorption and desorption of ^{14}C-labelled phenanthrene on soils, during 6 months of natural conditions. The results showed that phenanthrene was strongly adsorbed and poorly desorbed. Consequently, in the leachates of microlysimeters, phenanthrene represented only a limited proportion, from 20 to 40%. Leaching primarily concerned degradation products, maybe including traces of 9,10-phenanthrenedione, which was present in significant quantities in the soils in an extractable state. Lastly, the sequestration or formation of non-extractable residues can contribute in a very significant and quasi-final way to the exclusion of the pollutant from any other process which defines its behaviour in the soil. Amellal et al. (2006b) observed that when biological degradation occurs, there was an increased in sequestration and in leaching, as well as in the concentration in residues in pore water. The increased in availability for leaching was exclusively due to the formation of degradation products that were more soluble and/or less adsorbed than phenanthrene on the surface of the aggregates. In the same way, these compounds were responsible for the increased in the concentration in residues observed in the immobile water of the soil. The passage of phenanthrene degradation products into soil pore water occurs by diffusion due to the effect of a concentration gradient created by the degradation of phenanthrene (and the formation of degradation products) on the surface of the aggregates where it was initially adsorbed. In addition, this diffusion facilitates the sequestration of the compounds in the soil. They concluded that in soils contaminated by strongly adsorbed organic compounds, water pollution and sequestration are primarily controlled by degradation.

Many studies have been reported for the PAHs in soils in some countries (Wild et al. 1990; Weiz et al. 1994; Nam et al. 2003). It was reported that the typical endogenous PAHs in soils are in the range of 1–10 μg/kg (Edwards 1983), and most of them are derived from oil seeps, bitumens, coal, plant debris, forest and prairie fires (Yunker et al. 2002). PAHs with three rings or more tend to be very strongly adsorbed to the soil matrices (Knox et al. 1993), preferentially to small aggregates (<50 μm), which also contain the most humified organic matter (Quantin et al. 2005). Strong adsorption coupled with very low water solubility render PAH loss by leaching insignificant. Lichtfouse et al. (1997) reported that soil PAHs were mainly of pyrolytic origin, e.g. exogenous (airborne) contributions of fossil fuel combustion products and (old) vegetation fires, with a small contribution of uncombusted fossil fuels.

A number of studies have shown that PAHs adsorb very strongly to the surface of soil particles and organic matter, and are hence not readily available for biodegradation. For example, Fu et al. (1994) observed that 30 ± 50% of naphthalene, the most water-soluble and bioavailable PAH, resisted desorption from an Oklahoma soil consisting of 92% sand.

According to Vyskocil et al. (2000) the total PAH concentration in the soil from the kindergarten situated in the "polluted" area was twice higher than that in the "non-polluted" area (see Table 10).

Table 10. PAH concentrations in soil (Vyskocil et al. 2000)

Place	Total PAH[a] (ng/g)	Pyrene (ng/g)
Kindergarten in a polluted area	132	10.4
Kindergarten in a non polluted area	65	6.1

[a] Total of 12 PAH: phenanthrene, anthracene, fluoranthene, pyrene, benzo(a)anthracene, chrysene, benzo(b)fluoranthene, benzo(k)fluoranthene, benzo(a)pyrene, dibenzo(a,h)anthracene, benzo(ghi)perylene, indeno(1,2,3-c,d) pyrene

The highest concentration of PAHs was detected in the chemical area of Tarragona (476.2 ng/g), with values two- and four-times higher than those of the urban/residential (206.9 ng/g) and the petrochemical zones (119.7 ng/g), respectively. Finally, the lowest concentration corresponded to soils sampled in the unpolluted area (52.8 ng/g). Notwithstanding, the differences did not reach the level of statistical significance, for any of the individual PAHs, as well as for the sum of 16 PAHs and the sum of 7 carcinogenic PAHs. With respect to the total PAH sum, the concentration in soil samples from the chemical and urban areas decreased, while the levels corresponding to the petrochemical and unpolluted zones increased. Excepting those samples collected in the Southern complex, a similar tendency was noted for the levels of seven carcinogenic PAHs. However, the temporal variation was not statistically significant for any of both groups of organic pollutants here analyzed. When the hydrocarbons were individually assessed, it was found that the concentration of naphthalene increased significantly from 5.3 to 24.4 ng/g ($p < 0.05$) and from 3.7 to 16.6 ng/g ($p < 0.001$) in the chemical and petrochemical areas, respectively. In contrast, acenaphthylene levels significantly decreased in urban samples (from 23.4 to 3.0 ng/g, $p < 0.01$), while no significant changes were found for the remaining PAHs (Nadal et al. 2007).

Wilcke and Amelung (2000) reported that the total of 20 PAHs in the soils from North America Great Plain ranges from 60 to 180 μg/kg, while PAHs in the vegetable soils in Guangzhou (China) is higher than this range (Chen et al. 2005).

In another work PAH were present in the 0–10 cm soil from the Swiss soil monitoring network (NABO) observation sites in concentrations from 50 to 619 μg/kg dw. The highest concentrations were observed in urban areas (400–619 μg/kg dw), irrespective of the land utilization type (city park, forests, arable land). In rural areas, concentrations of 16 PAH in arable soils ranged from 60 to 145 μg/kg dw, in permanent and pasture grassland soils from 50 to 314 μg/kg dw, and in forest soils from 98 to 219 μg/kg dw. Median values obtained in these main compartments increased in the order of arable soils (66 μg/kg dw) < permanent and pasture grassland (142 μg/kg dw) < forests (164 μg/kg dw) < urban soils (451 μg/kg dw) (Bucheli et al. 2004). However, Capuano et al. (2005) reported that the values of PAH decrease with depth soil, for example, for sample 1 total PAH concentrations were 318,9 μg/kg dw (10 cm), 88,9 μg/kg dw (20 cm) and 74,1 μg/kg dw (30 cm). The PAH-concentrations in soil samples from site 1 are approximately two orders of magnitude higher than those in other soils in the vicinity of human activities (see Table 11).

According to Masih and Taneja (2006) the average concentration of total PAH in all samples (Agra—semi-arid region of India) was 12.1 μg/g and the range was from 3.1 to 28.5 μg/g. The total PAH concentrations were 13.72, 12.98, 9.37 and 6.73 μg/g at industrial, roadside, residential and agricultural sites, respectively. The industrial sites had the highest total PAH concentration followed by roadside, residential and agricultural site. High concentrations at industrial site can be due to the location of the site, which is well known for generator manufacturing, tanning and iron casting industries. The maximum concentrations of PAHs were found to be in winter season. The concentration of PAH decreased in the order chrysene > benzo(b)fluoranthene > fluoranthene (Masih and Taneja

2006). This might be due to industrial-oil burning, wood combustion and emission coming from diesel powered vehicles (Ravindra et al. 2001).

Table 11. Total concentrations of PAHs in soils (µg g dry wt.) in urban areas and in the vicinity of human (industrial) activities (after Bakker et al. 2000)

Location	Number of PAHS	Soil depth (cm)	∑PAH concentration in soil (µg g dry wt.)	Reference
Highway (CAN)	17	Surface	1.4	Wang and Meresz (1982)
Motorway (UK)	7	0–4	20 at 1 m	Butler et al. (1984)
			4–8 at 500 m	
Highway (USA)	14	0–5	3	Yang et al. (1991)
Urban (JAP)	8	0–3	1.3 ± 0.8	Spitzer and Kuwatsuka (1993)
Chemical plant (AUS)	18	0–5	0.3–79	Weiss et al. (1994)
Urban (CH)	16	0–20	11–12	Niederer et al. (1995)
Al plant (SK)	17	Surface[a]	40–200	Wickle et al. (1996)
Urban (UK)	16	0–10	2.7 ± 0.5	Meharge et al. (1998)
Polyprop. fire (UK)	16	0–10	12–18	Meharge et al. (1998)

[a] Organic surface layer

The total PAHs concentrations in soils from the paddy fields at Hunpu wastewater irrigation areas in Shenyang, China, were between 950 and 2,790 μg/kg. There was no regular pattern on the accumulation of PAHs in soil along the irrigation channel from the upper reach to the lower reach. For instance, total PAHs measured in soil at sites 1 and 2 were 2,760 and 2,000 μg/kg, respectively, but the content of total PAHs was 2,230 μg/kg in soil at site 6 about 25 km further downstream from site 1, which suggested that the channel played no role in the dilution and decontamination of the pollutants. Two reasons contributed to this phenomenon: one was the re-solubilizing of pollutants from the channel sediment, former study (Song et al. 2005) had showed that the total content of 16 PAHs in the channel sediment had reached up to 67 mg/kg; the other might be the possible additional unregulated discharge of sewages in the middle and/or lower reach of the channel, both of which would result in the elevated accumulation of pollutants in soils at the sites far from the pollution sources of the upper reach. This was consistent with communications with local farmers at the sampling sites. The total concentration of the 14 PAHs in soil collected from the wastewater irrigation region in the former eastern Berlin, ranged from 830 to 3,690 μg/kg; there were no significant differences between residual total PAHs for samples from adjacent sites (Song et al. 2006). For comparison, Chen et al. (2003) reported that the total PAHs were 3,000–5,000 μg/kg in the agricultural soil within the wastewater irrigation region in Tianjin, China. The background level of TPAHs in the unpolluted paddy soil in Shenyang, China, was 275–317 μg/kg (Song et al. 2002).

PAH contents ranged from 450 μg/kg of dry material at Honfleur to 5,650 μg/kg at Notre-Dame de Gravenchon (France) (Motelay-Massei et al. 2004). Maliszewska-Kordybach (1996) measured an average of 264 μg/kg (for 16 PAHs) in agricultural soils in Poland.

Aamot et al. (1996) found low values (144 μg/kg on average) in forested soils of Norway. However, in Estonian urbanized areas, the high levels reported were from 2,200 to 12,300 μg/kg (Trapido 1999). Mielke et al. (2001) found concentrations of 3,700 μg/kg in the urban centre of New Orleans (USA). The concentration range of 16 PAHs varied by over two orders of magnitude from 0.016 μg/kg in rural to 3.884 μg/kg in suburban soils of Beijing (China) (Ma et al. 2005). The concentrations of the total PAH at rural site were generally less than those of urban site of Tianjin, China. Total PAH concentrations ranged from 144 to 308 ng/m^3 and 239 to 537 ng/m^3 for Chenlou and Meteorological tower sampling, respectively (Wu et al. 2006). The total PAH at meteorological tower (urban) were about one or two order higher than those of Mumbai (21–42 ng/m^3; Venkataraman et al. 1999) and Kuala Lumpur (6.28 ± 4.35 ng/m^3; Omar et al. 2002), and were close to the concentration of Inchon (410 ng/m^3; Bae et al. 2002) in winter season. The average total PAH at

Chenlou was also much higher than that of Yangsuri rural (119 ng/m^3; Bae et al. 2002). The total PAH differences between Chenlou (rural) and meteorological tower (urban) were significant and similar results have been reported in Massachusetts (Allen et al. 1996), Kuala Lumpur (Omar et al. 2002), Birmingham (Smith and Harrison 1996) and Korea (Bae et al. 2002) with the urban and rural total PAH were 49.5, 6.28, 4.37, 410 and 1.94, 1.3, 1.13, 119 ng/m^3, respectively.

Crépineau et al. (2003) reported a study on the contamination of pastures, grass and soil samples by PAHs in the vicinity of a highway (70,000 vehicles per day). All samples have been collected from 10 m (d1), 50 m (d2), and 150 m (d3) perpendicular to a French highway and at a control site in a rural area away from nearby contaminating sources. Total PAH concentration were ranges from 767 ng/g dry weight to 3,989 ng/g dry weight, according to the matrix and the distance from the highway. The authors concluded that the distance was not a significant factor for PAH deposition on grass, while in soil it has an effect between d1 and d2 or d3. The total PAH concentration in highway samples was eight times higher than in control site samples for grass and seven to four times higher for soil. Fluoranthene, pyrene, and phenanthrene were the major PAHs in grass samples at the control site and the highway, but the concentrations were about five times higher near the highway. In soil samples collected near the highway, the values of concentrations between all compounds were not statistically different. PAH deposition on grass was linked to the physicochemical properties of the compounds, which lead to a specific distribution of each molecule (according to their volatility and the number of aromatic rings) while no specific behavior was revealed in soil. In another work (Bryselbout et al. 2000) results showed that the low-molecular weight PAHs such as naphthalene, fluorene, and anthracene increased from 2 to 6 m of a highway (Nancy, France). However, the concentrations of high-molecular weight PAHs such as benzo(a)pyrene and benzo(g,h,i)pyrene decreased from 2 to 6 m. According to these authors the results strengthen the hypothesis of a local distillation effects around the highway.

Table 12 shows a worldwide comparison of PAHs concentration (Masih and Taneja 2006). The PAH concentration in soil of industrial (13.72 μg/g) and roadside (12.98 μg/g) area of Agra is less than the concentration found in Austria/Germany (79.00/16.00 μg/g) and USA (58.60 μg/g), respectively, whereas residential (9.37 μg/g) and agricultural (6.73 μg/g) sites concentrations of PAH were found to be higher than in UK (4.20 μg/g) and Germany (1.90 μg/g), respectively. As evident from the Table 12 the concentrations measured in soils of various sites at Agra (industrial, roadside, residential and agricultural) show much difference between each other. In the present data, contamination in the urban industrial area appears to be two times higher than in agricultural areas; similar results have been reported in earlier studies (Tremolada et al. 1996; Wagrowski and Hites 1997).

Table 12. Soil PAH concentrations compiled from literature data (after Masih and Taneja 2006)

Study area	PAH concentration (μg/g)	Number of PAH	Reference
Agricultural (rural)			
Brazil	0.096	20	Wickle et al. (1999)
UK	0.19	12	Wild and Jones (1995)
Germany	1.90	06	Tebaay et al. (1993)
India	6.7	11	Masih and Taneja (2006)
Residential (urban)			
Bangkok	0.38	20	Wickle and Muller (1999)
Brazil	0.39	20	Wickle et al. (1999)
Germany	1.80	06	Tebaay et al. (1993)
UK	4.20	12	Wild and Jones (1995)
India	9.3	11	Masih and Taneja (2006)
Roadside (urban)			
Australia	3.30	14	Yang et al.
USA	58.68	14	Rogge et al. (1993)
India	12.9	14	Masih and Taneja (2006)
Industrial (urban)			
UK	4.50	12	Wild and Jones (1995)
Germany	16.0	06	Tebaay et al. (1993)
Austria	79.0	18	Weiss et al. (1994)
India	13.7	11	Masih and Taneja (2006)

8. SEDIMENTS

Sediments are deposited material consisting of organic matter in various stages of decomposition, particulate mineral matter and inorganic material of biogenic origin. They have been proven an efficient tool to identify environmental impacts (Evans et al. 1990). Due to the exposure time to industrial effluents, they are valid for long-term studies. Sediments constitute a pollutant trap and are an important factor to establish the assimilative capacity of the environment. Sediments may contain a high level of pollutants ready to pass on to the food chain or be mobilised by anthropogenic or natural means. Therefore, sediments can act as indicator of the relationship between natural and anthropogenic variables (Salomons 1995; Calmano et al. 1996).

PAHs entering the aquatic environment, due to their hydrophobicity, rapidly become associated with sediments and suspended particles. Sediment retention capacity may be related to physico-chemical properties such as grain size and organic matter. Previous results have demonstrated that the sorption of hydrocarbons is related to the organic matter of the sediments.

According to Viguri et al. (2002) the results of the distribution of PAHs in surface sediments of the Santander Bay, Cantabric Sea, (Spain), the concentration of PAHs were strongly dependent on the organic matter, with sediments below 2% of organic matter always below 10 μg Σ16 PAHs/g dw. The left side of the Bay, where the main urban and industrial activities take place, shows sediment samples with very high concentrations (355 μg/g dw) of the sum of the 16 PAHs. An important change in the sediments characterisation can be observed according to the industrial activities along the rivers in the estuaries. From heavily polluted sediments in the Boo estuary, moderately polluted sediments in the Solia and San Salvador estuary to slightly/non-polluted sediments in the Cubas estuary. This result allows us to consider that important changes in the PAHs accumulation in sediments take place depending on the direct influence of anthropogenic activities, mainly combustion processes of traffic and industrial activities.

9. OCCUPATIONAL EXPOSURE

An increased risk of cancer has been reported in occupations with heavy exposure to traffic related air pollution, such as professional drivers. PAHs are present in gasoline and diesel engine exhaust, thus making, motor vehicle emissions in urban areas a major source of exposure to these compounds (Fromme et al. 1998). Personal PAH exposure due to urban traffic has been the subject of several investigations, be it in the general population (Minoia et al. 1997) or in occupationally exposed individuals, such as policemen (Merlo et al. 1997; Perico et al. 2001) and newsagents (Piccardo et al. 2003). Menichini (1992) reported that the evaluate the daily personal exposure to benzo(a)pyrene is one of the most important of taxi drivers. Twenty-four hour personal exposures (both occupational and environmental) to benzo(a)pyrene of 15 taxi drivers working in Genoa was determined by Piccardo et al. (2004). Daily personal B(a)P exposures of taxi drivers and controls in Genoa and mean daily airborne B(a)P concentrations (ng/m^3) measured by three fixed samplers in different monitoring periods shown in Table 13. Table 13 also reports the daily BaP concentrations measured by three fixed PM_{10} monitoring stations during the same periods of personal exposure sampling. The mean BaP exposures measured at these different times were 1.4, 1.23 and 1.22 ng/m^3, respectively, values were significantly greater than the levels found in controls (0.16 ± 0.2 ng/m^3). Mean daily personal B(a)P exposures of taxi drivers were not statistically different from the mean daily airborne B(a)P concentrations measured by fixed samplers during the same sampling periods. However, they concluded that mean daily B(a)P exposures measured in taxi drivers during May–June 1998, (n = 9; 1.19 ± 0.7 ng/m^3) were significantly higher (P – 0.01) than the daily BaP exposures measured during the same sampling months in another occupational category, namely non-smoker newsagents (n = 16; 0.65 ± 0.25 ng/m^3) (Piccardo et al. 2003). Newsagents work 12 h shifts a day inside open newsstands commonly placed in heavily trafficked areas. Therefore, their lower mean BaP exposure would seem to confirm the importance of pollution levels usually found inside vehicles. Indeed, in Berlin Fromme et al. (1998) sampled B(a)P in the interior of a car in the breathing zone of passengers for 9 h (from 07:00 to 16:00 h under different meteorologic conditions). The mean values obtained during the two measurement periods (*summer* and winter) inside the car were *1.0* and 3.2 ng/m^3 for benzo(a)pyrene, *10.2* and 28.7 ng/m^3 for total-measured-PAHs, and in the subway *0.7* and 4.0 ng/m^3 for benzo(a)pyrene, *30.2* and 67.5 ng/m^3 for total PAHs. A comparison between subway and car exposures shows significantly higher concentration of PAHs in the subway train, which can be explained by relatively high concentrations of fluoranthene and pyrene in the subway. So far a satisfactory explanation has not been found, but one source might be the wooden railway ties which were formerly preserved with tar based products. In wintertime in both transportation systems the concentrations of benzo(a)pyrene are three to four times higher than in summer corresponding to the changing of the ambient air concentrations.

Mielke et al. (2004) were determined PAH concentrations of the inner-city and suburban census tracts. Table 14 list the results of the medians of each local site by percentiles. The largest medians are found along busy streets and the ranking for the remaining sites are foundations > residential streets > open areas of New Orleans. Note that the same overall trend exists for both PAHs and metals. The most pronounced trend is for PAHs. Busy streets have over twice the quantities of PAHs as foundation soils, the second largest PAH containing soil. Vehicle exhaust is the major source PAHs in residential communities. Note that as traffic volumes change from busy streets to residential streets, the amount of PAHs also changes in the direction expected (Mielke et al. 2004).

Table 13. Daily personal BaP exposures of taxi drivers and controls in Genoa and mean daily airborne BaP concentrations (ng/m^3) measured by three fixed samplers in different monitoring periods (after Piccardo et al. 2004)

	No. samples	Mean ± SD	Geometric mean	Min	Max
Personal exposure of taxi drivers					
May–July 1998	15	1.04 ± 0.66	0.85	0.17	2.66
February 1999	7	1.23 ± 0.50	1.16	0.72	2.27
June 1999	7	1.22 ± 1.10	0.84	0.22	2.82
Personal exposure of controls					
April 1999	5	0.16 ± 0.12	0.11	0.03	0.28
Fixed sampling stations					
May–July 1998	13	1.05 ± 0.5	0.78	0.28	1.9
February 1999	5	1.52 ± 0.6	1.44	0.95	2.4
June 1999	4	1.16 ± 0.2	1.15	1.0	1.4
April 1999	3	1.17 ± 0.4	1.12	0.8	1.6

From the huge number of studies on the PAH exposure at workplaces a small selection is collated in Table 15. which presents concentration of benzo(a)pyrene only. It shows that there are enormous variations even at similar or identical workplaces, but also that significant progress has been made by modernisation of the plants to reduce the emission of PAH (Jacob and Seidel 2002). This holds in particular for coke plants for which the limit values have been set to 5 mg B(a)P/m^3 for the oven platform and to 2 mg B(a)P/m^3 for other workplaces in Germany (TRK-Wert 1989), Poland (Srogi 2007) and 0.15 mg B(a)P/m^3 in France for the production of coke electrodes (Jacob and Seidel 2002).

It should be noted that, industrial contamination of workers of PAH has been found to occur by disposition from air or following contact with contaminated surfaces including clothing and tools. The studies indicate that total PAH in skin oil of roofing workers and road-paving crews may reach levels as high as 1,400 ng/cm^2 (Jongeneelen et al. 1988). Of concern in such individuals, PAH are highly lipophillic compounds and thus are readily absorbed through the skin (Van Rooij et al. 1993).

10. 1-HYDROXYPYRENE—BIOLOGICAL MONITORING

The assessment of environmental exposure to PAH requires the measurement of levels present in each pathway of possible contact. The biological monitoring of PAH enables the best way to assess the internal dose of PAH.

Table 14. Median PAH concentrations stratified by residential collection sites for the combined data of the inner-city and suburban census tracts (Mielke et al. 2004)

PAH	Busy streets	Residential streets	Open areas	Foundation samples
A	10	8	2	8
B	6	0	5	9
C	1	4	4	5
D	1	0	3	3
E	133	29	21	77
F	38	8	7	10
G	266	107	82	105
H	237	97	67	78
I	145	41	28	51
J	111	42	32	40
K	319	112	94	105
L	95	34	24	72
M	255	141	144	163
N	322	202	205	233
O	208	126	123	128
P	237	125	122	137
PAH	2,469	1,061	975	1,188
N	8	18	6	6

Units are ng/g. Key for polycyclic aromatic hydrocarbons: A, naphthalene; B, aenaphthylene; C, acenaphthene; D, fluorene; E, phenanthrene; F, anthracene; G, fluroanthene; H, pyrene; I, benz(a) anthracene; J, chrysene; K, benzo(b)fluoranthene; L, benzo(k)fluoranthene; M, benzo(a)pyrene; N, indeno[1,3,3-cd]pyrene; O, dibenz (a,h)anthracene; P, benzo(g,h,i)perylene; total PAHs

Table 15. Benzo[*a*]pyrene concentration at various workplaces

Workplace	Country	Benzo(a)piren (μg/m³)	References
Coke plants	Poland		Braszczyńska et al. (1975, 1976)
Oven platforms—before modernisation		99 (A), 112.3 (B), 112,4 (C)	
After modernisation		25.8 1.3 2.3	
Coke plants	Poland		Brzeźnicki (1995)
Coke charging man		25,7; 3.3	Braszczyńska (1989)
Coke guide man		4,5	
Coke-quenching man		0,4	
Coke man		2,6; 1.7	
Coke door man		6,5; 25.6	
Coke plants	Poland		Luks-Betlej and Bodzek (1992)
Battery			
Coke charging car		2.72	
Coke plants	Poland		Smolik (2001)
Battery top		5.54	
Battery side		3.11	
Battery bottom		0.65	
Coke plants	Poland		Braszczyńska et al. (1976)
Battery			
Coke charging system		273.2	
Coke plants	Sweden		Levin (1995)
Battery top[a]		16	
Battery top[b]		4.0	
Coal conveyer[c]		2.6	
Charging car		12	
Oven doors[a]		22	
Oven doors[b]		1.9	
Oven platforms	Sweden	9.4–13.5	Lindstedt and Sollenberg (1982)
Driver of filling container		4.7–17	
Aluminum plants—various workplaces	Norway	11.3–854	Bjørseth et al. (1978)
Coke plants	Finland		Pyy et al. (1997)
Gas worker		1.0; 2.2[d]	
On the bridge of the larry car		5.0	

[a] 20 cigarettes/day

[b] Non-smoking

[c] 15 cigarettes/day

[d] In 1994

1-Hydroxypyrene in human urine has been examined as a potential biomarker for the assessment of human exposure to PAHs (Viau et al. 1995). In general, background urinary 1-hydroxypyrene levels are influenced by factors such as diet, industrial pollution, and cigarette smoking. Most previous studies show that the concentrations of urinary 1-hydroxypyrene are correlated with cigarette smoking in a non-occupational population (Van Rooij et al. 1994; Levin 1995). However, other studies show a statistically significant correlation between the levels of 1-hydroxypyrene in human urine and those of pyrene and benzo(a)pyrene in ambient air (Jongeneelen et al. 1986). Smoking has little influence on this correlation when persons smoke less than 20 cigarettes daily in an environment with a relatively high baseline concentration of air-borne PAHs. These observations support the use of 1-hydroxypyrene as a potential biomarker of PAH exposure.

According to Li et al. (2000) the average concentrations of urinary 1-hydroxypyrene were 0.04 μmol/mol creatinine in non-smokers, and 0.20 μmol/mol creatinine in light smokers, 0.46 μmol/mol creatinine in medium smokers, and 1.16 μmol/mol creatinine in heavy smokers. The urinary 1-hydroxypyrene levels of light, medium and heavy smokers were 5, 11, and 28 times higher than those of non-smokers, respectively. The apparent values (from non-detectable to 3.03 μmol/mol creatinine) of urinary 1-hydroxypyrene found in this study are higher and vary greater than those previously reported: 0.07 ± 0.76 μmol/mol creatinine (Van Rooij et al. 1994). The dose-response correlation is closer than those previously reported (Van Rooij et al. 1994; Boogaard and van Sittert 1995; Göen et al. 1995; Levin 1995). Despite many factors such as sampling, analysis and air quality, the difference may be attributed to the population which may differ in personal smoking characteristics, smoking histories and diets.

It is well known that cigarette smoke contains a variety of potentially toxic components including PAHs. Pyrene from smoking is presumably the main source of 1-hydroxypyrene detected in the cigarette smokers' urine, particularly when compared with non-cigarette smokers. The background urinary 1-hydroxypyrene can be derived from a number of sources such as diets, ambient air, and second hand smoke. The baseline values of urinary 1-hydroxypyrene vary greatly among population subgroups, occupation and air quality (Van Rooij et al. 1994; Levin 1995) Li et al. (2000) found low amounts of 1-hydroxypyrene in non-smokers' urine and a clear difference in the excretion of urinary 1-hydroxypyrene between non-smokers and smokers who smoked one, two, or three packs of cigarettes per day. This study confirms that 1-hydroxypyrene is a potential biomarker of PAH exposure from cigarette smoking.

Hansen et al. (2005) concluded that children living in urban residences are more exposed to PAH than children living in rural residences. Time spent outdoors increased the excretion of 1-hydroxypyrene among urban children but less so among rural children. Higher concentrations of ambient air pollution in urban areas may explain this finding. No influence of environmental tobacco smoke, cooking habits, and heating facilities was detected. The present study illustrates the feasibility of obtaining urine samples from a relatively large population of children, in connection with environmental sampling. During the week, the children excreted on average 0.07 µmol urinary 1-hydroxypyrene per mol creatinine. Children living in urban residences excreted 0.02 µmol more 1-hydroxypyrene than children living in rural residences. This was confirmed in the multiple regression analysis showing a 29% higher excretion among urban children than rural children. Moreover, the regression analysis showed that for each hour per day spent outside the children excreted 58% more 1-hydroxypyrene in urine (Hansen et al. 2005).

It should be mentioned that Jacob and Seidel (2002) reviewed the PAH metabolites in human urine as the method of choice to determine occupational and/or environmental exposure of an individual to PAH, in particular, when multiple routes of exposure have to be taken into account. Also, studies using 1-hydroxypyrene or phenanthrene metabolites including its phenols and dihydrodiols were summarized in described work.

11. CONCLUSION

Polycyclic aromatic hydrocarbons (PAHs) comprise the largest class of chemical compounds known to be cancer causing agents. Some, while not carcinogenic, may act as synergists. Some of these PAHs are manufactured for research or are used in medicines, dyes, plastics and pesticides such as naphthalene found in mothballs. PAHs can also be found in coal tar, bitumen, crude oil, creosote and roofing tar.

Thus, potential exposures to chemicals may be assessed by testing contaminated soil, air or water for the chemicals of interest and estimating the degree of intake of each of these media into the human body. In conclusion, the reader is referred to the book (Lichtfouse et al. 2005) for more details associated with review on the actual knowledge about PAHs.

REFERENCES

1. Aamot E, Steinnes E, Schmid R (1996) Polycyclic aromatic hydrocarbons in Norwegian forest soils: impact of long range atmospheric transport. Environ Pollut 92(3):275–280
2. Allen JO, Dookeran NM, Smith KA, Sarofim AF, Taghizadeh K, Lafleur AL (1996) Measurement of PAHs associated with size-segregated atmospheric aerosols in Massachusetts. Environ Sci Technol 30:1023–1031
3. Amellal S, Boivin A, Perrin Ganier C, Schiavon M (2006a) High sorption of phenanthrene in agricultural soils. Agron Sustain Dev 26:99–106
4. Amellal S, Boivin A, Perrin Ganier C, Schiavon M (2006b) Effect of ageing on mobility and sequestration of phenanthrene in an agricultural soil. Agron Sustain Dev 26:269–275
5. Anyakora C, Ogbeche A, Palmer P, Coker H (2005) Determination of polynuclear aromatic hydrocarbons in marine samples of Siokolo Fishing Settlement. J Chromatogr Part A 1073:323–330
6. Ardito G, Badino G, Baker AJ, Orsi M, Ostacoli G, Parodi A (1999) PAH and heavy-metal biomonitoring along the A32 motorway (North West Italy). In: Proceedings of Fifth International Conference on the Biogeochemistry of Trace Elements. Vienna, Austria
7. ATSDR (1995) Toxicological profile for polycyclic aromatic hydrocarbons, US Department of Health and Human Services, Public Health Service, Agency for Toxic Substances and Disease Registry, Atlanta, GA
8. Bae SY, Yi SM, Kim YP (2002) Temporal and spatial variations of the particles size distribution of PAHs and their dry deposition fluxes in Korea. Atmos Environ 36:5491–5500
9. Baek SO, Goldstone ME, Kirk PWW, Lester JN, Pery R (1991) Concentrations of particulate and gaseous polycyclic aromatic hydrocarbons in London air following a reduction in the Lead content of petrol in the United Kingdom. Sci Total Environ 11:169–199
10. Baumann PC, Harshbarger JC (1995) Decline in liver neoplasms in wild brown bullhead catfish after coking plant closes and environmental PAHs plummet. Environ Health Perspect 103:168–170
11. Bakker MI, Casado B, Koerselman JW, Tolls J, Kollöffel C (2000) Polycyclic aromatic hydrocarbons in soil and plant samples from the vicinity of an oil refinery. Sci Total Environ 263:91–100
12. Binet S, Pfohl-Leszkowicz A, Brandt H, Lafontaine M, Castegnaro M (2002) Bitumen fumes: review of work on the potential risk to workers and the present knowledge on its origin. Sci Total Environ Stud 300:37–49
13. Bjørseth A, Bjørseth O, Fjeldstad PE (1978) Characterization of PAH in Work Atmospheres, Royal Norwegian Council for Scientific and Industrial Research, Oslo, Norway, Report no. 6
14. Bodzek D, Luks-Betlej K, Warzecha L (1993) Determination of particle-associated polycyclic aromatic hydrocarbons in ambient air samples from the Upper Silesia region of Poland. Atmos Environ 27A:759–764
15. Boffetta P, Jourenkova N, Gustavsson P (1997) Cancer risk from occupational and environmental exposure to polycyclic aromatic hydrocarbons. Cancer Causes Control 8:444–472
16. Boogaard PJ, van Sittert NJ (1995) Urinary 1-hydroxypyrene as biomarker of exposure to polycyclic aromatic hydrocarbons in workers in petrochemical industries: baseline values and dermal uptake. Sci Total Environ 163:203–209
17. Bouloubassi I, Saliot A (1991) Composition and sources of dissolved and particulate PAH from the Rhone delta (NW Mediterranean). Mar Pollut Bull 22:588–594
18. Braszczyńska Z (1989) Report CPBR 11.5.62. IMP, Sosnowiec (in Polish)
19. Braszczyńska Z, Linscheid D, Szyja W (1975) Monitoring of particle-associated polycyclic aromatic hydrocarbons at a coking plant. Med Pr 26:4–10
20. Braszczyńska Z, Linscheid D, Osińska R (1976) Estimation of a new technology in workplace. Med Pr 29:5–11

21. Broman D, Naef C, Zeburch Y (1991) Long-term high-and-low-volume air sampling of polychlorinated dibenzo-p-dioxins and dibenzofurans and polycyclic aromatic hydrocarbons along a transect from urban to remote areas on the Swedish Baltic Coast. Environ Sci Technol 25:1841–1850
22. Brorström-Lundén E, Löfgren C (1998) Atmospheric fluxes of persistent semivolatile organic pollutants to a forest ecological system at the Swedish west coast and accumulation in spruce needles. Environ Pollut 102:139–149
23. Brorström-Lundén, E, Skärby L (1984) Plants as monitoring samplers of airborne PAH. In: Versino B, Angeletti G (eds) Physico-chemical behaviour of atmospheric pollutants. Riedel, Dordrecht
24. Brunekreef B, Holgate ST (2002) Air pollution and health. Lancet 360(9341):1233–1242
25. Bryselbout C, Henner P, Carsignol J, Lichtfouse É (2000) Polycyclic aromatic hydrocarbons in highway plants and soils. Evidence for a local distillation effect. Analusis 28:32–35
26. Brzeźnicki S (1995) Coke plant. In: Wodniak W, Gromca P, Więcka E (eds) Monitoring of polycyclic aromatic hydrocarbons and asbestos in the industry. IMP Łódź
27. Bucheli TD, Blum F, Desaules A, Gustafsson Ö (2004) Polycyclic aromatic hydrocarbons, black carbon, and molecular markers in soils of Switzerland. Chemosphere 56:1061–1076
28. Burkhard LP, Lukasewycz MT (2000) Some bioaccumulation factors and biota-sediment accumulation factors for polycyclic aromatic hydrocarbons in lake trout. Environ Toxicol Chem 19:1427–1429
29. Butler JD, Butterworth V, Kellow SC, Robinson HG (1984) Some observations on the polycyclic aromatic hydrocarbon (PAH) content of surface soils in urban areas. Sci Total Environ 33:75–85
30. Butler JP, Post GB, Lioy PJ, Waldman JM, Greenberg A (1993) Assessment of carcinogenic risk from personal exposure to benzo(a)pyrene in the total human environmental exposure study (THEES). J Air Waste Manage Assoc 43:970–977
31. Bylina BG, Rakwic B, Pastuszka JS (2005) Assessment of exposure to traffic-related aerosol and to particle-associated PAHs in Gliwice, Poland. Polish J Environ Stud 14:117–123
32. Calmano W, Ahlf W, F€orstner U (1996) Sediment quality assessment: chemical and biological approaches. In: Calmano W, Förstner U (eds) Sediments and toxic substances. Springer, Berlin, pp 1–35
33. Capuano F, Cavalchi B, Martinelli G, Pecchini G, Renna E, Scaroni I, Bertacchi M, Bigliardi G (2005) Environmental prospection for PCDD/PCDF, PAH, PCB and heavy metals around the incinerator power plant of Reggio Emilia town (Northern Italy) and surrounding main roads. Chemosphere 58:1563–1569
34. Caricchia AM, Chiavarini S, Pezza M (1999) Polycyclic aromatic hydrocarbons in the urban atmospheric particulate matter in the city of Naples (Italy). Atmos Environ 33:3731–3738
35. Cavret S, Feidt C (2005) Intestinal metabolism of PAH: in vitro demonstration and study of its impact on PAH transfer through the intestinal epithelium. Environ Res 98:22–32
36. Chen J, Wang X, Tao S, Liu S, Zhang Z, Shen W, Qin B, Sun R, Zhang W (2003) Vertical distribution of polycyclic aromatic hydrocarbons in wastewater irrigated soil in Tianjin area. Urban Environ Urban Ecol 16(6):272–274
37. Chen L, Ran Y, Xing B, Mai B, He J, Wei X, Fu J, Sheng G (2005) Contents and sources of polycyclic aromatic hydrocarbons and organochlorine pesticides in vegetable soils of Guangzhou, China. Chemosphere 60:879–890
38. Chetwittayachan T, Shimazaki D, Yamamoto K (2002) A comparison of temporal variation of particle-bound polycyclic aromatic hydrocarbons (pPAHs) concentration in different urban environments: Tokyo, Japan, and Bangkok, Thailand. Atmos Environ 36:2027–2037
39. Cheung KC, Leung HM, Kong KY, Wong MH (2007) Residual levels of DDTs and PAHs in freshwater and marine fish from Hong Kong markets and their health risk assessment. Chemosphere 66:460–468

40. Chiang T-A, Wu P-F, Ko Y-Ch (1999) Identification of carcinogens in cooking oil fumes. Environ Res Sec A 81:18–22
41. Cormier SM, Lin EL, Fulk F, Subramanian B (2000) Estimation of exposure criteria values for biliary polycyclic aromatic hydrocarbons metabolite concentration in white suckers (Catostomus commersoni). Environ Toxicol Chem 19:1120–1126
42. Council Directive 80/778/EEC of 15 July 1980 Relating to the Quality of Water Intended for Human Consumption Official. J L 229, 30 August 1980, pp. 0011–0029 (http://www.europa.eu.int.)
43. Crépineau C, Rychen G, Feidt C, Le Roux Y, Lichtfouse É, Laurent F (2003) Contamination of pastures by polycyclic aromatic hydrocarbons (PAHs) in the vicinity of a highway. J Agric Food Chem 51:4841–4845
44. Dachs J, Bayona JM, Raoux C, Albaigẽs J (1997) Spatial, vertical distribution and budget of polycyclic aromatic hydrocarbons in the western Mediterranean seawater. Environ Sci Technol 31:682–688
45. Davis LC, Castro-Diaz S, Zhang Q, Erickson LE (2002) Benefits of vegetation for soils with organic contaminants. Crit Rev Plant Sci 21:457–491
46. De Nicola F, Maisto G, Prati MV, Alfani A (2005) Temporal variations in PAH concentrations in Quercus ilex L. (holm oak) leaves in an urban area. Chemosphere 61:432–440
47. Dyke PH, Foan C, Fiedler H (2003) PCB and PAH releases from power stations and waste incineration processes in the UK. Chemosphere 50:469–480
48. Edwards NTJ (1983) Polycyclic aromatic hydrocarbons (PAHs) in the terrestrial environment—a review. J Environ Qual 12:427–441
49. Elskus AA, Stegeman JJ (1989) Induced cytochrome P-450 in Fundulus heteroclitus associated with environmental contamination by polychlorinated biphenyls and polynuclear aromatic hydrocarbons. Mar Environ Res 27:31–45
50. Evans K, Gill R, Robotham P (1990) The PAH and organic content of sediment particle size fractions. Water Air Soil Pollut 51:13–31
51. Fang G-C, Wu YS, Chen MH, Ho TT, Huang SH, Rau JY (2004a) Polycyclic aromatic hydrocarbons study in Taichung, Taiwan, during 2002–2003. Atmos Environ. 38:3385–3391
52. Fang G-C, Chang KF, Lu C, Bai H (2004b) Estimation of PAHs dry deposition and BaP toxic equivalency factors (TEFs) study at Urban, Industry Park and rural sampling sites in central Taiwan, Taichung. Chemosphere 55:787–96
53. Fang G-C, Wu YS, Fu PC, Yang IL, Chen MH (2004c) Polycyclic aromatic hydrocarbons in the ambient air of suburban and industrial regions of central Taiwan. Chemosphere 54:443–52
54. Fang G-C, Wu YS, Fu PC, Chang Ch-N, Chen M-H (2005) Characteristic study of polycyclic aromatic hydrocarbons for fine and coarse particulates at Pastureland near Industrial Park sampling site of central Taiwan. Chemosphere 60:427–433
55. Fang G-C, Wu YS, Fu PC, Chen M-H, Chang, Ch-N, Ho T-T (2006) Characteristic of polycyclic aromatic hydrocarbon concentrations and source identification for fine and coarse particulates at Taichung Harbor near Taiwan Strait during 2004–2005. Sci Total Environ 366:729–738
56. Fiedler H, Cheung CK, Wong MH (2002) PCDD/PCDF, chlorinated pesticides and PAH in Chinese teas. Chemosphere 46:1429–1433
57. Fismes J, Perrin-Ganier C, Empereur-Bissonnet P, Morel JL (2002) Soil-to-root transfer and translocation of polycyclic aromatic hydrocarbons by vegetables grown on industrial contaminated soils. J Environ Qual 31:1649–1656
58. Franzaring J (1997) Temperature and concentration effects in biomonitoring of organic air pollutants. Environ Monit Assess 46:209–220
59. Fromme H, Oddoy A, Piloty M, Krause M, Lahrz T (1998) Polycyclic aromatic hydrocarbons (PAH) and diesel engine emission (elemental carbon) inside a car and subway train. Sci Total Environ 217:165–173
60. Fu G, Kan AT, Tomson M (1994) Adsorption and desorption hysteresis of PAHs in surface sediment. Environ Toxicol Chem 13:1559–1567

61. Fuentes-Rios D, Orrego R, Rudolph A, Mendoza G, Gavilán JF, Barra R (2005) EROD activity and biliary fluorescence in Schroederichthys chilensis (Guichenot 1848): biomarkers of PAH exposure in coastal environments of the South Pacific Ocean. Chemosphere 61:192–199
62. Gill RA, Jackson RB (2000) Global patterns of root turnover for terrestrial ecosystems. New Phytol 147:13–31
63. Göen Th, Göundel J, Schaller K-H, Angerer J (1995) The elimination f 1-hydroxypyrene in the urine of the general population and workers with different occupational exposures to PAH. Sci Total Environ 163:195–201
64. Grayston SJ, Vaughan D, Jones D (1996) Rhizosphere carbon flow in trees, in comparison with annual plants: the importance of root exudation and its impact on microbial activity and nutrient availability. Appl Soil Ecol 5:29–56
65. Grova N, Laurent C, Feidt C, Rychen G, Laurent F (2000) Gas chromatography-mass spectrometry of polycyclic aromatic hydrocarbons in grass and milk from urban and rural farms. Eur J Mass Spectrom 6:457–460
66. Grova N, Rychen G, Monteau F, LeBizec B, Feidit C (2006) Effect of oral exposure to polycyclic aromatic hydrocarbons on goat's milk contamination. Agron Sustain Dev 26:195–199
67. Guerrin MR (1993) Environment tobacco smoke exposure assessment. In: Tobacco smoke. Springer, New York
68. Guo H, Lee SC, Ho KF, Wang XM, Zou SC (2003) Particle-associated polycyclic aromatic hydrocarbons in urban air of Hong Kong. Atmos Environ 37:5307–5317
69. Halsall CJ, Coleman PJ, Davis BJ et al (1994) Polycyclic aromatic hydrocarbons in UK urban air. Environ Sci Technol 28:2380–2389
70. Hansen ÅM, Raaschou-Nielsen O, Knudsen LE (2005) Urinary 1-hydroxypyrene in children living in city and rural residences in Denmark. Sci Total Environ 347:98–105
71. Henner P, Schiavon M, Druelle V, Lichtfouse E (1999) Phytotoxicity of ancient gaswork soils. Effect of polycyclic aromatic hydrocarbons (PAHs) on plant germination. Org Geochem 30:963–969
72. Hoek G, Brunekreef B, Goldbohm S, Fischer, van den Brandt P, Piet A (2002) Association between mortality and indicators of traffic related air pollution in the Netherlands: a cohort study. Lancet 360(9341):1203–1209
73. Howsam M, Jones KC, Ineson P (2000) PAHs associated with the leaves of three deciduous tree species. I: concentrations and profiles. Environ Pollut 108:413–424
74. Howsam M, Jones KC, Ineson P (2001) PAHs associated with the leaves of three deciduous tree species. II: uptake during a growing season. Chemosphere 44:155–164
75. Hyötyläinen T, Oikari A (2004) Bioaccumulation of PAHs from creosote-contaminated sediment in a laboratory-exposed freshwater oligochaete, Lumbriculus variegates. Chemosphere 57:159–164
76. IARC (1992) IARC Working Group on the Evaluation on Carcinogenic Risk to Humans: Occupational Exposures to Mists and Vapors from Strong Inorganic Acids and Other Industrial Chemicals, vol 54. IARC, Lyon
77. IARC (International Agency for Research on Cancer) (1983) IARC monographs on the evaluation of the carcinogenic risk of chemicals to human. Polynuclear aromatic compounds, Part I, Chemical, environmental, and experimental data. World Health Organization, Geneva, Switzerland
78. Jacob J, Seidel A (2002) Biomonitoring of polycyclic aromatic hydrocarbons in human urine. J Chromatogr Part B 778:31–48
79. JEA (Japanese Environmental Agency) (2000) Report on auto-related environmental taxes; utilization of auto-related taxes in environmental policy. http://www.eic.or.jp/eanet/en/index.html. JEA, Tokyo
80. Jongeneelen FJ (1994) Biological monitoring of environmental exposure to polycyclic aromatic hydrocarbons; 1-hydroxypyrene in urine of people. Toxicol Lett 72:205–212
81. Jongeneelen FJ, Bos RP, Anzion RBM, Theuws JLG, Henderson PT (1986) Biological monitoring of polycyclic aromatic hydrocarbons metabolites in urine. Scand J Work Environ Health 12:137–143

82. Jongeneelen FJ, Sheepers PTJ, Groenendijk A, Van Aerts LAGJM, Anzion RBM, Ros RP, Veenstra SJ (1988) Airborne concentrations, skin contamination, urinary metabolite excretion of polycyclic aromatic hydrocarbons among paving workers exposed to coal tar derived road tars. Am Ind Hyg Assoc 49:600–607
83. Jouraeva VA, Johnson DL, Hassett JP, Nowak DJ (2002) Differences in accumulation of PAHs and metals on the leaves of Tilia xeuchlora and Pyrus calleryana. Environ Pollut 120:331–338
84. Kanoh T, Fukuda M, Onozuka H, Kinouchi T, Ohnishi Y (1993) Urinary 1-hydroxypyrene as a marker of exposure to polycyclic aromatic hydrocarbons in environment. Environ Res 62:230–245
85. King AJ, Readman JW, Zhou JL (2004) Determination of polycyclic aromatic hydrocarbons in water by solid-phase microextraction–gas chromatography–mass spectrometry. Anal Chem Acta 523:259–267
86. Kipopoulou AM, Manoli E, Samara C (1999) Bioconcentration of polycyclic aromatic hydrocarbons in vegetables grown in an industrial area. Environ Pollut 106:369–380
87. Kishikawa N, Wada M, Kuroda N, Akiyama S, Nakashima N (2003) Determination of polycyclic aromatic hydrocarbons in milk sample by high-performance liquid chromatography with fluorescence detection. J Chromatogr Part B 789:257–264
88. Kiss G, Varga-Punchony Z, Rohrbacher G, Hlavay J (1998) Distribution of polycyclic aromatic hydrocarbons on atmospheric aerosol particles of different sizes. Atmos Res 46:253–261
89. Knox RC, Sabatini DA, Canter LW (1993) Subsurface transport and fate processes. Lewis Publishing, Boca Raton
90. Kong KY, Cheung KC, Wong CRC, Wong MH (2005) Residues of DDTs, PAHs and some heavy metals in fish (Tilapia) collected from Hong Kong and mainland China. J Environ Sci Heal 40:1–11
91. Kulkarni P, Venkataraman C (2000) Atmospheric polycyclic aromatic hydrocarbons in Mumbai, India. Atmos Environ 34:2785–2790
92. Kuo C-Y, Cheng Y-W, Chen Y-W, Lee H (1998) Correlation between the amounts of polycyclic aromatic hydrocarbons and mutagenicity of airborne particulate samples from Taichung City, Taiwan. Environ Res Sec A 78:43–49
93. Larsson B, Sahlberg G (1982) Polycyclic aromatic hydrocarbons in lettuce. Influence of a highway and an aluminium smelter. In: Cooke M, Dennis AJ, Fisher GL (eds) Polynuclear aromatic hydrocarbons: physical and biological chemistry. Springer, New York, pp 417–426
94. Laurent C, Feidt C, Lichtfouse É, Grova N, Laurent F, Rychen G (2001) Milk-blood transfer of 14C-tagged polycyclic aromatic hydrocarbons (PAHs) in pigs. J Agric Food Chem 49:2493–2496
95. Laurent C, Feidt C, Grova N, Mpassi D, Lichtfouse É, Laurent F, Rychen G (2002) Portal absorption of 14C after ingestion of spiked milk with 14C-phenanthrene, 14C-benzo[a]pyrene or 14C-TCDD in growing pigs. Chemosphere 48:843–848
96. Lehndorff E, Schwark L (2004) Biomonitoring of air quality in the Cologne Conurbation using pine needles as a passive sampler—Part II: polycyclic aromatic hydrocarbons (PAH). Atmos Environ 38:3793–3808
97. Levin JO (1995) First international workshop on hydroxypyrene as a biomarker for PAH exposure in man—summary and conclusions. Sci Total Environ 163:165–168
98. Li CK, Kamens RM (1993) The use of polycyclic aromatic hydrocarbons as source signatures in receptor modeling. Atmos Environ 27:523–532
99. Li CS, Ro YS (2000) Indoor characteristics of polycyclic aromatic hydrocarbons in the urban atmosphere of Taipei. Atmos Environ 34:611–620
100. Li H, Krieger RI, Li QX (2000) Improved HPLC method for analysis of 1-hydroxypyrene in human urine specimens of cigarette smokers. Sci Total Environ 257:147–153
101. Lichtfouse É, Budzinski H, Garrigues P, Eglinton TI (1997) Ancient polycyclic aromatic hydrocarbons in modern soils: 13C, 14C and biomarker evidence. Org Geochem 26:353–359

102. Lichtfouse É, Apitz S, Nanny M (1999) The biogeochemistry of polycyclic aromatic hydrocarbons. Org Geochem 30:873–969
103. Lichtfouse É, Schwarzbauer J, Robert D (eds) (2005) Environmental chemistry. Green chemistry and pollutants in ecosystems, 36–41 chaps. Springer, Berlin
104. Lin TC, Chang FH, Hsieh JH, Chao HR, Chao MR (2002) Characteristics of polycyclic aromatic hydrocarbons and total suspended particulate in indoor and outdoor atmosphere of a Taiwanese temple. J Hazard Mater A 95:1–12
105. Lindstedt G, Sollenberg J (1982) Polycyclic aromatic hydrocarbons in the occupational environment. Scand Work Environ Health 8:1–19
106. Liu YJ, Zhu LZ, Shen XY (2001) Polycyclic aromatic hydrocarbons (PAHs) in indoor and outdoor air of Hangzhou, China. Environ Sci Technol 35:840–844
107. Liu YN, Tao S, Dou H, Zhang TW, Zhang XL, Dawson R (2007) Exposure of traffic police to Polycyclic aromatic hydrocarbons in Beijing, China. Chemosphere 66:1922–1928
108. Lodovici M, Dolara P, Taiti S et al (1994) Polynuclear aromatic hydrocarbons in the leaves of the evergreen tree (Laurus nobilis). Sci Total Environ 153:61–68
109. Lodovici M, Dolara P, Casalini C, Ciappellano S, Testolin G (1995) Polycyclic aromatic hydrocarbon contamination in the Italian diet. Food Addit Contam 12:703–712
110. Lodovici M, Akpan V, Casalini C, Zappa C, Dolara P (1998) Polycyclic aromatic hydrocarbon in Laurus nobolis leaves as a measure of air pollution in urban and rural sites of Tuscany. Chemosphere 6:1703–1712
111. Luks-Betlej K, Bodzek D (1992) Monitoring of exposure to polycyclic aromatic hydrocarbons at a coking plant. Med Pr 43:515–521
112. Ma LL, Chu SG, Wang XT, Cheng HX, Liu XF, Xu XB (2005) Polycyclic aromatic hydrocarbons in the surface soils from outskirts of Beijing, China. Chemosphere 58:1355–1363
113. Magee BR, Lion LW, Lemley AT (1991) Transport of dissolved organic macromolecules and their effect on the transport of phenanthrene in porous media. Environ Sci Technol 25:323–331
114. Maliszewska-Kordybach B (1996) Polycyclic aromatic hydrocarbons in agricultural soils in Poland: preliminary proposals for criteria to evaluate the level of soil contamination. Appl Geochem 11(1–2):121–127
115. Manoli E, Samara C (1999) Polycyclic aromatic hydrocarbons in natural waters: sources, occurrence and analysis. Tredns Anal Chem 18:717–428
116. Manoli E, Voutsa D, Samara C (2002) Chemical characterization and source identification/apportionment of fine and coarse air particles in Thessaloniki, Greece. Atmos Environ 36:949–961
117. Masih A, Taneja A (2006) Polycyclic aromatic hydrocarbons (PAHs) concentrations and related carcinogenic potencies in soil at a semi-arid region of India. Chemosphere 65:449–456
118. McDonald SJ, Kennicutt MC, Liu H, Safe SH (1995) Assessing aromatic hydrocarbon exposure in Antarctic fish captured near Palmer and McMurdo stations Antarctica. Arch Environ Contam Toxicol 29:232–240
119. Meador JP, Stein JE, Reichert WL, Varanasi U (1995) Bioaccumulation of polycyclic aromatic hydrocarbons by marine organisms. Rev Environ Contam Toxicol 143:79–165
120. Meharg AA, Wright J, Dyke H, Osborn D (1998) Polycyclic aromatic hydrocarbon (PAH) dispersion and deposition to vegetation and soil following a large scale chemical fire. Environ Pollut 99:29–36
121. Menichini E (1992) Urban air pollution by polycyclic aromatic hydrocarbons: levels and source of variability. Sci Total Environ 116:109–135
122. Menzie CA, Potocki BB, Santodonato J (1992) Exposure to carcinogenic PAHs in the environment. Environ Sci Technol 26:1278–1284
123. Merlo F, Bolognesi C, Peluso M, Valerio F, Abbondandolo A, Puntoni R (1997) Airborne levels of polycyclic aromatic hydrocarbons: 32P-Postlabeling DNA adducts and micronuclei in white blood cells from traffic police workers and urban residents. J Environ Pathol Toxicol Oncol 16:157–162

124. Meudec A, Dussauze J, Deslandes E, Poupart N (2006) Evidence for bioaccumulation of PAHs within internal shoot tissues by a halophytic plant artificially exposed to petroleum-polluted sediments. Chemosphere 65:474–481
125. Mielke HW, Wang G, Gonzales CR, Le B, Quach VN, Mielke PW (2001) PAH and metal mixtures in New Orleans soils and sediments. Sci Tot Environ 281:217–227
126. Mielke HW, Wang G, Gonzales, ChR, Powell ET, Le B, Quach VN (2004) PAHs and metals in the soils of inner-city and suburban New Orleans, Louisiana, USA. Environ Toxicol Pharm 18:243–247
127. Miguel AH, Kirchestetter TW, Harely RA, Hering AV (1998) On-road emissions of particulate polycyclic aromatic hydrocarbons and black carbon from gasoline and diesel vehicles. Environ Sci Technol 32:450–455
128. Minoia C, Magnaghi S, Micoli G, Fiorentino ML, Turci R, Angeleri S, Berri A (1997) Determination of environmental reference concentration of six PAHs in urban areas (Pavia, Italy). Sci Total Environ 198:33–41
129. Moret S, Conte LS (2000) Polycyclic aromatic hydrocarbons in edible fats and oils: occurrence and analytical methods. J Chromatogr 882:245–253
130. Motelay-Massei A, Ollivion D, Garban B, Teil MJ, Blanchard M, Chevreuil M (2004) Distribution and spatial trends of PAHs and PCBs in soils in the Seine River basin, France. Chemosphere 55:555–565
131. Mueller KE, Shann JR (2006) PAH dissipation in spiked soil: Impacts of bioavailability, microbial activity, and trees. Chemosphere 64:1006–1014
132. Müller JF, Hawker DW, Connel DW (1998) Polycyclic aromatic hydrocarbons in the atmospheric environment of Brisbane, Australia. Chemosphere 37:1369–1383
133. Müller JF, Hawker DW, McLachlan MS, Connel DW (2001) PAHs, PCDD/Fs, PCBs and HCB levels from Brisbane, Australia. Chemosphere 43:507–515
134. Nadal M, Schuhmacher M, Domingo JL (2007) Levels of metals, PCBs, PCNs and PAHs in soils of a highly industrialized chemical/petrochemical area: temporal trend. Chemosphere 66:267–276
135. Nakajima D, Yoshida Y, Suzuki S (1995) Seasonal changes in the concentration of polycyclic aromatic hydrocarbons in Azalea leaves and relationship to atmospheric concentrations. Chemosphere 30:409–418
136. Nam JJ, Song BH, Eom KC, Lee SH, Smith A (2003) Distribution of polycyclic aromatic hydrocarbons in agricultural soils in South Korea. Chemosphere 50:1281–1289
137. Niederer M, Maschka-Selig A, Hohl C (1995) Monitoring polycyclic aromatic hydrocarbons (PAHs) and heavy metals in urban soil, compost and vegetation. Environ Sci Pollut Res 2:83–89
138. Niu J, Chen J, Martens D, Quan X, Yang F, Kettrup A, Schramm K-W (2003) Photolysis of polycyclic aromatic hydrocarbons adsorbed on spruce [Picea abies (L.) Karst] needles under sunlight irradiation. Environ Pollut 123:39–45
139. Omar NYMJ, Abas MRB, Ketuly KA, Tahir NM (2002) Concentrations of PAHs in atmospheric particles (PM-10) and roadside soil particles collected in Kuala Lumpur, Malaysia. Atmos Environ 36:247–254
140. Pandit GG, Srivastava PK, Mohan Rao AM (2001) Monitoring of indoor volatile organic compounds and polycyclic aromatic hydrocarbons arising from kerosene cooking fuel. Chemosphere 279:159–165
141. Panshin SY, Hites RA (1994) Atmospheric concentrations of polychlorinated biphenyls in Bloomington, Indiana. Environ Sci Technol 28:2008–2013
142. Papageorgopoulou A, Manoli E, Touloumi E, Samara C (1999) Polycyclic aromatic hydrocarbons in the ambient air of Greek towns in relation to other atmospheric pollutants. Chemosphere 39:2183–2199
143. Park SS, Kim YJ, Kang CH (2002) Atmospheric polycyclic aromatic hydrocarbons in Seoul, Korea. Atmos Environ 36:2917–2924
144. Perico A, Gottardi M, Boddi V, Bavazzano P, Lanxciotti E (2001) Assessment of exposure to Polycyclic Aromatic Hydrocarbons in police in Florence, Italy, throught personal air sampling and biological monitoring of the urinary metabolite hydroxypyrene. Arch Environ Health 56:512–513

145. Perugini M, Visciano P, Giammarino A, Manera M, Di Nardo W, Amorena M (2007) Polycyclic aromatic hydrocarbons in marine organisms from the Adriatic Sea, Italy. Chemosphere 66:1904–1910
146. Phillips DH (1999) Polycyclic aromatic hydrocarbons in the diet. Mutat Res Genet Toxicol Environ Mutagen 443:139–147
147. Piccardo MT, Stella A, Redaelli A, Minoia C, Valerio F (2003) Newsagents personal exposures to benzo(a)pyrene in Genoa, Italy. Atmos Environ 37:603–613
148. Piccardo MT, Stella A, Redaelli A, Balducci D, Coradeghini R, Minoia C, Valerio F (2004) Personal daily exposures to benzo(a)pyrene of taxi drivers in Genoa, Italy. Sci Total Environ 330:39–45
149. Pyy L, Mäkelä M, Hakal E, Kakko K, Lapinlampi T, Lisko A, Yrjänheikki E, Vähäkangas K (1997) Ambient and biological monitoring of exposure to polycyclic aromatic hydrocarbons at a coking plant. Sci Total Environ 199:151–158
150. Quantin C, Joner EJ, Portal JM, Berthelin J (2005) PAH dissipation in a contaminated river sediment under oxic and anoxic conditions. Environ Pollut 134:315–322
151. Raiyani CV, Shah JA (1993) Levels of PAHs in ambient environment of Ahmedabad city. Indian J Environ Protect 13:206–215
152. Rauret G, Llauradó M, Tent J, Rigol A, Alegre LH, Utrillas MJ (1994) Deposition on holm oak leaf surface of accidentally released radionuclides. Sci Total Environ 157:7–16
153. Ravindra, Mittal AK, Grieken RV (2001) Health risk suspended particulate matter with special reference to PAHs: a review. Rev Environ Health 16:169–189
154. Reeves WR, McDonald TJ, Cizmas L, Donnelly KC (2004) Partitioning and desorption behavior of polycyclic aromatic hydrocarbons from disparate sources. Sci Total Environ 332:183–192
155. Rehwagen M, Müller A, Massolo L, Herbath O, Ronco A (2005) Polycyclic aromatic hydrocarbons associated with particles in ambient air from urban and industrial areas. Sci Total Environ 348:199–210
156. Reynolds WJ, Feist SW, Jones GJ, Lyons BP, Sheahan DA, Stentiford GD (2003) Comparison of biomarker and pathological responses in flounder (Platichthys flesus L.) induced by ingested polycyclic aromatic hydrocarbon (PAH) contamination. Chemosphere 52:1135–1145
157. Ringuette S, Germain A, Gonthier C, Perron F (1993) Presence of PAHs in the Canadian Environment-An Overview. (Priority substances list supporting document no. 2) Canadian Environmental Protection Act. Environment Canada, Québec Region
158. Rockens E, Dumollin J, Matheeussen C (2000) PM10 dust and chemical characterisation of aerosols in Flanders, Belgium. Adv Air Pollut 8:699–707
159. Rogge WF, Hildemann LM, Mazurck MA (1993) Sources of the fine organic aerosol 3. Road dust, tire debris, and organometallic brake lining dust: roads as sources and sinks. Environ Sci Technol 27:1892–1904
160. Sakai R, Siegmann HC, Sato H, Voorhees AS (2002) Particulate matter and particle-attached polycyclic aromatic hydrocarbons in the indoor and outdoor air of Tokyo measured with personal monitors. Environ Res Sec A 89:66–71
161. Salomons W (1995) Long-term strategies for handling contaminated sites and large-scale areas. In: Salomons W, Stigliani W (eds) Biogeodynamics of pollutants in soils and sediments. Springer, Berlin, pp 1–30
162. Santasiero A, Settimo G, Cappiello G, Viviano G, Dell'Andrea E, Gentilini L (2005) Urban crematoria pollution related to the management of the deceased. Microchem J 79:307–317
163. Schreiber L, Schönherr J (1992) Uptake of organic chemicals in conifer needles: surface adsorption and permeability of cuticles. Environ Sci Technol 26:153–159
164. Seigmann K, Scherrer L, Siegmann HC (1999) Physical and chemical properties of airborne nanoscales and how to measure the impact on human health. J Mol Struct (Teochem) 458:191–201
165. Sharma H, Jain VK, Khan ZH (2007) Characterization and source identification of polycyclic aromatic hydrocarbons (PAHs) in the urban environment of Delhi. Chemosphere 66:302–310

166. Sheu HL, Lee WJ, Lin SJ, Fang G-C, Chang HC, You WC (1997) Particle bound PAH content in ambient air. Environ Pollut 96:369–82
167. Simonich SL, Hites RA (1994) Vegetation-atmosphere partitioning of polycyclic aromatic hydrocarbons. Environ Sci Technol 28:939–943
168. Simonich SL, Hites RA (1995) Organic pollutant accumulation in vegetation. Environ Sci Technol 29:2905–2914
169. Simpson SL, Burston VL, Jolley DF, Chau K (2006) Application of surrogate methods for assessing the bioavailability of PAHs in sediments to a sediment ingesting bivalve. Chemosphere 65:2401–2410
170. Smith DJT, Harrison RM (1996) Concentrations, trends and vehicle source profile of polynuclear aromatic hydrocarbon concentrations in the UK atmosphere. Atmos Environ 30:2513–2525
171. Smolik E (2001) Exposure to PAHs at a coking plant. Med Pr 3:77–81
172. Song YF, Jing X, Fleischmanna S, Wilke BM (2002) Comparative study of extraction methods for the determination of PAHs from soils and sediments. Chemosphere 48:993–1001
173. Song YF, Zhou QX, Gong P, Sun TH (2005) Ecotoxicity of soils contaminated with industrial and domestic wastewater in western Shenyang China. Sci China Ser C Life Sci 48 (Suppl):48–56
174. Song YF, Wilke B-M, Song YX, Gong P, Zhou QX, Yang GF (2006) Polycyclic aromatic hydrocarbons (PAHs), polychlorinated biphenyls (PCBs) and heavy metals (HMs) as well as their genotoxicity in soil after long-term wastewater irrigation. Chemosphere 65:1859–1868
175. Spitzer T, Kuwatsuka S (1993) Residue levels of polynuclear aromatic compounds in urban surface soil from Japan. J Chromatogr 643:305–317
176. Srogi K (2007) Human exposure to polycyclic aromatic hydrocarbons (PAHs) from coke plant. Air Protec Waste Probl 2:45–51
177. Sverdrup LE, Krogh PH, Nielsen T, Kjaer C, Stenersen J (2003) Toxicity of eight polycyclic aromatic compounds to red clover (Trifolium pratense), ryegrass (Lolium perenne), and mustard (Sinapsis alba). Chemosphere 53:993–1003
178. Tebaay RH, Welp G, Brummer GW (1993) Gehalt an Polycyclischen Aromatischen Kohlenwasserstoffen (PAK) und deren Verteilungsmuster in unterschiedlivh belasteten Boden. Z Pflanzernernaehr Bodenkd 156:1–10
179. Thomann RV, Komlos J (1999) Model of biota-sediment accumulation factor for polycyclic aromatic hydrocarbons. Environ Toxicol Chem 18:1060–1068
180. TMG (Tokyo Metropolitan Government) (2000) Say no! to dirty diesel vehicles. http://www.kankyo.metro.tokyo.jp/dno/ENG/index.ht. TMG, Tokyo
181. Tolosa I, de Mora S, Sheikholeslami MR, Villeneuve J-P, Bartocci J, Cattini C (2004) Aliphatic and aromatic hydrocarbons in coastal Caspian Sea sediments. Mar Pollut Bull 48:44–60
182. Trapido M (1999) Polycyclic aromatic hydrocarbons in Estonian soil: contamination and profiles. Environ Pollut 105:67–74
183. Tremolada P, Burnutt V, Calamari D, Jones KC (1996) Spatial distribution of PAHs in the UK atmosphere using pine needles. Environ Sci Technol 30:3570–3577
184. TRK-Wert for benzo[a]pyrene in Bundesarbeitsblatt 10, German Federal Department for Worker Safety, Dortmund, 1989, p 58
185. Tsai PJ, Shieh HY, Lee WJ, Lai SO (2002) Characterization of PAHs in the atmosphere of carbon black manufacturing workplaces. J Hazard Mater A91:25–42
186. US EPA (1999) Compendium of methods for the determination of toxic organic compounds in ambient air. Environmental Protection Agency, US Federal Register, Compendium Method TO-13A, pp 1–42
187. Van der Oost R, Heida H, Opperhuizen A, Vermeulen NPE (1991) Bioaccumulation of organic micropollutants in different aquatic organisms: sublethal toxic effects on fish. In: Mayes MA, Baron MG (eds) Aquatic Toxicology and Risk Assessment ASTM STP 1124, vol 14, American Society for Testing and Materials, Philadelphia, pp 166–180

188. Van Rooij JGM, De Ros JHC, Bodelier-Bade MM, Jongeneelen FJ (1993) Absorption of polycyclic aromatic hydrocarbons through skin human: differences between anatomical sites and individuals. J Toxicol Environ Health 38:355–368
189. Van Rooij JGM, Veeger MMS, Bodelier-Bade MM, Scheepers PTJ, Jongeneelen FJ (1994) Smoking and dietary intake of polycyclic aromatic hydrocarbons as sources of interindividual variability in the baseline excretion of 1-hydroxypyrene in urine. Int Arch Occup Environ Health 66:55–65
190. Van der Oost R, van Gastel L, Worst D, Hanraads M, Satumalay K, van Schooten FJ, Heida H, Vermeulen NPE (1994) Biochemical markers in feral roach (Rutilus rutilus) in relation to the bioaccumulation of organic trace pollutants. Chemosphere 29:801–/817
191. Van der Oost R, Beyer J, Vermeulen NPE (2003) Fish bioaccumulation and biomarkers in environmental risk assessment: a review. Environ Toxicol Pharm 13:57–149
192. Venkataraman C, Lyons JM, Fiedlander S (1994) Size distribution of aromatic hydrocarbons and elemental carbon, sampling measurement methods and source characterization. Environ Sci Technol 28:535–562
193. Venkataraman C, Thomas S, Kulkarni P (1999) Size distributions of polycyclic aromatic hydrocarbons–gas/particle partitioning to urban aerosols. J Aerosol Sci 30:759–770
194. Verweij F, Booij K, Satumalay K, van der Molen N, van der Oost R (2004) Assessment of bioavailable PAH, PCB and OCP concentrations in water, using semipermeable membrane devices (SPMDs), sediments and caged carp. Chemosphere 54:1675–1689
195. Viau C, Carrier G, Vyckocil A, Dodd C (1995) Urinary excretion kinetics of 1-hydroxypyrene in volunteers exposed to pyrene by the oral and dermal route. Sci Total Environ 163:179–186
196. Viguli J, Verde J, Irabien A (2002) Environmental assessment of polycyclic aromatic hydrocarbons (PAHs) in surface sediments of the Santander Bay, Northern Spain. Chemosphere 48:157–165
197. Vyskocil A, Fiala Z, Fialova D, Krajak V, Viau C (1997) Environmental exposure to polycyclic aromatic hydrocarbons in Czech Republic. Hum Exp Toxicol 16:589–597
198. Vyskocil A, Fiala Z, Chénier V, Krajak V, Ettlerova E, Bukac J, Viau C, Emminger S (2000) Assessment of multipathway exposure of small children to PAH. Environ Toxicol Pharm 8:111–118
199. Wagrowski DM, Hites RA (1997) Polycyclic aromatic hydrocarbon accumulation in urban, suburban, and rural vegetation. Environ Sci Technol 31:279–282
200. Wang DT, Meresz O (1982) Occurrence and potential uptake of polynuclear aromatic hydrocarbons of highway traffic origin by proximally grown food crops. In: Cooke M, Dennis AJ, Fisher GL (eds) Polynuclear aromatic hydrocarbons: physical and biological chemistry. Springer, New York, pp 885–896
201. Wang XL, Tao S, Dawson RW, Xu FL (2002) Characterizing and comparing risks of polycyclic aromatic hydrocarbons in a Tianjin wastewater-irrigated area. Environ Res 90:201–206
202. Wania F, Haugen JE, Lei YD, Mackay D (1998) Temperature dependence of atmospheric concentrations of semivolatile organic compounds. Environ Sci Technol 32:1013–1021
203. Watts WA, Ballestero TP, Gardner KH (2006) Uptake of polycyclic aromatic hydrocarbons (PAHs) in salt marsh plants Spartina alterniflora grown in contaminated sediments. Chemosphere 62:1253–1260
204. Weiss P, Riss A, Gschmeidler E, Schentz H (1994) Investigation of heavy metal, PAHs, PCB patters and PCDD/F profiles of soil samples from an industrialized urban area (Linz, Upper Austria) with multivariate statistical methods. Chemosphere 29:2223–2236
205. WHO (1987) Polynuclear aromatic hydrocarbons (PAHs). In: Air quality guidelines for Europe. Copenhagen, World Health Organization Regional Office for Europe, pp 105–117
206. WHO (1998) Guidelines for drinking-water quality, 2nd edn. World Health Organization, Geneva, pp 123–152
207. Wilcke W, Muller S (1999) Polycyclic aromatic hydrocarbons in hydromorphic soils of the tropical metropolis Bangkok. Geoderma 91:297–309
208. Wilcke W, Zech W, Kobza J (1996) PAH-pools in soils along a PAH-deposition gradient. Environ Pollut 92:307–313

209. Wilcke W, Lillienfein J, do Carmo Lima S (1999) Contamination of highly weathered urban soils in Uberlandia, Brazil. J Plant Nutr Soil Sci 162:539–548
210. Wilcke W, Amelung W (2000) Persistent organic pollutants in native grassland soils along a climosequence in North America. Soil Sci Soc Am J 64:2140–2148
211. Wild SR, Jones KC (1995) Polynuclear aromatic hydrocarbons in the UK environment: a preliminary source inventory and budget. Environ Pollut 88:91–108
212. Wild SR, Waterhouse KS, McGrath SP, Jones KC (1990) Organic contaminants in an agricultural soil with a known history of sewage sludge amendments: polycyclic aromatic hydrocarbons. Environ Sci Technol 24:1706–1711
213. Wu SP, Tao S, Liu WX (2006) Particle size distributions of polycyclic aromatic hydrocarbons in rural and urban atmosphere of Tianjin, China. Chemosphere 62:357–367
214. Yang SYN, Connell DW, Hawker DW (1991) PAHs in air, soil, and vegetation in the vicinity of an urban roadway. Sci Total Environ 66:127–136
215. Yang HH, Lai SO, Hsieh LT, Hsueh HJ, Chi TW (2002) Profiles of PAH emission from steel and iron industries. Chemosphere 48:1061–1074
216. Yang HH, Hsieh LT, Liu HS, Mi HH (2005) Polycyclic aromatic hydrocarbon emissions from motorcycles. Atmos Environ 39:17–25
217. Yunker MB, Macdonald RW, Vingarzan R, Mitchell RH, Goyette D, Sylvestre S (2002) PAHs in the Fraser River basin: a critical appraisal of PAH ratios as indicators of PAH source and composition. Org Geochem 33:489–515
218. Zhao Z-H, Quan W-Y, Tian D-H (1992) The relationship between polynuclear aromatic hydrocarbons in ambient air and 1-hydroxypyrene in human urine. J Environ Sci Health A 27:1949–1955
219. Zhou J, Wang T, Huang Y, Mao T, Zhong N (2005) Size distribution of polycyclic aromatic hydrocarbons in urban and suburban sites of Beijing, China. Chemosphere 61:792–799
220. Zhu L, Wang J (2003) Sources and patterns of polycyclic aromatic hydrocarbons pollution in kitchen air, China. Chemosphere 50:611–618
221. Zhu LZ, Amagai Y, Takahashi T (1997) High sensitive automatic analysis of polycyclic aromatic hydrocarbons in indoor and outdoor air. Talanta 45:113–118

CHAPTER 2

Temperature Effects on Zn(II) Toxicity to Metabolisms of Polyphosphate Accumulating Organisms in Aerobic and Anaerobic Conditions

Tsai YP*, lu MS, Lin JW, Chou CC and HU LY

Department of Civil Engineering, National Chi-Nan University, Taiwan, Republic of China

ABSTRACT

This study explores the influence of temperature on the tolerance of polyphosphate accumulating organisms (PAO_S) to Zn(II) in enhanced biological phosphorus removal. The results show anaerobic and aerobic metabolisms of PAO_S decreased with increasing Zn(II) concentration varying between 0-2 mg L^{-1} and temperature indeed affected inhibitive degrees. Furthermore, Zn(II) is more toxic to anaerobic poly-P hydrolysis and glycogen degradation at 10°C than at 20 and 30°C. For anaerobic polyhydroxyalkanoate (PHA) synthesis, Zn(II) had a highly inhibitive effect at 10°C too. The inhibitions of PAO_S aerobically taking up phosphorus, degrading PHA and replenishing glycogen in the presence of Zn(II) were amplified at 10°C. The metabolism of aerobic PHA degradation at 10°C was completely terminated and the aerobic replenishment of glycogen at 10° C was also completely terminated at 0.1 mg L^{-1} of Zn(II) due to the complete inhibition of aerobic PHA metabolism, which provides the required reducing power for synthesizing glycogen.

Keywords: Polyhydroxyalkanoates (PHAs); Zinc; Phosphorus accumulating organisms (PAO_s); Temperature; Phosphorus removal

1. INTRODUCTION

Though enhanced biological phosphorus removal (EBPR) is a standard technology for removing phosphorus from wastewater, the responsible organisms and relevant mechanisms are complex and not well-understood [1]. Substrate type has been identified as an important factor affecting the competition between polyphosphate accumulating organisms (PAO_s) and glycogen accumulating organisms (GAOs), which are usually dominant in a failed EBPR [2]. In addition, temperature is also considered an important factor in practice governing the efficiency of phosphate removal in EBPR because the population competition between PAO_s and GAO_s is significantly Ref. [3] affected by water temperature. Ref. [4] found that the predominant microbial population changed from PAO_s to GAO_s when temperature gradually rose from 20°C to 35°C. Similarly [5] found PAO_swere the dominant microorganisms at low temperature (10°C), regardless of the influent carbon source or pH. However [6] found conflicting results. Heavy metal ions have also been reported to be able to inhibit biological reactions responsible for carbon, nitrogen and phosphate removals. Ref. [7] found that the presence of Cd (II) significantly changed the structure of microbial population and was toxic to the bacterial community in nutrient removal sludge, especially to the bacteria in the Beta proteobacteria. Monitoring of a full-scale advanced municipal wastewater treatment plant showed the P removal efficiency decreased dramatically after tin (Sn) levels in the solids fraction of the mixed liquid suspended solids (MLSS) exceeded 4 μgSn L^{-1}[8]. A similar study by [9] showed 2 mg L^{-1} of Cd(II) took the lead in affecting biological phosphate removal. At the level of 5 mg L^{-1} of Cd (II), the total nitrogen removal and nitrification efficiency substantially declined and the denitrification rate was inhibited by about 61%. At the same time, the inhibition percentages of anaerobic release, anoxic and aerobic uptake rates of phosphate were about 76, 64, and 90%, respectively Ref. [10] further conducted a series of batch experiments to identify the toxic effects of Pb(II), Ni(II) and Cd(II) on nitrification and denitrification for nutrient removal sludges. Although both the inhibition kinetics of Ni(II) and Cd(II) followed first-order models, no significant inhibition to nitrification was observed for Pb(II) case, even when the Pb (II) concentration was as high as 40 mg L^{-1}. In contrast, 80% inhibition of nitrification was observed when Ni

(II) or Cd (II) concentration was higher than 5 or 25 mg L^{-1}, respectively. No synergistic effect upon adding different heavy metals at one time was observed for nitrification. However, a significant synergistic effect was found for denitrification. A similar study was conducted by [11] who investigated the toxic effects of Cu (II), Zn(II) and Cr(VI) on nitrification and denitrification in sequencing batch reactor (SBR) sludge. Ref. [12] also found the nitrifiers in SBR sludge could withstand more metal ion toxicity than those in anaerobic-anoxic-oxic (A_2O) sludge.

The mechanisms of biological phosphorus removal are very complex, especially for the competition of PAO_s and GAO_s. Many factors can potentially affect this competition, including pH, temperature, carbon sources, metal ions, sludge retention time (SRT) etc. The influences of those factors on phosphorus removal are probably not independent. Interaction might exist between one factor and the others, for example, the toxic metal effect is possibly subject to pH, temperature, SRT, and carbon source etc. Ref. [13] investigated the shock load effect of Cu (II) on PAO_s behaviors with respect to the transformations of poly-P, intracellular polyhydroxyalkanoates (PHA_s), and glycogen. They found Cu (II) inhibitory effect was reduced by increasing the pH to 7.6, revealing that pH was able to influence the endurance of PAO_s biomass to Cu (II) and there was an interaction between pH and toxic effect. Ref. [14] further experimentally demonstrated that PAO_s loss of the ability to anaerobically synthesize PHA_s and to aerobically take up phosphate under the presence of Cu (II) was due to inhibition of the enzyme activities of acetyl-CoA synthases (ACS) and of polyphosphate kinase (PPK), respectively. The apparent maximum specific activity (V_{max}) of ACS decreased with increasing Cu (II) concentration, revealing that Cu (II) has a higher affinity for free ACS than for ACScoenzyme A complex. A similar result Ref. [15] was found in PPK activity tests, showing that Cu (II) more easily bound to free PPK than to PPK-Adenosine triphosphate complex.

Investigated the effect of Pb (II) on metabolisms of PAO_s and GAO_s fed with acetic acid or with glucose, respectively, as their sole carbon source. Results from trials with a high concentration of acetic acid showed there was an abnormal aerobic phosphate release. Ref. [16] found that the effect of Cu (II) on phosphate removal was significant and the SRT was an important factor influencing the endurance of PAOs to Cu (II). The PHA storage in anaerobic stage for the sludge at 10d SRT was less influenced by Cu (II) than storage at 5d or 15d SRT_s. The reaction of PAO_s to aerobically take up phosphate for the sludge at 5d or 15d SRT almost ceased at 2 mg Cu L^{-1}, whereas PAO_s in the sludge at 10d SRT retained half of their ability to take up phosphate.

To date, however, little is known about how water temperature affects the endurance of PAOs biomass to metal ion toxicity. Therefore, this study examines the shock load effect of heavy metal (Zn^{2+}) on the behavior of PAOs biomass in activated sludges at different temperature levels (10, 20 and 30°C) with respect to the anaerobic phosphate release, the aerobic/anoxic phosphate uptake, the transformations of poly-P and intracellular PHA_S, as well as the utilization and replenishment of glycogen.

2. MATERIALS AND METHODS

2.1 Sbr pilot-plant

An SBR pilot-plant was used to chronically acclimatize the activated sludge without the addition of metal ions. The system had a total reactor volume of 160 L, base volume of 80 L, and effective volume of 148.5 L. It was operated at 2 cycles per day with 12 hours per cycle. The phases and durations in a cycle followed the sequence: anaerobic 150 min, aerobic 330 min, anoxic 180 min, re-aeration 10 min, sedimentation 30 min, and decantation 20 min. The SBR system was fed with a synthetic wastewater containing milk, KH_2PO_4, Urea, $FeCl_3$, CH_3COOH, Glucose, and NH_4Cl. The pH of the synthetic wastewater was adjusted to 6.8-7.2 with 6N NaOH. The pH in the aerobic phase of SBR system was automatically maintained at 7.2 ± 0.2 by adding a $NaOH/NaHCO_3$solution, and the dissolved oxygen (DO) concentration in the aerobic phase was also controlled at 2.0 ± 0.3mgL^{-1}. The SBR system was acclimatized at the condition of 10d SRT. The batch experiments described below were carried out when the SBR system had reached a steady state condition. Other conditions are described in Ref. [15].

2.2 Batch experiments

The batch test was carried out in a 2 L acrylic, magnetically stirred reaction vessel. The sludge for the batch test was taken from the SBR system at the end of the aerobic phase. Before starting a batch test, the sludge was pre-treated according to Ref. [15]. A required amount of $ZnCl_2$ solution was then added into the batch reactor and the pH was adjusted to 7.5. The concentration of acetate in the batch reactor equaled 100 mg L^{-1} as chemical oxygen demand (COD). The mixed liquid in the batch reactor bubbling with nitrogen gas was then anaerobically incubated at a constant temperature in a water-bath tank for 120 min. Later, the mixed liquid was aerated for 240 min to examine the aerobic phosphate uptake of PAO_S.

2.3 Analytical methods

The liquid samples taken from the batch reactor were immediately filtered through Millipore filter (0.45 μm pore size) and then COD, total phosphorus (TP), PO_4^{3-}-P, MLSS and mixed liquor volatile suspended solid (MLVSS) were analyzed in accordance with Standard Methods [17]. The PHA concentration was measured according to [18]. The intracellular glycogen was analyzed using the method described in [19]. The concentration of Zn^{2+} was analyzed by an Atomic Absorption Spectrometry device (Shimadzu AA-6200), and the averaged value of duplicate analysis was used in this study.

2.4 Data analysis

To compare variation in treatments, analysis of variance (ANOVA) F-test was applied to test whether the temperature, zinc concentration and their interaction significantly influenced the metabolisms of PAO_s in the anaerobic and aerobic stages. If significant factors (e.g., temperature and/or zinc concentration) were found, then the least significant difference test (LSD) was further applied to establish which particular levels of a factor were different. A paired sample t-test was also used to determine if means of interested variables were not statistically equal. All tests were performed at a significance level (p-value) of 0.05, using standard statistical software (SPSS release 10.1.0, SPSS Inc., Chicago, IL).

3. RESULTS AND DISCUSSION

The average water qualities and operational conditions of the SBR system under steady state conditions are shown in **TABLE 1**. The removal rates of phosphate and total phosphorus were 99% and 98%, respectively. The data show the sludge used in batch tests was good with high carbon, nitrogen, and phosphate removal efficiencies. Before batch tests, a full cycle examination of the SBR system was conducted and the result showed the sludge in the SBR system could release 30 mgl^{-1} of total phosphorus after 150 minutes in the anaerobic stage, indicating the predominant population of sludge in SBR system was PAO_S, rather than GAO_S. The 4, 6-diamidino-2-phenylindole (DAPI) method was also used to reconfirm this issue in the study (data not shown).

Table 1: Influent and effluent water qualities of the SBR pilot-scale system at steady state (n>8).

Water Quality Parameters	Influent (mgl^{-1})	Effluent (mgl^{-1})	Removal (%)	Operational Parameters
Total COD	357 ± 30.5	17.5 ± 6.62	95	
Soluble COD	309 ± 25.3	9.68 ± 5.23	96	
Total P	6.46 ± 0.44	0.13 ± 0.1	98	
PO_4^{3-}-P	6.20 ± 0.36	0.09 ± 0.25	99	
Total nitrogen	51.8 ± 14.	11.9 ± 1.2	74	
NH^{4+}-N	30.0 ± 6.03	ND	100	
NO_x^--N	2.39 ± 0.86	11.5 ± 0.83		
Suspended solids		8.39 ± 3.28		
MLSS (mgl)				1366 ± 59
SRT (d)				10
MLVSS (mgl^{-1})				1151 ± 53
Organic loading (g COD g^{-1}VSS d^{-1})				0.29 ± 0.03

3.1 Background study - temperature effects on PAO metabolisms in the absence of Zn(II)

To understand the effect of temperature on PAO metabolisms in the absence of Zn(II), two to four times experiments were conducted as background study. The averages and standard deviations of these data were further used to statistically compare their differences.

PAO ANAEROBIC METABOLISMS IN THE ABSENCE OF ZN(II): The experimental results in **FIGURE 1(A)** show the anaerobic specific substrate utilization rates (SSUR) of PAOs at the temperatures of 10, 20, and 30°C were 17.7 ± 1.2, 19.7 ± 6.2, and 23.8 ± 3.6 mgCOD g^{-1}VSS h^{-1}, respectively. Statistical analysis showed the temperature effect on anaerobic substrate utilization of microorganisms was insignificant (p=0.494).

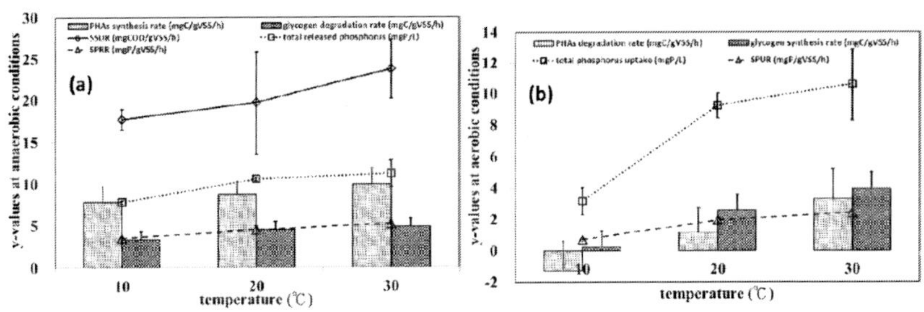

Figure 1: Temperature effects on (a) Anaerobic, and (b) Aerobic metabolisms of PAOs in the absence of Zn(II).

FIGURE 1(A) also shows the total phosphorus releases (TPR_S) and the specific phosphorus release rates (SPRRs) for PAOs at different temperature conditions. The PAOs released 7.8 ± 0.2, 10.6 ± 0.2, and 11.2 ± 1.6 mg L^{-1} of phosphorus at 10, 20, and 30°C, respectively. The corresponding SPRRs were 3.5 ± 0.4, 4.5 ± 0.5, and 5.2 ± 0.1 mgP g^{-1}VSS h^{-1}. Statistical analysis showed temperature variation significantly affected anaerobic TPRs and SPRRs (P=0.01 and 0.002, respectively) and the metabolic activity of PAO_S releasing phosphorus in anaerobic condition increased with increasing temperature.

From an energy viewpoint, PAO_S are primarily characterized by their ability to anaerobically take up organic substrates and store them by utilizing energy from the hydrolysis of stored poly-P without consuming any electron acceptors. When organic substrates are taken up anaerobically, they are usually converted to PHA and stored. The anaerobic conversion of acetate to PHA requires a reducing power because PHA is a more reduced compound than acetate. Only 30% of the reducing power is supplied by the citric acid (TCA) cycle, which functions under anaerobic conditions to oxidize a portion of the acetate to CO_2 and generate the reducing power in the form of nicotinamide adenine dinucleotide hydrogen (NADH). The remaining 70% of the reducing power is supplied by anaerobic degradation of intracellularly stored glycogen to acetyl-coA, together with its partial oxidation to CO_2 [20]. The metabolic function of glycogen degradation, therefore, appears to be essential for anaerobic substrate uptake, PHA synthesis, and PAO proliferation in the EBPR process [21].

In **FIGURE 1(A)** it shows the anaerobic PHAs synthesis rates at 10, 20, and 30°C were 7.9 ± 1.9, 8.8 ± 1.6, and 10.0 ± 1.9 mg g^{-1}VSS h^{-1}, respectively. **FIGURE 1(A)** also shows the anaerobic glycogen degradation rates at 10, 20, and 30°C were 3.3 ± 0.7, 4.5 ± 0.3, and 4.9 ± 1.4 mg g^{-1}VSS h^{-1}, respectively. Statistical analysis shows the temperature effects on both rates were insignificant ($P=0.40$ and 0.21, respectively). The results are similar to that of SSUR, indicating that the anaerobic metabolisms of PAO_s taking up substrate and consuming glycogen to supply reducing power to convert acetate to PHA are not statistically influenced by temperature variation.

PAO AEROBIC METABOLISMS IN THE ABSENCE OF ZN(II): In **FIGURE 1(B)** it shows that the total phosphorus uptakes (TPU_s) and the specific phosphorus uptake rates ($SPUR_s$) for PAOs at different temperature conditions. PAOs aerobically took up 3.2 ± 0.9, 9.2 ± 0.8 and 10.6 ± 2.3 mg L^{-1} of phosphorus at 10, 20, and 30°C, respectively. The corresponding SPURs were 0.7 ± 0.1, 2.0 ± 0.2, and 2.4 ± 0.2 mgP g^{-1}VSS h^{-1}. Statistical analysis indicates temperature variation significantly affected aerobic TPU_s and $SPUR_s$($P<0.001$ and <0.001, respectively) and the aerobic TPU_s and $SPUR_s$ increased with increasing temperature.

In **FIGURE 1(B)** it also shows the PHA_s energy consumption and glycogen replenishment at different temperature conditions. The PHA degradation rates were -1.3 ± 0.4, 1.2 ± 0.3, and 3.3 ± 0.6 mg g^{-1}VSS h^{-1} for 10, 20, and 30°C, respectively. At the same time, the glycogen replenishment rates were 0.2 ± 0.6, 2.6 ± 0.7, and 4.0 ± 0.6 mg g^{-1}VSS h^{-1} for 10, 20, and 30°C, respectively. It is noticeable that PAO_s might have been unable to use PHA_s as a carbon source when the temperature was as low as 10°C. At the same time, the mechanism for PAO_s aerobically replenishing glycogen was also inhibited due to the low temperature. Both PHA degradation rates and glycogen synthesis rates in the aerobic stage were significantly influenced by temperature variation in that both rates increased with increasing temperature ($p<0.001$).

TEMPERATURE EFFECTS ON THE ENDURANCE OF PAO_S TO ZN(II) TOXICITY: To understand how the temperature variation influences Zn(II) toxicity to PAO metabolisms, $ZnCl_2$ solution was added to the batch reactor at different temperature conditions. The inhibition kinetic concept proposed in [8] was used to compare Zn(II) toxic intensity according to inhibition constants (r_i) which are shown in the regression equation in each graph. The larger the constant value is, the larger the metabolic inhibition.

3.2 Anaerobic metabolisms of PAOs in the presence of Zn(II)

In **FIGURE 2(A)** it shows the results of microorganisms anaerobically taking up substrate at different temperature conditions in the presence of Zn (II). At 10°C, the $SSUR_s$ of microorganisms at 0, 0.1, 0.3, 0.5, 1.0, and 2.0 mg L^{-1} of Zn(II) are 17.7, 16.9, 11.9, 11.6, 9.9, and 6.4 mgCOD g^{-1}VSS h^{-1}, respectively. At 20°C, the corresponding SSURs are 23.7, 21.3, 19.4, 18.4, 17.8, and 15.2 mgCOD g^{-1}VSS h^{-1}. At 30°C, the corresponding SSUR are 28.0, 21.3, 18.4, 11.8, 11.8, and 10.0 mgCOD g^{-1}VSS h^{-1}. The SSUR for microorganisms at anaerobic conditions decreased with increasing Zn(II) concentration, regardless of temperature. The inhibition constants (ri) at temperatures 10, 20, and 30°C are 1.64, 1.94, and 3.46 L mg^{-1}, respectively, indicating Zn(II) is more toxic to substrate uptake of microorganisms at higher temperature condition (30°C) than at lower conditions (10 and 20°C).

In **FIGURE 2(B)** it shows the results of PAOs anaerobically releasing phosphorus at different temperature conditions in the presence of Zn(II). At 10°C, the SPRRs of PAOs at 0, 0.1, 0.3, 0.5, 1.0, and 2.0 mg L^{-1} of Zn(II) are 4.0, 3.0, 2.4, 2.1, 1.8, and 1.3 mgP g^{-1}VSS h^{-1}, respectively. At 20°C, the corresponding SPRRs are 4.8, 4.8, 4.3, 4.0, 3.1, and 2.9 mgP g^{-1}VSS h^{-1}. At 30°C, the corresponding SPRRs are 6.7, 5.7, 5.3, 4.9, 4.2, and 3.8 mgP g^{-1}VSS h^{-1}. It can be seen that the hydrolysis of stored poly-P in sludge at anaerobic conditions decreased with increasing Zn(II) concentration, regardless of temperature. The inhibition constants at temperatures 10, 20, and 30°C were 3.01, 1.08, and 1.95 L mg^{-1}, respectively, indicating Zn(II) is more toxic to poly-P hydrolysis at 10°C than at 20 or 30°C.

In **FIGURE 2(C)** it shows the results of PAO_S anaerobically synthesizing PHA_S at different temperature conditions. At 10°C, the PHA synthesis rates of PAO_S at 0, 0.1, 0.3, 0.5, 1.0, and 2.0 mg L^{-1} of Zn(II) were 12.3, 8.5, 6.8, 6.8, 5.2, and 2.5 mg g^{-1}VSS h^{-1}, respectively. At 20°C, the PHA synthesis rates at 0, 0.1, 0.3, 0.5, 1.0, and 2.0 mg L^{-1} of Zn(II) were 13.9, 13.7, 10.4, 9.6, 9.6, and 6.2 mg g^{-1}VSS h^{-1}, respectively. At 30°C, the PHA synthesis rates at 0, 0.1, 0.3, 0.5, 1.0, and 2.0 mg L^{-1} of Zn(II) were 16.2, 14.2, 10.4, 7.9, 6.1, and 4.1 mg g^{-1}VSS h^{-1}, respectively. It can be seen that the PHA synthesis rates for PAO_S at anaerobic conditions decreased with increasing Zn(II) concentration, regardless of temperature. The inhibition constants at temperature 10, 20, and 30°C are 1.94, 1.31, and 2.17 L mg^{-1}, respectively, indicating Zn(II) is more toxic to PHA synthesis at 10 and 30°C than at 20°C.

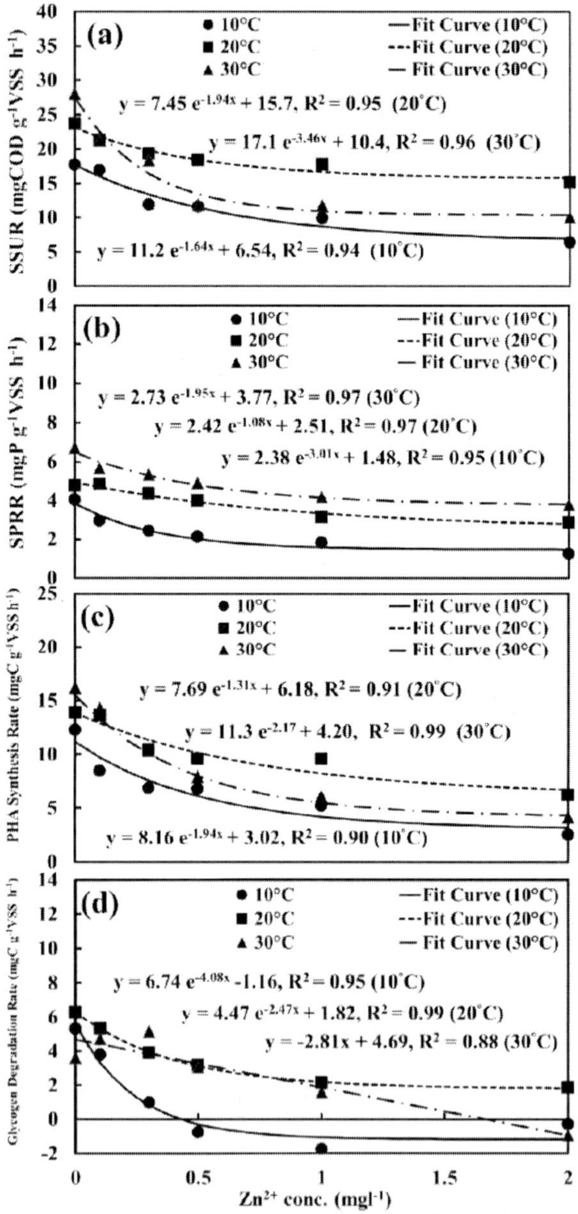

Figure 2: Temperature effects on Anaerobic metabolisms of PAOs in the presence of Zn(II).

FIGURE 2(D) shows the results of PAO_s anaerobically degrading glycogen for energy requirement at different temperature conditions. At 10°C, the glycogen degradation rates of PAOs at 0, 0.1, 0.3, 0.5, 1.0, and 2.0 mg L^{-1} of Zn(II) were 5.3, 3.8, 1.0, -0.8, -1.8, and -0.3 mg g^{-1}VSS h^{-1}, respectively. At 20°C, the glycogen degradation rates at 0, 0.1, 0.3, 0.5, 1.0, and 2.0 mg L^{-1} of Zn(II) were 6.3, 5.3, 3.9, 3.2, 2.1, and 1.9 mg g^{-1} VSS h^{-1}, respectively. At 30°C, the glycogen degradation rates at 0, 0.1, 0.3, 0.5, 1.0, and 2.0 mg L^{-1} of Zn(II) were 3.6, 4.7, 5.1, 3.0, 1.6, and -0.9 mg g^{-1} VSS h^{-1}, respectively. It can be seen that the glycogen consumption rates for PAOs at anaerobic conditions also decreased with increasing Zn(II) concentration, regardless of temperature. The inhibition constants at temperature 10 and 20°C are 4.08 and 2.47 L mg^{-1}, respectively, but the data at 30°C have a better fit with a linear regression instead of the first order model. It is noteworthy that the metabolism of glycogen degradation was completely terminated at 10°C when more than 0.5 mg L^{-1} of zinc ions was added, indicating Zn(II) is more toxic to the glycogen metabolism at 10°C than at 20 or 30°C.

In this study, both the metabolic rates of anaerobic glycogen degradation and PHAs synthesis were reduced with increasing Zn(II) concentration, revealing that the presence of Zn(II) caused a reduction in the degradation of glycogen for production of NADH, which is essential for PHA synthesis.

AEROBIC METABOLISMS OF PAOS IN THE PRESENCE OF ZN(II): In **FIGURE 3(A)** it shows the results of PAO_s aerobically taking up phosphorus at different temperature conditions. At 10°C the SPUR levels of PAOs at 0, 0.1, 0.3, 0.5, 1.0, and 2.0 mg L^{-1} of Zn(II) were 0.9, 0.4, 0.2, -0.04, -0.2, and -0.2 mgP g^{-1}VSS h^{-1}, respectively. At 20°C, the SPUR levels at 0, 0.1, 0.3, 0.5, 1.0, and 2.0 mg L^{-1} of Zn(II) were 2.0, 1.8, 1.0, 0.6, 0.3, and -0.2 mgP g^{-1}VSS h^{-1}, respectively. At 30°C, the SPUR levels at 0, 0.1, 0.3, 0.5, 1.0, and 2.0 mg $L^{-1}$of Zn(II) were 3.2, 2.5, 1.3, 0.6, 0.1, and -0.4 mgP g^{-1}VSS h^{-1}, respectively. The phosphorus uptake activity for PAOs at aerobic conditions decreased with increasing Zn(II) concentration and was completely inhibited at 2.0 mg L^{-1} of Zn (II), regardless of temperature. Furthermore, the inhibition by Zn(II) was accelerated at 10°C, since PAO_s had completely stopped taking up phosphorus at 0.5 mg L^{-1} of Zn (II). The inhibition constants at temperatures 10, 20, and 30°C were 3.94, 1.84, and 2.43 L mg^{-1}, respectively, indicating Zn(II) is more toxic to phosphorus uptake by PAO_s at 10°C than at 20 or 30°C.

In **FIGURE 3(B)** it shows the results of PAOs aerobically consuming PHAs at different temperature conditions. At 10°C, the PHA degradation rates of PAOs at 0, 0.1, 0.3, 0.5, 1.0, and 2.0 mg L^{-1} of Zn(II) were -1.0, -0.6, -2.0, -3.4, -1.8, and -2.3 mg g^{-1}VSS h^{-1}, respectively. At 20°C, the PHA degradation rates at 0, 0.1, 0.3, 0.5, 1.0, and 2.0 mg L^{-1} of Zn(II) were 2.7, 2.3, 1.6, 0.4, 0.1, and -1.8 mg g^{-1}VSS h^{-1}, respectively. At 30°C, the PHA degradation rates at 0, 0.1, 0.3, 0.5, 1.0, and 2.0 mg L^{-1} of Zn(II) were 5.7, 4.1, 3.6, 2.1, 1.6, and 0.02 mg

$g^{-1}VSS\ h^{-1}$, respectively. The PHA degradation rates for PAOs at 20 and 30°C decreased with increasing Zn(II) concentration, regardless of temperature. The corresponding inhibition constants were 0.70, and 0.46 L mg^{-1}, respectively, whereas the aerobic PHA metabolism at 10°C completely stopped, regardless of Zn(II) concentration. Moreover, the PHA metabolisms at 20 and 30°C also stopped at 2.0 mg L^{-1} of Zn (II).

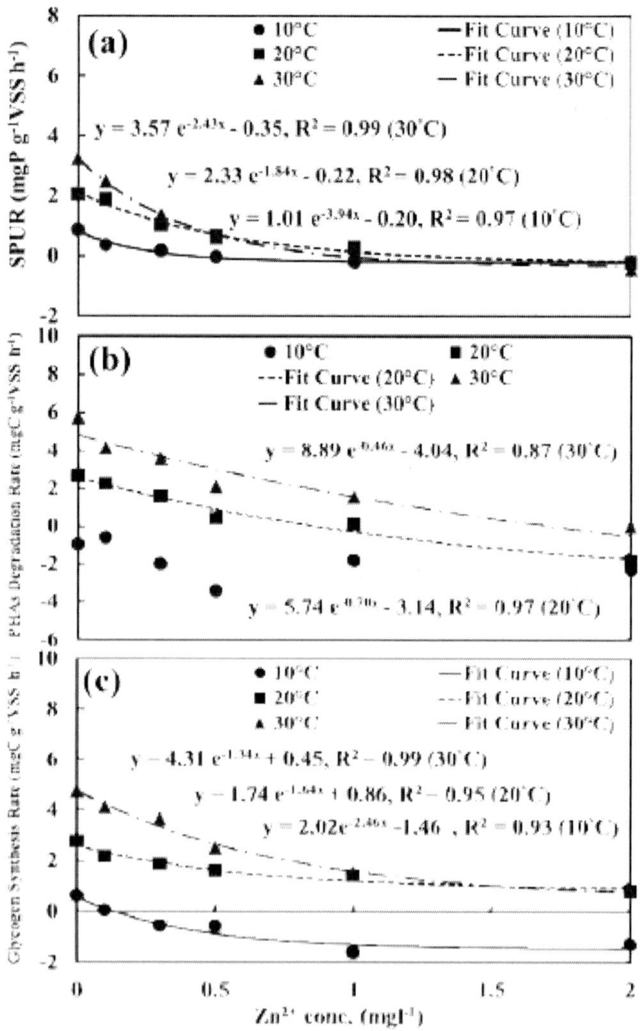

Figure 3: Temperature effects on Aerobic metabolisms of PAOs in the presence of Zn(II).

In **FIGURE 3(C)** it shows the results of PAO_S aerobically replenishing intracellular glycogen at different temperature conditions. At 10°C, the glycogen synthesis rates of PAO_S at 0, 0.1, 0.3, 0.5, 1.0, and 2.0 mg L^{-1} of Zn(II) were 0.6, 0.03, -0.6, -0.6, -1.6, and -1.3 mg g^{-1}VSS h^{-1}, respectively. It is noteworthy that the aerobic glycogen metabolism was completely inhibited at 0.1 mg L^{-1} of Zn(II) due to the entire inhibition of aerobic PHAs metabolism, which provides the required reducing power for synthesizing glycogen. At 20°C, the glycogen synthesis rates at 0, 0.1, 0.3, 0.5, 1.0, and 2.0 mg L^{-1} of Zn(II) were 2.8, 2.2, 1.9, 1.6, 1.4, and 0.8 mg g^{-1}VSS h^{-1}, respectively. At 30°C, the glycogen synthesis rates at 0, 0.1, 0.3, 0.5, 1.0, and 2.0 mg L^{-1} of Zn(II) were 4.7, 4.1, 3.6, 2.5, 1.5, and 0.8 mg g^{-1}VSS h^{-1}, respectively. The glycogen synthesis rates for PAOs at aerobic conditions decreased with increasing Zn(II) concentration, regardless of temperature. The inhibition constants at temperatures 10, 20, and 30°C were 2.46, 1.64, and 1.34 L mg^{-1}, respectively, indicating Zn(II) is more toxic to aerobic glycogen metabolism for PAOs at 10°C than at 20 or 30°C.

INHIBITION PERCENTAGES FOR METABOLISMS OF PAOS IN THE PRESENCE OF ZN(II): To understand how the Zn(II) toxicity influences PAO_S metabolisms, inhibition percentages at different Zn(II) concentration and temperature conditions were calculated and are shown in **FIGURE 4**.

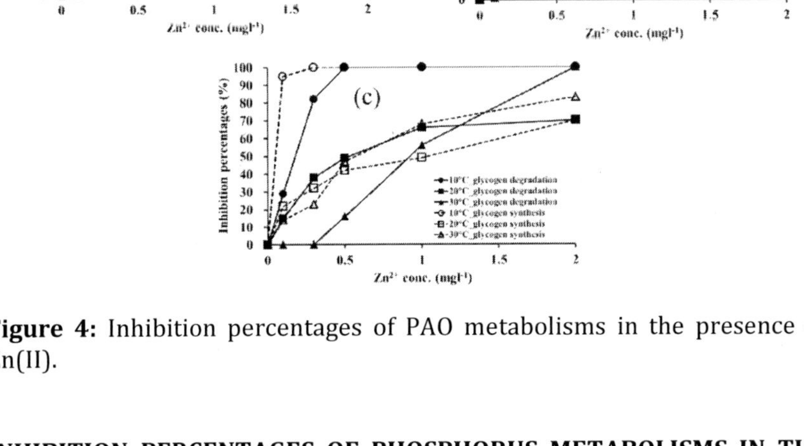

Figure 4: Inhibition percentages of PAO metabolisms in the presence of Zn(II).

INHIBITION PERCENTAGES OF PHOSPHORUS METABOLISMS IN THE PRESENCE OF ZN(II): In **FIGURE 4(A)** it shows the inhibition percentages of PAOs anaerobically releasing phosphorus and aerobically taking up phosphorus at different Zn(II) concentrations and temperature conditions. The order of the magnitude of inhibition percentages at different temperature is 10°C >30°C >20°C for both SPRR and SPUR cases. Clearly, Zn(II) toxicity to the PAO activities of releasing and taking up phosphorus was magnified at the low temperature condition of 10°C Furthermore, the inhibition percentages of SPUR by Zn(II) were significantly larger than those of SPRR, regardless of temperature. For example, regardless of temperature, the inhibition percentage of SPURs was nearly 100% at 2 mg L^{-1} of Zn(II), whereas only 69, 40 and 44% of SPRR inhibition were found at 10, 20 and 30°C, respectively.

INHIBITION PERCENTAGES OF PHA METABOLISMS IN THE PRESENCE OF ZN(II): In **FIGURE 4(B)** it shows the inhibition percentages of PHA metabolisms at different Zn(II) concentration and temperature conditions. The order of magnitude for inhibition percentages at different temperatures was 10°C 20°C >30°C or the case of aerobic PHA degradation and 10°C ≈ 30°C >20°C for the case of anaerobic PHAs synthesis. This indicates that, similar to phosphorus metabolisms, the 10°C low temperature condition also magnified the toxicity of Zn(II) to PHA metabolisms. Moreover, similar to phosphorus metabolisms, the inhibition percentages of aerobic PHA degradation by Zn(II) were significantly larger than those of anaerobic PHA synthesis, regardless of temperature. It is noteworthy that the metabolism of aerobic PHAs degradation completely stopped at the low temperature condition (10°C), regardless of Zn(II) concentration, whereas completely stopped at 2 mg L^{-1} of Zn(II) for 20 and 30°C. At the same time, only 80, 55, and 75% of the inhibition of anaerobic PHA synthesis were found at 10, 20 and 30°C, respectively.

INHIBITION PERCENTAGES OF GLYCOGEN METABOLISMS IN THE PRESENCE OF ZN(II): In **FIGURE 4(C)** it shows the inhibition percentages of glycogen metabolisms at different Zn(II) concentration and temperature conditions. The order of the magnitude of inhibition percentages at different temperature is 10°C >30°C >20°C for aerobic glycogen synthesis case and 10°C >20°C >30°C for anaerobic glycogen degradation case. Similar to phosphorus and PHA metabolisms, the 10°C low temperature condition also magnified the toxicity of Zn(II) to glycogen metabolisms. However, unlike phosphorus and PHAs metabolisms, the inhibition percentages of aerobic glycogens synthesis by Zn(II) are not clearly larger than those of anaerobic glycogens degradation. It is noteworthy that only 0.3-0.5 mg L^{-1} of Zn(II) completely terminated the glycogen metabolisms at low temperature condition (10 °C).

Distinguishing the individual degrees of Zn(II) toxicity to anaerobic and aerobic metabolisms from **FIGURES 4(A)** and **4(B)**, it can be seen that the inhibition percentages by Zn(II) with respect to phosphorus and PHA metabolisms at the aerobic stage were larger than those at the anaerobic stage. This seemingly indicates Zn(II) is more toxic to the aerobic metabolism than anaerobic metabolism. However, it is difficult to draw an exact conclusion with respect to this issue because Zn(II) was added at the beginning of the anaerobic stage, which was subsequently followed by the aerobic stage in a batch test of this study. Thus an interesting problem arises as to whether or not the toxic effect of Zn(II) that began at the aerobic stage included the toxic effect at the previous anaerobic stage. In other words, does the high metabolic inhibition at aerobic stage come from the direct interference of Zn(II) in aerobic metabolisms? Or, does a partial inhibition come from the indirect interference in previous anaerobic metabolisms linked to subsequent aerobic metabolisms? To verify this issue, we conducted two simultaneous series of batch tests at 20°C, one added Zn(II) at the beginning of the anaerobic stage and the other not until the end of the anaerobic stage (i.e., the beginning of the aerobic stage). We focused on the differences of phosphorus and PHA metabolisms.

INHIBITION SOURCES FOR PHOSPHORUS METABOLISMS: In **FIGURE 5(A)** it shows the inhibition percentages of PAO_s anaerobically releasing phosphorus (SPRR) and aerobically taking up phosphorus (SPUR) at different Zn(II) occurring times and concentrations. The difference of inhibition percentages of $SPUR_s$ between Zn(II) initially presented at the anaerobic stage and at the aerobic stage is defined as the inhibition percentage only contributed by Zn(II) toxicity at the anaerobic stage. At the same time, the inhibition percentages of SPUR while Zn(II) initially presented at the aerobic stage is defined as the inhibition percentage only contributed by Zn(II) toxicity at the aerobic stage. The same concept is used for the SPRR, and then the inhibition percentages of SPRR contributed only from Zn(II) toxicity at the anaerobic stage could be obtained. All results are shown in **FIGURE 5(B)**. The inhibitions of SPRR originated from Zn(II) toxicity at the anaerobic stage for 0.5, 1.0, and 2.0 mg L^{-1} of Zn(II) are low (10, 5, and 14%, respectively). The inhibitions of SPUR that originated from Zn(II) toxicity at the anaerobic stage for 0.5, 1.0, and 2.0 mg L^{-1} of Zn(II) only are 2, 2, and 11%, respectively. However, the inhibitions of SPUR that originated from Zn(II) toxicity at the aerobic stage for 0.5, 1.0, and 2.0 mg L^{-1} of Zn(II) are 42, 80, and 89%, respectively. Consequently, it is clear the majority of metabolic inhibition (>90% of total inhibitions) for PAOs aerobically taking up phosphorus came from the direct interference of Zn(II) in aerobic metabolisms, regardless of Zn(II) concentration. Only a small amount of the inhibitions (<10% of total inhibitions) came from the indirect interference in previous anaerobic metabolisms linked to subsequent aerobic metabolisms.

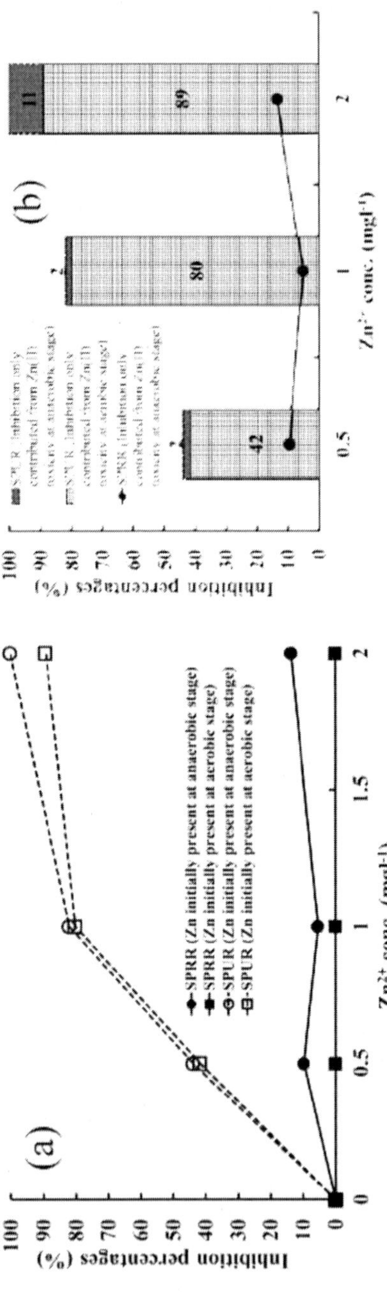

Figure 5: Contribution of inhibition percentages for Phosphorus metabolisms of PAOs in the presence of Zn(II).

INHIBITION SOURCES FOR PHA METABOLISMS: In **FIGURE 6(A)** it shows the inhibition percentages of PAOs anaerobically synthesizing PHAs and aerobically degrading PHAs at different Zn(II) times of occurrence and concentrations. The same calculations for phosphorus metabolisms are applied to PHA metabolisms, and then the inhibition percentages of PHA degradation contributed only by Zn(II) toxicity at the anaerobic and aerobic stages, as well as the inhibition percentages of PHA synthesis contributed only by Zn(II) toxicity at the anaerobic stage could be obtained, as shown in **FIGURE 6(B)**. Unlike SPRR, the inhibitions of PHA synthesis originated from Zn(II) toxicity at the anaerobic stage for 0.5, 1.0, and 2.0 mg L^{-1} of Zn(II) are significant (23, 30, and 67%, respectively). The inhibitions of PHA degradation originated from Zn(II) toxicity at the anaerobic stage for 0.5, 1.0, and 2.0 mg $L^{-1}$of Zn(II) are 14, 16, and 60%, respectively. And, the inhibitions of PHA degradation originated from Zn(II) toxicity at the aerobic stage for 0.5, 1.0, and 2.0 mg L^{-1} of Zn(II) are 25, 22, and 38%, respectively. Like phosphorus metabolisms, the majority of metabolic inhibition for PAOs aerobically degrading PHAs came from the direct interference of Zn(II) in aerobic metabolisms when 0.5 or 1.0 mg L^{-1} of Zn(II) was added (64 and 58% of total inhibitions, respectively). Only less than half inhibitions came from the indirect interference in previous anaerobic metabolisms linked to subsequent aerobic metabolisms. On the contrary, the major inhibition of aerobic PHA metabolisms came from the indirect interference in previous anaerobic metabolisms (61% of total inhibitions) while 2.0 mg L^{-1} of Zn(II) was added. Only 39% of total inhibitions came from the direct toxic effects to aerobic PHA metabolisms. This amplification trend of indirect interference followed by increasing Zn(II) concentration for aerobic PHA degradation was consistent with the trend of direct interference for anaerobic PHA synthesis. This indicates that the degree to which PAO_S lose their ability to utilize PHAs is dependent on how much PHA was synthesized in the previous anaerobic stage under the presence of Zn (II). This is reasonable because the source of PHAs providing for PAO use at the aerobic stage were those PHAs synthesized at the previous anaerobic stage.

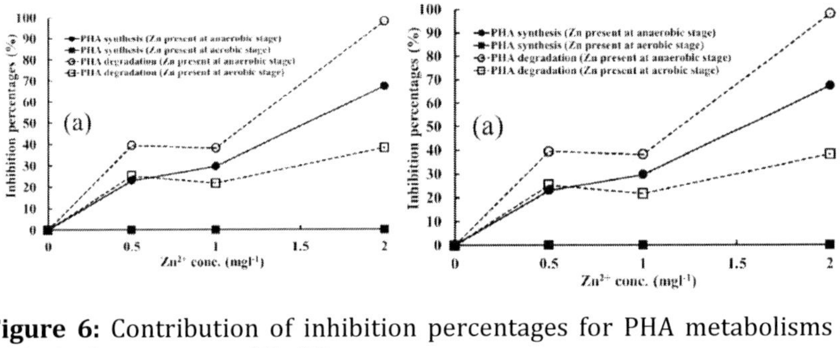

Figure 6: Contribution of inhibition percentages for PHA metabolisms of PAOs in the presence of Zn(II).

CONCLUSIONS

Anaerobic and aerobic metabolisms of PAO_s decrease with increasing Zn(II) concentration varying between 0-2 mg L^{-1} and temperature variation indeed affects inhibitive degrees. The majority of metabolic inhibition (>90% of total inhibitions) for PAO_saerobically taking up phosphorus came from the direct interference of Zn(II) in aerobic metabolisms. Only a small amount of the inhibitions (<10% of total inhibitions) came from the indirect interference in previous anaerobic metabolisms linked to subsequent aerobic metabolisms. The results are helpful to understanding the reasons of the failure of EBPR systems under the presence of toxic substances and temperature variation.

ACKNOWLEDGEMENTS

This study was supported by grants received from the National Science Council, Taiwan, Republic of China (NSC 99-2918-I-260-001 and NSC 100-2221-E-260- 002-MY3).

REFERENCES

1. Gebremariam SY, Beutel MW, Christian D, Hess TF (2011) Research advances and challenges in the microbiology of enhanced biological phosphorus removal - A critical review. Water Environ Res 83: 195-219.
2. You SJ, Tsai YP, Cho BC, Chou YH (2011) Metabolic influence of lead on polyhydroxyalkanoates (PHA) production and phosphate uptake in activated sludge fed with glucose or acetic acid as carbon source. Bioresour Technol 102: 8165-8170.
3. Oehmena A, Lemosa PC, Carvalhoa G, Yuanb Z, Kellerb J, et al. (2007) Advances in enhanced biological phosphorus removal: From micro to macro scale. Water Res 41: 2271-2300.
4. Panswad T, Doungchai A, Anotai J (2003) Temperature effect on microbial community of enhanced biological phosphorus removal system. Water Res 37: 409-415.
5. Lopez-Vazquez CM, Oehmen A, Hooijmans CM, Brdjanovic D, Gijzen HJ, et al. (2009) Modeling the PAO-GAO competition: effects of carbon source, pH and temperature. Water Res 43: 450-462.
6. Ren N, Kang H, Wang X, Li N (2011) Short-term effect of temperature variation on the competition between PAOs and GAOs during acclimation period of an EBPR system. Frontiers of Environmental Science & Engineering in China 5: 227.
7. Tsai YP, You SJ, Pai TY, Chen KW (2005) Effect of cadmium on composition and diversity of bacterial communities in activated sludges. Int Biodeter& Biodegr 55: 285-291.
8. Rayne S, Carey S, Forest K (2005) Evidence for tin inhibition of enhanced biological phosphorus removal at a municipal wastewater treatment plant. J Environ Sci Health A Tox Hazard Subst Environ. Eng 40: 535-551.

9. Tsai YP, You SJ, Pai TY, Chen KW (2006) Effect of Cd(II) on different bacteria species present in a single sludge activated sludge process for carbon and nutrient removal. J Environ Eng ASCE 132: 173-180.
10. You SJ, Tsai YP, Huang RY (2009) Effect of heavy metals on nitrification performance in different activated sludge processes. J Hazard Mater 165: 987-994.
11. Tsai YP, Chou CC, Lin JW, Lu MS, Chen JK (2013). Metal ions toxicities to nitrification and denitrification of a sequential batch reactor. J Environ Protect 4: 10-15.
12. You SJ, Tsai YP, Huang RY (2009) Effects of heavy metals on the specific ammonia and nitrate uptake rates in activated sludge. Environ Eng Sci 26: 1207-1215.
13. Wang Y, Ren Z, Jiang F, Geng J, He W, et al. (2011). Effect of copper ion on the anaerobic and aerobic metabolism of phosphorus accumulating organisms linked to intracellular storage compounds. J Hazard Mater 186: 313-319.
14. Tsai YP, Tzeng HF, Lin JW, Lu MS, Lin JY (2013) Verification of enzymes deterioration due to Cu(II) presence in an enhanced biological phosphorus removal system. Chemosphere 91: 602-607.
15. You SJ, Tsai YP, Cho BC, Chou YH (2011) Metabolic influence of lead on polyhydroxyalkanoates (PHA) production and phosphate uptake in activated sludge fed with glucose or acetic acid as carbon source. Bioresour Technol 102: 8165-8170.
16. Tsai YP, Chen HT (2011) Influence of sludge retention time on tolerance of copper toxicity for polyphosphate accumulating organisms linked to polyhydroxyalkanoates metabolism and phosphate removal. Bioresour. Technol 102: 11043-11047.
17. APHA (1999) Standard Methods for the Examination of Water and Wastewater. 20th edn. American Public Health Association/American Water Works Association/American Water Environment Federation, Washington, DC.
18. Chua CF, Yuang HC, Yos J (1998) Phosphorus and polyhydroxyalkanoates variation in a combined process with activated sludge and biofilm. Water Sci. Technol 37: 593-597.
19. Sudiana IM, Mino TH, Satoh K, Nakamura, Matsuo T (1999) Metabolism of enhanced biological phosphorus removal and non-enhanced biological phosphorus removal sludge with acetate and glucose as carbon source. Water Sci. Technol 39: 29-35.
20. Mino T, Van Loosdrecht MCM, Heijnen JJ (1998) Microbiology and biochemistry of the enhanced biological phosphate removal process. Water Res 32: 3193-3207.
21. Pereira H, Lemos PC, Reis MAM, Crespo JPSG, Carrondo MJT, et al. (1996) Model for carbon metabolism in biological phosphorus removal processes based on in vivo 13C-NMR labeling experiments. Water Res 30: 2128-2138.

CHAPTER 3

Mineral Chemistry and Thermobarometry of the Volcanic Rocks in Torud, Iran

Elham Shah Hosseini[1], Afshin Ashja Ardalan[2], Mohammad Hashem Emami[3], Mohammad Hossein Razavi[4]

[1]*Department of Geology, North Tehran branch, Islamic Azad University, Tehran, Iran*

[2]*Department of Geology, North Tehran Branch, Islamic Azad University, Tehran, Iran*

[3]*Department of Geology, Eslamshahr Branch, Islamic Azad University, Eslamshahr, Iran*

[4]*Department of Geology, Faculty of Sciences, Kharazmi University, Tehran, Iran*

ABSTRACT

This paper elucidates the compositional studies on clinopyroxene, plagioclase of basalts to andesitic rocks of Torud area to understand the geotectonic and geothermobarometry conditions. Early Eocene-Oligocene calc-alkaline volcanic rocks are exposed around Torud in the Central Iranian zone. Volcanic rocks consist of basaltic, andesite basalt, Tracyandesite, and andesite. Minerals in the volcanic rocks exhibit degrees of disequilibrium features. Plagioclase as dominant mineral in these rocks generally displays oscillatory zoning. Mineral chemistry studies show that clinopyroxenes in the volcanic rocks are diopside, augite and plotted in medium pressure field. The clinopyroxene composition yields the crystallization temperatures 900°C - 1000°C. The mineral composition indicates that these rocks are formed in a tensional environment.

Keywords: Volcanic Rocks, Mineral Chemistry, clinopyroxenes, plagioclase, Torud

1. INTRODUCTION

Central Iran is one of the main and most complex geological zones of Iran (Figure 1). The oldest metamorphism (Precambrian) and the youngest semi-active volcanism occurred in this zone. In fact, this region is the oldest micro-continent plate in Iran which has suffered various geological processes. One of the most important and interesting geological events in Iran is the occurrence of orogenic movements comparable with the Katangan in Gondwanaland and Baikalian in the Eurasian continent. In Central East Iran the Microplate was more or less connected to the southern rim of Eurasia during the Paleozoic and early Mesozoic. A significant separation from Eurasia took place during the Jurassic (Soffel and Forster, 1984) [1] . As a result of Neo-Tethys subduction beneath the Central Iran Microplate (CIM) and following the collision of Iranian and Afro-Arabian plates, various structural zones developed in Iran (Alavi, 2004, Nadimi, 2007, Sengör and Yilmaz, 1981 and Ahmadzadeh, Jahangiri, Lentz and Mojtahedi, 2010) [2] - [5] . However, peak of this subduction-related calc-alkaline magmatic activity is thought to have been in the Eocene age (Stocklin, 1973, Farhoudi, 1978, Emami, 1981, Jahangiri, 2007, Berberian and King, 1981, Dupuy and Dostal, 1984, Azizi and Jahangiri, 2008, Jung, Küsten and Tarkian, M., 1976) [6] -[13] . The area under studied is located in the 115 km south of Damghan city and 120 km north southwest of Shahrood with 70 square kilometers. And longitudes of 54°30' and 54°45' and latitudes of 35°15' and 35°30' (Figure 1) are part of Troud 1:250000 scale map of Central Iran micro plate.

2. GEOLOGY

Volcanic rocks and volcanic—sedimentary rocks of Middle to Upper Eocene in Torud area are composed of lava alternate with basic-intermediate pyroclastic rocks and they are deposited with the sedimentary layers (siltstone, nummulities bearing limestone) in a shallow marine environment. Combined spectrum of volcanic rocks consisting of basalt, andesite, trachyandesite, andesite and dacite were obtained by fractional crystallization and eventually contamination process, and clinopyroxene, amphibole and plagioclase, are the major minerals in the rocks. Porphyritic to mga-Porphyritic textures microlithic groundmass are the nature of these rocks. These features are along with its alkaline nature, and represent rocks formed in a back-arc tectonic position of the Middle Eocene—upper area.

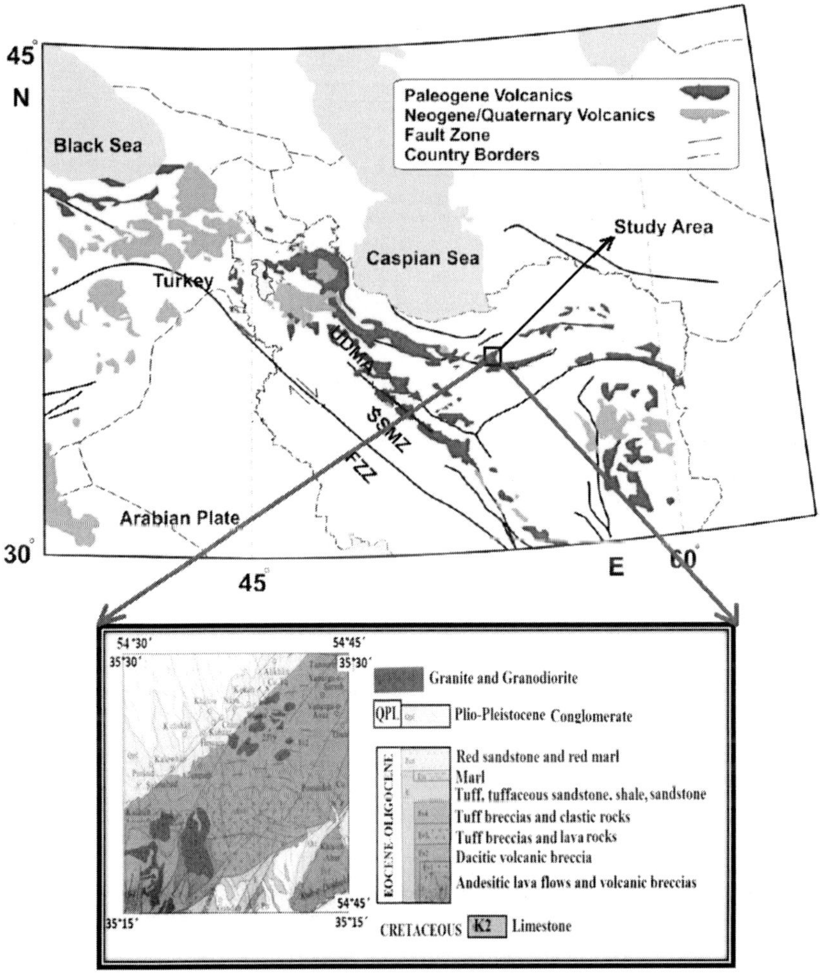

Figure 1. Geographical location of the study area of (1:250,000) Troud map (Housh- mandzadeh, Alavi, Right Pvr, 1357) [14] .

3. ANALYTICAL METHODS

In order to achieve the aims of this work, at the first field surveying and sampling was done, then thin and 38 thin polished sections were prepared.

About the rocks of area, 5 rocks include andesitic basalt, basaltic 39 andesite and andesite were selected for microprobe analysis and their olivine, clinopyroxene and 40 plagioclase minerals by using XGT-7200 micro-XRF analyzer—Horiba with voltage of 50 kv and a 41 current of 3 mA in Kansaran Binaloud Laboratory were analyzed. A total of 80 points was analyzed 42 which results of this analysis are given in Tables 1-3.

Table 1. EPMA data of representative plagioclases from basalts to andesite composition of Torud area.

	PL				
	E28-b-1	E28-b-2	E75-An-1	E75-An-2	E65-An-1
Element	Mass (%)	Mass (%)	Mass (%)	Mass (%)	Mass (%)
Na_2O	0.01	4.37	6.64	4.09	5.45
Al_2O_3	19.12	23.38	22.18	23.58	22.3
MgO	18.31	0.09	0.49	0.81	0.5
CaO	9.45	12.53	4.02	5.95	5.4
P_2O_5	0	0	0.04	0.01	0.01
Fe_2O_3	5.96	0.77	0.7	0.79	0.7
SiO_2	45.82	57.57	64.48	63.33	64.5
K_2O	0.16	1.07	1.26	1.16	1.2
Cr_2O_3	0	0	0	0	0
Rb_2O	0	0.01	0	0.01	0.01
ZrO_2	0.01	0.01	0.01	0.01	0.01
SO_3	0.01	0.02	0	0.04	0.04
SrO	0.01	0.07	0.09	0.09	0.09
MnO_2	0.93	0.06	0.03	0.05	0.03
TiO_2	0.07	0.03	0.05	0.04	0.04
V_2O_5	0.05	0.02	0	0	0
ZnO	0.06	0	0	0.03	0.03
Total	99.97	100	99.99	99.99	100.31
SiO[2]	1.84	2.31	2.58	2.55	2.57
TiO[2]	0.00	0.00	0.00	0.00	0.00
Al[2]O[3]	1.02	1.25	1.18	1.27	1.18
Cr	0.00	0.00	0.00	0.00	0.00
FeO	0.48	0.06	0.06	0.06	0.06
MnO	0.08	0.01	0.00	0.00	0.02
MgO	1.47	0.01	0.04	0.02	0.04
CaO	0.76	1.00	0.32	0.48	0.43
Na[2]O	0.00	0.70	1.06	0.66	0.87
K[2]O	0.03	0.17	0.20	0.19	0.19
Total	5.66	5.50	5.46	5.23	5.36
Or	3.269	9.141	67.003	49.726	58.29
Ab	0.204	37.334	20.283	36.17	28.88
An	96.527	53.524	12.714	14.103	12.83

Table 2. Type of plagioclase using Spreadsheet.

Samples	Or	Ab	An	Sum
E69-BA-1	3.56	0.45	95.9	100
E69-BA-2	3.67	0.86	95.47	100
E28-BA-1	3.269	0.204	96.52	100
E28-BA-2	9.141	37.334	53.52	99.99
E115-Ba-1	2.75	3.66	93.59	100
E115-Ba-2	2.87	0.65	96.48	100
E94-GA-1	57.203	15.352	27.445	100
E94-GA-2	68.928	16.519	14.553	100
E75-AN-1	67.003	20.283	12.714	100
E75-AN-2	49.726	36.17	14.103	99.99
E65-An-1	58.288	28.877	12.834	100

Table 3. EPMA data of representative Clinopyroxene from basalts to andesite composition of Torud area.

CPX	E38-Ba-1	E38-Ba-2	E38-Ba-3	E6-BA	E12-AN-1	E12-AN-2	E12-AN-3	E21-AN-1	E21-AN-2	E21-AN-3
Element	Mass (%)	Mass (%)	Mass (%)	Mass (%)	Mass (%)	Mass (%)	Mass (%)	Mass (%)	Mass (%)	Mass (%)
Na_2O	0.03	0.21	1.38	1.22	0.1	0.36	0.11	0.25	0.2	0.178
Al_2O_3	2.67	2.25	5.3	5.4	2.19	1.84	1.83	2.23	2.14	1.35
MgO	17.56	14.65	12.31	12.46	14.55	14.22	14.01	14.25	14.8	15.2
CaO	18.46	18.97	16.64	17.2	19.91	18.99	20.06	20.36	21	20.12
P_2O_5	0.51	0	0	0	0.01	0	0	0	0	0
Fe_2O_3	8.22	9.85	10.43	11.72	8.7	7.9	8.18	8.5	7.5	7.5
SiO_2	52.04	53	51.91	51.03	53.51	55.88	55.14	53.25	52.63	55.34
K2O	0.02	0.01	0.26	0.28	0.02	0	0	0.12	0.19	0.14
Cr_2O_3	0	0	0.11	0.11	0	0	0.03	0.03	0.03	0.03
Rb_2O	0	0.01	0.01	0.01	0	0	0	0	0	0
ZrO_2	0.02	0.02	0.01	0.01	0.01	0.02	0.02	0.02	0	0
SO_3	0.01	0.01	0.04	0.04	0.05	0.01	0	0.04	0	0
SrO	0.01	0.01	0.01	0.01	0.01	0	0	0	0	0
MnO_2	0.31	0.38	0.33	0.35	0.44	0.42	0.55	0.4	0.5	0.5
TiO_2	0.51	0.57	0.71	0.75	0.49	0.33	0	0.45	0.44	0.5
V_2O_5	0.08	0.06	0.08	0.08	0	0.04	0.06	0.05	0.05	0.05
ZnO	0	0	0	0	0	0	0	0	0	0
Total	100.43	112.31	99.53	100.67	99.99	100.01	99.99	99.95	99.48	100.908
SiO[2]	1.919	1.928	1.966	1.90	2.025	2.039	1.977	0.983	1.01	2.014
TiO[2]	0.014	0.020	0.016	0.02	0.000	0.009	0.014	0.008	0.00	0.011
Al[2]O[3]	0.116	0.232	0.098	0.24	0.079	0.079	0.095	0.066	0.05	0.058
Cr	0.000	0.003	0.000	0.00	0.001	0.002	0.000	0.000	0.00	0.001
FeO	0.253	0.324	0.305	0.36	0.251	0.241	0.269	0.305	0.25	0.228
MnO	0.010	0.010	0.012	0.01	0.017	0.013	0.014	0.012	0.02	0.015
MgO	0.966	0.682	0.810	0.69	0.767	0.773	0.801	0.810	0.77	0.825
CaO	0.729	0.662	0.754	0.69	0.789	0.742	0.788	0.754	0.79	0.784
Na[2]O	0.002	0.132	0.015	0.09	0.008	0.025	0.007	0.030	0.02	0.013
K[2]O	0.001	0.012	0.000	0.01	0.000	0.000	0.001	0.001	0.00	0.006
Total	4.011	4.006	3.977	4.01	3.938	3.924	3.966	2.969	2.91	3.955
Wo	37.20	36.59	39.71	37.27	42.98	41.24	41.88	42.74	41.24	42.97
En	49.30	37.67	42.68	37.58	41.77	42.97	42.59	41.63	42.97	42.97
Fs	13.39	18.45	16.82	20.37	14.82	14.37	15.14	14.68	14.37	42.97
Ac	0.11	7.28	0.80	4.78	0.43	1.41	0.38	0.95	1.15	0.67

4. DISCUSSION

4.1. Plagioclases

Plagioclase (PLAG) is largely unaltered in rocks from this site and shows a distinct difference in habit related to the type of occurrence and the magmatic affinity of the host lava. It occurs either as phenocrysts or microlites. Plagioclase compositions determined by microprobe are listed in Table 1. The overall and range is 97% - 30%, so the majority of plagioclase crystals are oligoclase, andesine, labradorite, by townite and anorthite in composition between basalt and esite to andesite composition. PLAG plotting in the ternary classification diagram Ab-An-Or (Figure 2) shows that there may be some overlap between phenocrysts of each magma compositions. The petrographic evidence that the PLAG was the liquidus phase is confirmed by the occurrence of individuals relatively rich in anorthite component (up to 95%). The most An-rich plagioclase phenocrysts occur in the Basalt rocks, while the most An-poor crystals are from the Andesitic rocks (Figure 2). Of albite, anorthite and alkali feldspar component is determined by the file Spreadsheet (Table 2).

The textural and compositional disequilibrium evidences in plagioclase are important. Sieve texture and zonation of plagioclase is noticeable in these volcanic rocks (Figure 3). This may be interpreted as resulting from either partial dissolution during the magma mixing process (Feeley and Dungan, 1996) [15] , or a decompression effect (Nelson and Montana, 1992) [16] .

4.2. Clinopyroxene

Clinopyroxene is one of the most abundant mineral phases in the mafic-intermadiate, calc-alkaline rocks from study areas. It occurs as large (up to 4 mm) isolated phenocrysts, glomerocrysts, microphenocrysts, and as a groundmass phase. The compositions of clinopyroxenes are predominantly augite and diopside according to the classification (Morimoto, 1989) [17] (Table 3) (Figure 4). Reverse zoning of phenocrysts can be explained by a number of processes, including: 1) decompression during magma ascent (Kontak, et al., 1984) [18] ; 2) more oxidizing conditions during later stages of crystallization (Luhr and Carmicheal, 1980) [19] and (Grunder and Mahood, 1988) [20] ; or 3) magma mixing (Nixon and Pearce, 1987) [21] . Sector zoning also occurs in some samples and can be interpreted to have originated from rapid crystal growth (Morrice and Gill, 1986) [22] . Of Anstatie Frosilit , Actinolite and Wollastonite component is determined by the file Spreadsheet (Table 4).

Experimental and petrological studies have demonstrated that partitioning of minor elements (e.g., Ti and Al) in pyroxene is strongly growth-rate dependent, although equilibrium partitioning of Ca, Mg and Fe may not be seriously affected by cooling rates (Gamble and Taylor 1980) [23] and hence increase in Al/Ti ratios even in liq- uids fractionating plagioclase (Table 2 and Table 3). On the pyroxene classification of (Morimoto, 1988) [17] the pyroxenes plot in the augite field approaching diopside compositions, similar to pyroxenes from other orogenic volcanic rocks (Ewart, 1979) [24] (Figure 4).

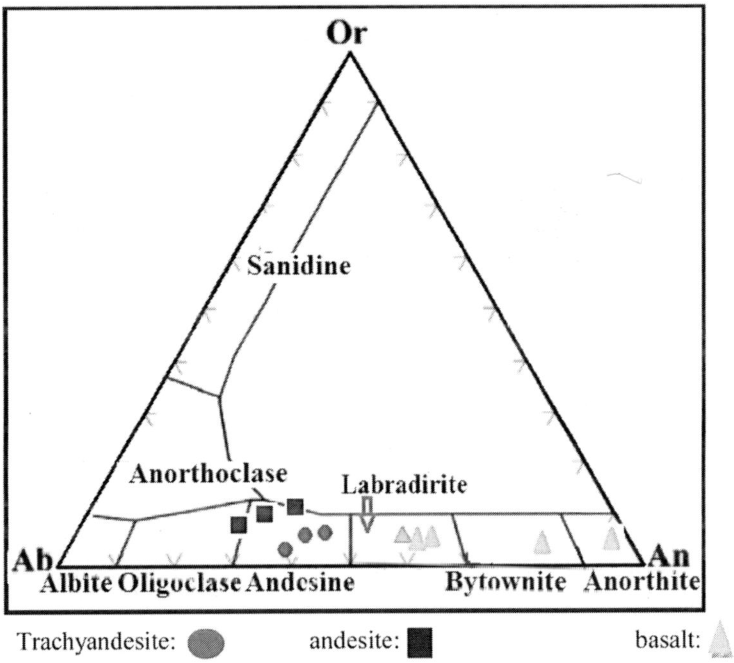

Figure 2. Ternary classification diagram of Ab-An-Or for felds pars (Deer, et al., 1997) [25] .

4.3. Evaluation of oxygen fugacity

Oxygen fugacity plays an important role in changing of the liquidus temperature, melt and crystals composition magmatic processes controlling, crystallization sequence and types of crystallized minerals (France, et al., 2010) [26] , (Kilinc, et al., 1983) [27] , (Moretti, 2005) [28] , (Botcharnikov, et al., 2005) [29] . Using Al^{IV} + Na vs. Al^{IV} + 2Ti + Cr diagram which depend on the amount of 3-valent iron in pyroxenes, we can get oxygen fugacity. The diagram is set based on the aluminum balance in the tetrahedral position with and Cr^{3+} in the octahedral position. The Fe^{3+} in pyroxenes can be displaced 3-valence elements such as Al^{VI}, Ti and Cr in the octahedral position. In the other hand, Fe^{3+} in pyroxenes depends on the amount of Al^{VI} which means it depends on the aluminum balance in tetrahedral and octahedral position. Figure 5 shows that the pyroxenes which crystallized at high oxygen fugacity, has been situated above the line of Fe^{3+}. Furthermore, (Papike and Cameron, 1976) [30] have mentioned the distances of the samples from the Fe^{3+} line and noted that further distances of the samples from this line were indicating more oxygen frugalities in their geological setting. In this diagram some samples are located above the line of Fe^{3+} line and some are below that (Figure 5).

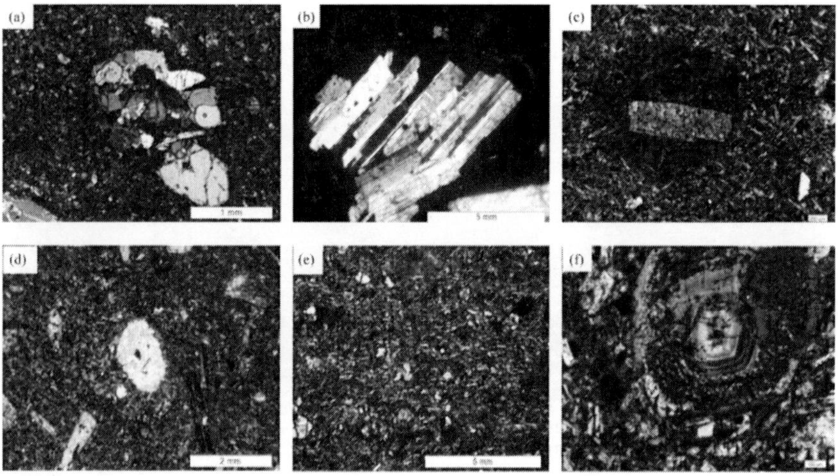

Figure 3. Photomicrographs of thin sections of lavas: (a) pyroxen phenocrysts in a andesite (XPL); (b) plagioclase phe- nocrysts in a basaltic andesite (XPL); (c) clinopyrpxene with microlitic texture in andesite basalt (XPL); (d) amphibole crystals extracted from a cavity filled by calcite in the andesite; (e) the mikrolite rounded olivine and plagioclase with clino-pyroxene phenocrysts of the lavas tracks mikrolite andesitic basalt and porphyritic texture create Stream; and (f) plagioclase phenocrysts with reaction and sieve-textured, resorption-generated zone in mixed lavas (XPL).

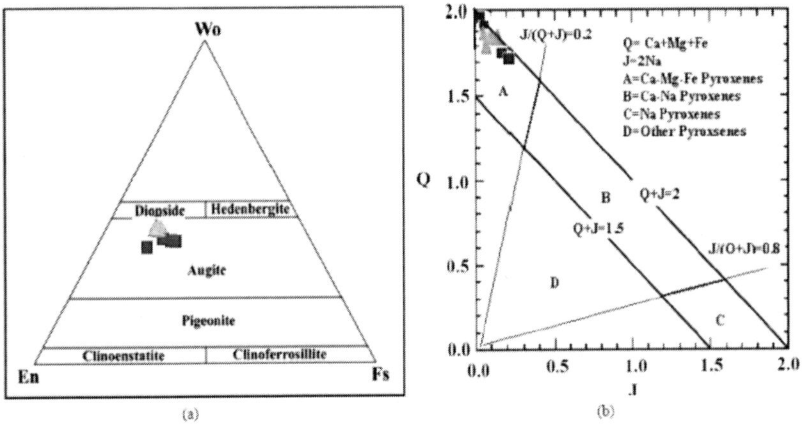

Figure 4. Composition of clinopyroxenes from Tafresh volcanic rocks are plotted in (a) (Deer, et al., 1997) [25] , and (b) the Q-J clinopyroxene classification diagram (Morimoto, 1988) [17] .

Table 4. Type of clinopyroxen using Spreadsheet.

Samples	Wo	En	Fs	Ac	Sum
E38-BA-1	37.20	49.30	13.39	0.11	100.00
E38-BA-2	36.59	37.67	18.45	7.28	100.00
E38-BA-3	39.71	42.68	16.82	0.80	100.00
E6-BA-1	37.27	37.58	20.37	4.78	100.00
E12-AN-1	42.98	41.77	14.82	0.43	100.00
E12-AN-2	41.24	42.97	14.37	1.41	100.00
E12-AN-3	41.88	42.59	15.14	0.38	100.00
E21-An-1	42.74	41.63	14.68	0.95	100.00
E21-AN-2	43.55	42.71	13.00	0.75	100.00
E21-AN-3	42.00	44.15	13.18	0.67	100.00

5. TECTONIC SETTING

Pyroxene composition depends on the chemical composition and tectonic setting of the host lava which can be used widely to determine geological setting of the rocks (Schweitzer, et al., 1979) [32] , (Kushiro, 1960) [33] (Hout, et al., 2002) [34] (Figure 6). Ti + Cr vs. Ca and TiO_2 vs. Na_2O + CaO diagrams define the pyroxenes which generated by tholeiitic and calc-alkaline magmas from those of alkaline. Plotting data on Ti + Cr vs. Ca and TiO_2 vs. Na_2O + CaO diagrams show that the samples are subalkaline to alkaline tendency (Leterrier, et al., 1982) [35] . Diagrams (Figure 6) indicate the extensional setting of the area when the rocks are formed.

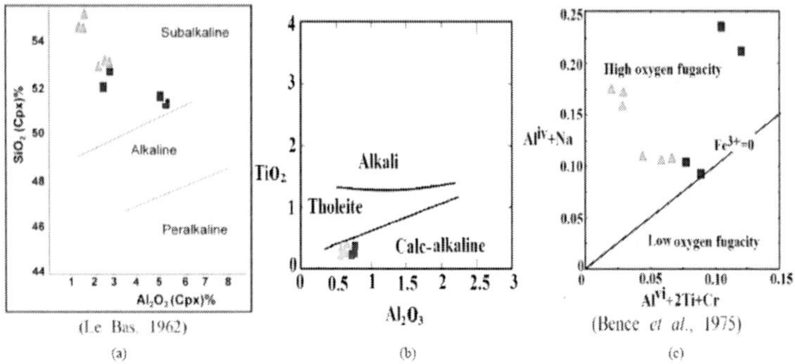

Figure 5. (a) TiO_2 vs. Al_2O_3 binary diagram after (Le Bas, 1962) [31] , for pyroxene; (b) SiO_2 vs. Al_2O_3 binary diagram after (Le Bas, 1962) for pyroxene; (c) Al^{VI} + Na vs. Al^{IV} + 2Ti + Cr diagram for Oxygen fugacity estimation of the cli- nopyroxenes of in volcanic rocks (Schweitzer, et al., 1979) [32] .

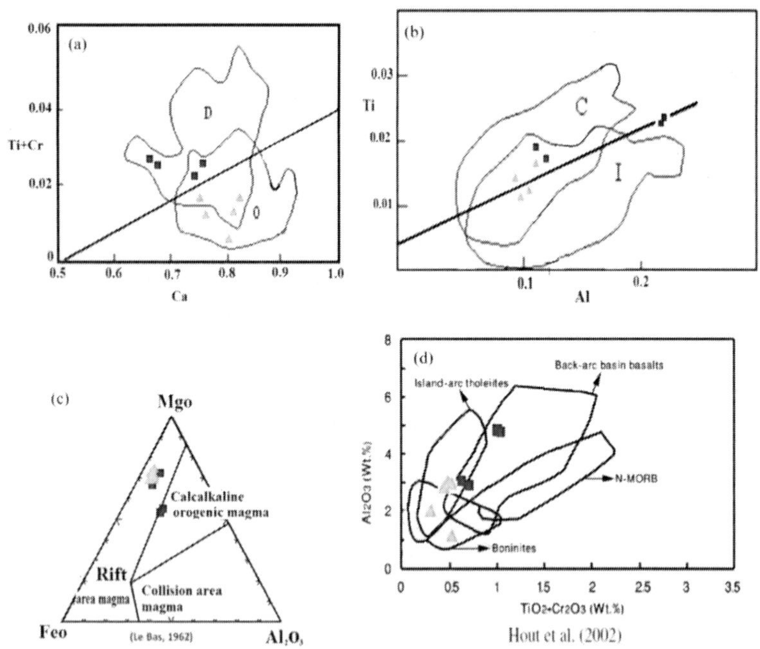

Figure 6. (a) TiO_2-Na_2O + CaO diagram; (b) Ti + Cr-Ca diagram (Leterrier, et al., 1982) [35] ; (c) diagram Tectonic setting (Hout, et al., 2002) [34]; (d) diagram Tectonic setting (Lebas, 1962) [31] . Abbrivation: O—Volcanic arc Basalt; D—MORB and other tholeiites of exten- tional setting; I—Island arc tholeiitic; C—Calc-alkaline basalt..

6. BAROMETERS

The ratios Al^{VI}/Al^{IV}, Ti + Al^{IV}/Si and $TiO_2/(Mg + Mg + Fe)$ in pyroxene could be used as a pressure gauge. The diagram Al^{VI}/Al^{IV} samples are in the range of moderate pressures (Figure 7(a)). Also, the structure of clinopyroxene, chrome is in equilibrium with AlVi, The ratio of Cr*100/Cr + Al^{VI}pressure is directly related to the pyroxene (Nimis and Taylor , 2000) [36] . The aluminum content in clinopyroxene at high pressure to the Reaction $NaAlSi_3O_8 = NaAlSi_2O_6 + SiO_2$ and low-pressure reaction is controlled $CaAl_2Si_2O_6 = CaAl_2SiO_6 + SiO_2$ (Green and Ringwood, 1967) [37] . The first reaction in the depth of about 120 km (containing garnet peridotite), and the second reaction occurs at depths less than 40 km . To determine the depth of the magma reservoir of aluminum in the pyroxene structure is used. Researchers, including (Helz, 1983) [38] have emphasized that the distribution of aluminum in the octahedral positions of tetragonal and clinopyroxene suitable criteria for estimating the water content of the magma and the pressure of the environment is the formation of igneous rocks. With this model, the pyroxene crystallized at a pressure of 5 kbar and water content of the magma is less than 10% (Figure 7(b)).

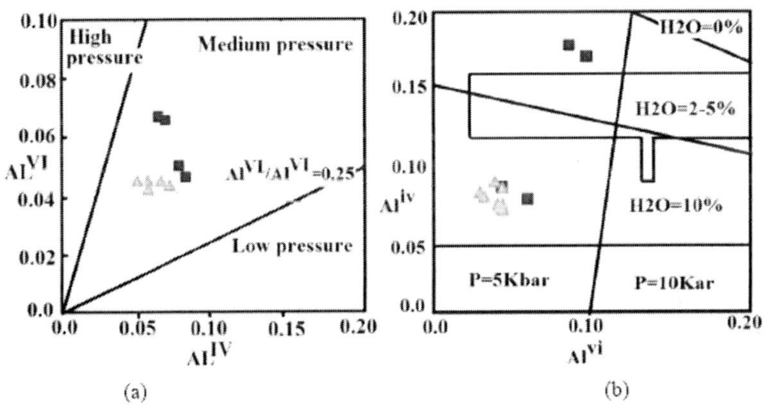

Figure 7. (a) Al^{VI} vs. Al^{IV} diagram (Aoki and Shiba, 1973) [39] , for pressure estimation of the clinopyroxenes of in volcanic rocks. HP = High-pressure field, MP = Medium-pressure field, LP = Low-pressure field. (b) diagram (Helz, 1973) [38] , for pressure estimation of the clinopyroxenes of in volcanic rocks.

7. CONCLUSIONS

- The composition of feldspars and clinopyroxenes determined by XPMA in Torud intrusive bodies is consistent with the results obtained from determination of extinction angles in microscopic thin-sections.
- Mineral types and P-T conditions of magma during crystallization can be calculated using Excel spreadsheet programs.
- The composition of clinopyroxenes, magmatic series and tectonic environments can be determined using XPMA data.
- The compositions of clinopyroxenes are predominantly augite and diopside according to the classification of Morimoto in 1989.
- According to the distribution of aluminum in clinopyroxenes, these minerals have been formed at <5 Kb pressure and water content between 5 to 10 percent.

Elham Shah Hosseini, Afshin Ashja Ardalan, Mohammad Hashem Emami, Mohammad Hossein Razavi · The mineral compositions of the rocks display an alkaline nature which indicates a tensional setting.

REFERENCES

1. Soffel, H. and Forster , H.G. (1984) Polar Wander Depth of the Central East Iran Microplate Including New Results. Neues Jahrbuch für Geologie und Paläontologie, 198, 165-172.
2. Alavi, M. (2004) Structures of the Zagros Fold-Thrust Belt in Iran. American Journal of Science, 307, 1064-1095.
3. Nadimi, A. (2007) Evolution of the Central Iranian Basement. Gondwana Research, 12, 324-333.
4. Celâl ^Sengör, A.M. and Yilmaz, Y. (1981) Tethyan Evolution of Turkey: A Plat Tectonic Approach. Tectonophysics, 75, 181-241.
5. Ahmadzadeh, G., Jahangiri, A., Lentz, D. and Mojtahedi, M. (2010) Petrogenesis of Plio-Quaternary Post-Collisional Ultrapotassic Volcanism in NW of Marand, NW Iran. Journal of Asian Earth Sciences, 39, 37-50.
6. Stocklin, Y. (1973) Basic Geological Study of Central Lut, East of Iran. Report 22-F, Institute of Geology and Mining Publications, Tehran, 56.
7. Farhoudi, G. (1978) A Comparison of Zagros Geology to Island Arcs. Journal of Geology, 86, 325-334.
8. Emami, M.H. (1981) Geologie de la Region de Qom. Aran (Iran) Contribution an l'etude Dynamique Etgeochemique du Volcanisme Tertiaire I 'Iran Central. These doctoratw Earth Grenoble, Vol. 21, 489 p.
9. Jahangiri, A. (2007) Post-Collisional Miocene Adakitic Volcanism in NW Iran: Geochemical and Geodynamic Implications. Journal of Asian Earth Sciences, 30, 433-447.
10. Berberian, M. and King, G.C.P. (1981) Towards a Paleogeography and Tectonic Evolution of Iran. Canadian Journal of Earth Sciences, 18, 210-256.

11. Dupuy, C. and Dostal, J. (1984) Trace Element Geochemistry of Some Continental Tholeiites. Earth and Planetary Science Letters, 67, 61-69.
12. Azizi, H. and Jahangiri, A. (2008) Cretaceous Subduction-Related Volcanism in the Northern Sanandaj-Sirjan Zone, Iran. Journal of Geodynamics, 45, 178-190.
13. Jung , D., Küsten, M. and Tarkian, M. (1976) Post-Mesozoic Volcanism in Iran and Its Relation to the Subduction of the Afro-Arabian under the Eurasian Plate. In: Pilger, A. and Rosler, A., Eds., Afar between Continental and Oceanic Rifting, E. Schweizerbart'scheVerlagsbuchhandlung, Stuttgart, 175-181.
14. Houshmandzadeh, A . S . , Alavi, M. and Pvr, R. (1357) Geological Evolution of the Phenomenon TORUD (from the Precambrian to the Present Covenant). Geological Survey, 138 p.
15. Feeley, T.C. and Dungan, M.A. (1996) Compositional and Dynamic Controls on Mafic-Silicic Magma Interactions at Continental Arc Volcanoes; Evidence from Cordon El Guadal , Tatara-San Pedro Complex , Chile. Journal of petrology, 37, 1547-1577.
16. Nelson , S.T. and Montana, A. (1992) Sieved Textured Plagioclase in Volcanic Rocks Produced by Rapid Decompression. American Mineralogist, 77, 1242-1249.
17. Morimoto, N. (1988) The Nomenclature of Pyroxenes. Mineralogical Magazine, 52, 425-433.
18. Kontak, D.J. , Clarck, A.H. and Pearce, T.H. (1984) Recognition of Simple and Complex Zoning in Olivine and Orthopyroxene Phenocrysts Using Laser Interference Microscopy. Mineralogical Magazine, 48, 547-550.
19. Luhr, J.F. and Carmichael , I.S.E. (1980) The Colima Volcanic Complex, Mexico. I. Post-Caldera and Esites from Volcan Colima. Contributions to Mineralogy Petrology, 71, 343-372.
20. Grunder, A.L. and Mahood, G.A. (1988) Physical and Chemical Models of Zoned Silicic Magmas: The Loma Seca Tuff and Calabozos Caldera, Southern Andes. Journal of Petrology, 29, 831 867.
21. Nixon , G.T. and Pearce, T.H. (1987) Laser-Interferometry of Oscillatory Zoning in Plagioclase: The Record of Magma Mixing and Phenocryst Recycling in Calc-Alkaline Magma Chambers, Iztaccihuat Volcano, Mexico. American Mineralogist, 72, 1144-1162.
22. Morrice, M.G. and Gill, J.B. (1986) Spatial Patterns in the Mineralogy of Island Arc Magma Series: Sangihe Arc. Indonesia. Journal of Volcanology and Geothermal Research, 29, 311-353.
23. Gamble, R.P. and Taylor , L.A. (1980) Crystal/Liquid Partitioning Augite: Effects of Cooling Rate. Earth and Planetary Science Letters, 47, 21-33.
24. Ewart, A. (1979) A Review of the Mineralogy and Chemistry of Tertiary—Recent Dacitic, Latitic, Rhyolitic, and Related Salic Volcanic Rocks. In: Barker, F., Ed., Trondhjemites, Dacites and Related Rocks, Elsevier, Amsterdam, 13- 111.
25. Deer, W.A. , Howie, R.A. and Zussman, J. (1997) Single-Chain Silicates. Geological Society of London, London.
26. France , L., Koepke, J., Ildefonse, B., Cichy, S.B. and Deschamps, F. (2010) Hydrous Partial Melting in the Sheeted Dike Complex at Fast Spreading Ridges: Experimental and Natural Observations. Contributions to Mineralogy and Pe- trology, 160, 683-704.
27. Kilinc, A., Carmichael , I., Rivers, M. and Sack, R. (1983) The Ferric-Ferrous Ratio of Natural Silicate Liquids Equilibrated in Air. Contributions to Mineralogy and Petrology, 83, 136-140.
28. Moretti, R. (2005) Polymerisation, Basicity, Oxidation State and Their Role in Ionic Modelling of Silicate Melts. Annals of Geophysics, 56, 340-368.
29. Botcharnikov, R., Koepke, J., Holtz, F., McCammon, C. and Wilke, M. (2005) The Effect of Water Activity on the Oxidation and Structural State of Fe in a Ferro-Basaltic Melt. Geochimica et Cosmochimica Acta, 69, 5071-5085.
30. Papike, J.J. and Cameron , M. (1976) Crystal Chemistry of Silicate Minerals of Geophysical Interest. Reviews of Geophysics, 14, 37-80.
31. Le Bas , M.J. , Le Maitre , R.W. , Streckeisen, A. and Zanettin, B. (1986) A Chemical Classification of Volcanic Rocks Based on the Total Alkali-Silica Diagram. Journal of Petrology, 27, 745-750.

32. Schweitzer , E.L. , Papike, J.J. and Bence, A.E. (1979) Statistical Analysis of Clinopyroxenes from Deepsea Basalts. American Mineralogist, 64, 502-513.
33. Kushiro, I. (1960) Si-Al Relation in Clinopyroxenes from Igneous Rocks. American Journal of Science, 258, 548-554.
34. Hout, F., Hébert, R., Varfalvy, V., Beaudoin, G., Wang, C.S., Liu, Z.F., Cotten, J. and Dostal, J. (2002) The Beimarang Melange (Southern Tibet) brings Additional Constraints in Assessing the Origin, Metamorphic Evolution and Obduction Processes of the Yarlung Zangbo Ophiolite. Journal of Asian Earth Sciences, 21, 307-322.
35. Leterrier, J., Maury , R.C. , Thonon, P., Girard , D. and Marchal, M. (1982) Clinopyroxene Composition as a Method of Identification of the Magmatic Affinities of Paleo-Volcanic Series. Earth and Planetary Science Letters, 59, 139-154.
36. Nimis, P. and Taylor , W.R. (2000) Single Clinopyroxene Thermobarometery for Garnet Peridotites. Part I. Calibration and Testing of the Cr-in-Cpx Barometer and an Enstitite-in-Cpx Thermometer. Contributions to Mineralogy and Petrology, 139, 541-554.
37. Green, D .H . and Ringwood, A .E. (1967) An Experimental Investigation of the Gabbro to Eclogite Transformation and Its Petrological Applications. Geochimica et Cosmochimica Acta, 31, 767-833.
38. Helz, R.T. (1973) Phase Relations of Basalts in Their Melting Ranges at P_{H_2O} = 5 kb as a Function of Oxygen Fugacity, Part I. Mafic Phases. Journal of Petrology, 14, 249-302.
39. Aoki, K.I. and Shiba, I. (1973) Pyroxenes from Lherzolite Inclusions of Itinome-Gata, Japan. Lithos, 6, 41-51.

CHAPTER 4

Heavy metal chemistry in soils received long-term application of organic wastes

Soon-Ik Kwon[1], Yeon-Ah Jang[1], Kye-Hoon Kim[2], Goo-Bok Jung[1], Min-Kyeong Kim[1], Hae Hwang[1], Mi-Jin Chae[1], Seung-Chang Hong[1], Kyu-Ho So[1], Sun-Gang Yun[1], Kwon-Rae Kim[3]*

[1]*National Academy of Agricultural Science (NAAS), RDA, Suwon, South Korea;*

[2]*The University of Seoul, Seoul, South Korea*

[3]*Gyeongnam National University of Science and Technology, Jinju, South Korea*

ABSTRACT

This study was carried out to understand the long-term effect of organic waste treatment on the fate of heavy metals originated from the organic wastes, together with examination of changes in soil properties. For this, the soils received three different organic wastes (municipal sewage sludge, alcohol fermentation processing sludge, pig manure compost) in three different rates (12.5, 25, 50 ton/ha/yr) for 7 years (1994 - 2000) were used. To see the long-term effect, plant growth study and soil examination were conducted twice in 2000 and 2010, respectively. There was no additional treatment of organic wastes for post ten years after ceasing organic waste treatment for seven years. Soil examination conducted in 2010 showed decreases in soil pH, EC, total nitrogen, organic matter, available phosphorus, exchangeable cations and heavy metal contents in all soils received organic wastes compared to the results obtained in 2000. Speciation of heavy metals in soil through sequential extraction showed that organically bound Cu was the dominant species in all treatment and exchangeable Cu was increased in the plots treated with municipal sewage sludge and alcohol fermentation processing sludge. organically bound Ni increased from 25% - 30% to 32% - 45% in 2010 in all treatment while Pb showed increase in carbonate form in all treatments. Zn existed mainly as sulfide and residual forms, showing increases in organically bound form in all treatment during post ten years.

Keywords: Heavy metal; Long-term application; Organic wastes; Sequential extraction

1. INTRODUCTION

Generation of wastes along with industrialization and urbanization of the society has been elevated and treatment of the wastes in safe manner is challenging task of human for the sustainable life in this planet. Recently, social agreement for reusing the wastes as resources has been made and many researches were conducted. In particular, organic wastes defined as wastes containing >40% organic materials such as municipal solid wastes and livestock wastes have been studied intensively to validate whether they are able to be used as alternative nutrient sources for crop plants [1]. Urine and fecal-, food-, and municipal solid-wastes are included and the amount of organic wastes generated has been increasing due to ever increased population. To date, landfill, incineration, dumping to the ocean has been mainly used to treat these wastes in addition to recycling them as resources [2] but as dumping to the ocean is banded from 2012, all the wastes must be treated in the land. In this circumstance, application of the organic wastes to agricultural land has been attracted to the governors as alternative way to treat wastes, in a mean while the soil physicochemical properties can be improved due to the organic substance and nutrients embed in the organic wastes. Epstein et al. (1976) and Smith (1996) reported that organic wastes treatment to soils induced soil amending effects by increasing water holding capacity, porosity, soil pH, cation exchange capacity (CEC), and soil aggregate, and by decreasing soil bulk density [3,4]. Despite such beneficial effects of the organic wastes to soils, actual treatment of the organic wastes has been limited mainly due to possibly contained hazardous substances such as heavy metals [5]. Nicholson et al. (1997) and Zhao et al. (1997) also reported that the elevated concentrations of heavy metals in the organic wastes are limiting factors for recycling the organic wastes in agricultural soils [6,7]. The heavy metal originated from the organic wastes can be accumulated in the soils eventually resulting in detrimental effects on crop plant performance and human health. Therefore, excessive application of the organic wastes was alerted and many researches were conducted to estimate the dose amount of organic waste to agricultural soils to avoid the accumulation of heavy metals in soils [8,9]. Also, to minimize the input of heavy metals to agricultural soils through organic wastes application, the standard guideline values of those metals in organic wastes has been legislated in Korean Fertilizer Management Act [10]. Despite these efforts to retard metal accumulation in soils, gradual increase of metal concentration in soils may occur by the delivered heavy metal ions through organic wastes even if the concentration of heavy metals in those organic wastes are below the guideline values. Therefore better understanding of the fate and chemistry of metal ions in soils originated from the organic wastes is required for developing advanced management protocol.

The current study was conducted to understand the changes in soil properties and heavy metal chemistry in field soils received long-term application of three different organic wastes (municipal sewage sludge, alcohol fermentation processing sludge, and pig manure compost).

2. METHODS AND MATERIALS

2.1. Treatment of Organic Wastes

For organic wastes treatment, 30 field plots ($1 \times 1 \times 1$ m^3) were constructed using concrete retaining walls and each plot was filled with cultivated upland soil after homogenization. The soils used for experimental plot preparation was Samgak series (coarse loamy, mesic family of Typic Distrochrepts) collected in Yeogi mountain, Suwon. The soils collected were sieved through 5 mm sieve prior to be used for filling the prepared plots. The soil was sandy loam with pH 4.85 containing around 1% organic matter. The concentrations of total nitrogen, available phosphorous and heavy metals were lower than the average values of the Korean cultivated upland soils.

Treatment of organic wastes to each plot soil was initiated in 1994 at three different rates of 12.5, 25, and 50 ton/ha/year and continued for another successive seven years till 2000 at the same treatment rate. The introduced organic wastes were municipal sewage sludge (MSS), alcohol fermentation processing sludge (FS), and pig manure compost (PMC). For comparison, chemical fertilizer treatment plot (234 kg N/ha, 51 kg P_2O_5/ha, and 81 kg K_2O/ha) were included and all the treatments were carried out in triplicates including control (no application of organic wastes and chemical fertilizer).

The chemical properties of organic wastes used were summarized in **Table 1**.

Table 1. Selected physicochemical properties and trace metal concentrations of the organic wastes used in this study.

	MSS	AFS	PMC
pH (1:5)	6.7	7.1	8.1
EC (dS/m)	42.6	51.8	60.5
Organic matter (g/kg)	559	788	707
Total nitrogen (g/kg)	34.3	75.1	26.0
P_2O_5 (g/kg)	4.61	40.8	60.7
Cations ($cmol_c$/kg)			
Ca	0.38	0.15	0.35
K	0.13	0.15	0.36
Mg	0.02	0.25	0.14
Trace metal (mg/kg)			
Cu	134.4	177.9	576.8
Pb	19.7	15.1	3.7
Zn	792.9	363.2	782.7

MSS = Municipal Sewage Sludge; AFS = Alcohol Fermentation Processing Sludge AFS; PMC = Pig Manure Compost.

Following the application for seven years, the soils in the plots were allowed to be managed by the conventional cultivation manner for ten years by 2010 without any further organic wastes applications. During this period, chemical fertilizers were applied for crop plant cultivations (bean, Chinese cabbage, reddish etc.).

2.2. Soil Sampling and Analysis

Soil sampling was carried out in March 1994 before application of organic wastes and in October 2000 and 2010 after harvesting crop plant. Five discrete soil samples collected using ø 5 cm augur sampler from each plot were composited to make a representative soil sample for each plot. The collected soils were thoroughly mixed, air-dried, sieved using 2 mm stainless steel sieve.

Soil pH and EC were determined on a 1:5 soil:water (w/v) suspension using a combination pH-EC meter (Thermo Orion 4 Star Plus, USA) following 1 h equilibration. Soil organic matter content was determined using the Walkely-Black method while organic matter contents of the organic waste were determined using loss on ignition at 550°C [11]. Total-N in soil and organic wastes were determined by the Kjeldahl digestion method followed by titration using 0.5 N $FeSO_4$ [12]. Soil texture was determined using a micro-pipette method [13]. Cation exchange capacity and exchangeable cations in soils were determined using a 1 M CH_3COOH (pH 7.0) extraction method [14].

Heavy metal concentrations in soils and organic wastes were determined by inductively coupled plasma-optical emission spectrometer (ICP-OES; GBC Integra XL, Australia) following acidic digestion. While soils (0.5 g, <2 mm) were digested with aqua regia (10 ml) using a micro-wave digestion oven (MARS5, CEM), organic wastes (0.5 g) were digested with concentrated HNO_3 (15 ml) using a block digester (Buchi digest system K-437, Buchi labortechnik, Switzerland) [15]. Each batch of digested soil samples also included a certified reference material (Montana Soil SRM2711, National Institute of Standards and Technology) and blanks to validate the digestion procedure. Each batch of digested organic waste samples also included blanks and spiked samples at a rate of 5% to ensure quality control. The digested suspensions of soils and organic wastes were filtered through 0.45 μm disposable filters and trace metal concentrations in filtrate determined using ICP. Hereafter, aqua regia acid extractable concentrations are referred to simply as the total metal concentration in the soil. Available and/or soluble metal pools in the soil were assessed using the 0.1 N HCl extraction method [16]. For this, soil (10 g, <2 mm) was shaken for one hour with 0.1 N HCl (50 mL) in a 250 mL Erlenmeyer flask. The soil suspension was subsequently filtered through 0.45 μm disposable filters and the trace metal concentrations in the filtrate analyzed using ICP.

In order to examine the heavy metal fraction existing in different forms in soils, sequential extraction was performed using the extraction method summarized in Table 2 [17]. The heavy metal fractions included soluble-(F1), exchangeable-(F2), organically bound-(F3), carbonate-(F4), and sulfide/residual-forms (F5). The extracted solution was filtered through

0.45 µm disposable filters followed by analysis of heavy metals in the filtrate using ICP.

3. RESULTS AND DISCUSSION

3.1. Changes in Soil Properties

Soil pH of all the experimental plots determined ten years after ceasing organic waste application ranged 3.6 - 4.9 being declined compared to those determined in 2000 (**Table 3**). Among the organic wastes, FS appeared to decline soil pH the most at 3.6 - 4.2. PMC treated soil showed increased soil pH up to 6.0 - 7.0 during the period when the soil received the annual application of PMC from 1994 to 2000 and then the pH declined back to below 5.0 with no application of PMC for post ten years. Soil pH is one of the most important environmental factors governing heavy metal chemistry in soils and hence it was expected that the heavy metals dynamics would be affected by the changes of soil pH along with different organic wastes treatment.

Table 2. Chemical extraction scheme for heavy metal fractionation in soils.

Fraction	Extractant	Shaking time	Metal species
F1	H_2O	30 min	Water soluble
F2	0.5 M KNO_3	16 hr	Exchangeable
F3	0.5 M NaOH	16 hr	Organically bound
F4	0.05 M EDTA	6 hr.	Carbonate
F5	4 M HNO_3	16 hr (at 80°C)	Sulfide & Residual

Organic waste treatment for seven years increased electrical conductivity (EC) of the soils and the increased magnitude was proportional with the application amount appeared the biggest increase occurred with 50 ton/ha/yr treatment. For instance, 50 ton/ha/yr FS treatment showed EC value at 3.39 dS/m in 2000 and this declined down to 1.05 dS/m in 2010. Like FS treated soils, decline of EC in 2010 compared to the values in 2000 was common for all organic waste treated soils. Despite decline of EC with time, the relatively higher EC shown with FS treatment implied that depending on the organic wastes, the application of excessive amount of organic wastes should be avoided not to experience salt damages on crop plants.

Along with organic waste treatment, the contents of organic matter, total-N, available P_2O_5 was increased and then decreased during another ten years without any more input of organic wastes. In particular, MSS and PMC increased available P at the range of 558 - 1500 mg/kg and 594 - 1919 mg/kg, respectively and the PMC treated soil appeared to be declined this value down to 356 - 680 mg/kg while MSS treated soil showed limited decline down to 426 - 1237 mg/kg. This result indicated that the accumulation of available P in soil might be occurred with long-term application of organic waste if the application amount of organic wastes is determined based on the nitrogen concentration embedded in organic wastes, which is generally performed. Therefore, the application amount of organic wastes such as PMC and MSS should be determined based on the available P contents and any insufficient nutrients should be supplied with chemical fertilizers. The previous study conducted by Kim et al. (2000) supported this explanation [18]. In their study, total-N and available P increased with increase in compost application amount and from this result, they concluded that P accumulation would be the detrimental effect with agricultural use of compost and hence countermeasure is required. In comparison with NPK treatment, exchangeable cations were not affected much with organic waste treatment except PMC which increased Ca accumulation levels in soils.

3.2. Changes in Heavy Metal Contents

In general, total concentration of heavy metals in soils with all organic waste treatments decreased during post ten years without any more input of organic wastes compared to those in 2000 (**Table 4**). The general decrease of total heavy metal concentration during ten years may contribute to metal uptake by crop plants and leaching through soil profile [19,20]. Unlike total metal concentration, changes in the 0.1 N HCl extractable concentration were dependant on the type of heavy metals and organic wastes. There was only slight decline in 0.1 N HCl extractable Cu in the soils received PMC and even increased extractable Cu was observed with 50 ton/ ha/yr PMC treatment despite the biggest decline in total Cu observed with PMC treated soil at 50 ton/ha/yr. This might be due to Cu property in regard with total organic carbon (TOC) in soil solution. Cu extractability can be increased with elevated TOC in soil solution and TOC concentration can be increased with successive crop cultivation [20]. Total Ni and Pb concentrations decreased by 1 - 3 mg/kg during post ten years with all treatments as application amount increased while no significant changes in extractable Ni and Pb was observed. This little decline in total concentration and no changes in extractable concentration in Ni and Pb reflected the lower mobility of these heavy metals in soils. Both total and extractable Zn concentration determined in 2000 and 2010 were higher in PMC treated soils than any other organic wastes treated soils even though Zn concentration in the examined organic wastes were in the order of MSS > PMC > FS. This implied that there would be differences in extractability and mobility of Zn depending on the type of organic wastes.

Table 3. Chemical properties in 2000 and 2010 of the soil treated in this experiment.

Year	Treatment	$pH_{1:5}$	$EC_{1:5}$	T-N	OM	AV.-P_2O_5	Exchangeable Ca	Exchangeable K	Exchangeable Mg
			dS/m	g/kg		mg/kg	$cmol_c$/kg		
2000	NPK	5.0	1.22	1.3	15.6	343	0.74	0.56	0.10
	MSS 12.5[a]	5.4	0.38	1.2	19.2	558	1.32	0.08	0.16
	MSS 25	4.8	0.68	1.4	23.5	907	0.79	0.09	0.12
	MSS 50	4.5	1.18	1.8	25.6	1500	0.69	0.10	0.12
	FS 12.5	4.5	1.01	2.6	23.9	366	0.49	0.09	0.30
	FS 25	4.2	1.62	4.3	39.5	495	0.52	0.17	0.43
	FS 50	4.4	3.39	8.5	51.3	765	0.84	0.50	1.00
	PMC 12.5	6.5	0.39	1.5	21.3	594	2.43	0.30	0.58
	PMC 25	6.6	0.62	2.1	25.2	1261	3.03	0.57	0.97
	PMC 50	6.7	1.00	3.5	45.6	1919	4.01	1.19	1.60
2010	NPK	4.7	0.45	0.9	9.3	253	1.18	0.08	0.22
	MSS 12.5	4.4	0.50	1.0	11.7	426	0.81	0.05	0.19
	MSS 25	4.1	0.49	1.2	12.6	807	0.45	0.03	0.10
	MSS 50	3.9	0.63	1.4	14.9	1237	0.33	0.02	0.10
	FS 12.5	4.2	0.59	1.8	12.6	328	0.56	0.04	0.13
	FS 25	3.9	0.71	2.8	15.6	414	0.48	0.04	0.12
	FS 50	3.6	1.05	4.5	23.1	466	0.35	0.03	0.12
	PMC 12.5	4.6	0.45	1.3	9.4	356	1.65	0.08	0.24
	PMC 25	4.7	0.44	1.9	11.2	481	2.21	0.10	0.28
	PMC 50	4.9	0.41	2.3	13.9	680	3.59	0.14	0.41

NPK: Chemical fertilizer treatment; MSS: Municipal sewage sludge; FS: Alcohol fermentation processing sludge; PMC: Pig manure compost; a: ton/ha/yr.

3.3. Heavy Metal Speciation

For copper, organically bound form (F3) was the main species existed in the soils determined both in 2000 and 2010 with all treatments (**Figure 1**). This was in good agreement with the general consensus that Cu has high affinity to bound on the organic substances [17]. The soils received PMC for seven years showed no changes in the fractions of exchangeable (F2), organically bound (F3), and carbonate bound (F4) ten years after ceasing the PMC application compared to the values in 2000. In contrast, FS treated soils appeared to decline in the fraction of F3 and F4 and to increase in the fraction of F2. This implied that decrease of F3 and F4 induced the increased F2 fraction of Cu in soils.

Table 4. Heavy metal concentrations of the experimental soils determined in 2000 and 2010.

Year	Treatment	Total				0.1 N HCl extractable			
		Cu	Ni	Pb	Zn	Cu	Ni	Pb	Zn
		mg·kg^{-1}				mg·kg^{-1}			
2000	NPK	13.7	12.9	25.4	95.4	1.4	0.3	4.8	2.7
	MSS 12.5[a]	19.7	13.7	26.7	112.9	5.3	0.2	6.4	26.4
	MSS 25	26.9	14.7	29.2	118.0	8.0	0.3	7.5	29.5
	MSS 50	38.9	14.9	32.8	127.4	10.8	0.3	7.2	27.1
	FS 12.5	23.6	13.1	23.4	76.9	5.9	0.3	4.2	5.1
	FS 25	33.7	14.1	25.8	84.2	9.7	0.4	3.8	5.7
	FS 50	51.9	14.8	28.7	84.7	16.6	1.0	1.8	12.4
	PMC 12.5	34.4	12.0	24.0	105.3	10.0	0.2	4.6	25.3
	PMC 25	55.4	12.4	24.4	138.7	13.5	0.4	3.7	57.0
	PMC 50	108.2	14.2	26.0	214.3	13.9	0.6	2.2	126.1
2010	NPK	12.3	12.2	22.0	77.5	2.0	0.4	5.1	3.3
	MSS 12.5	17.1	10.6	24.7	89.3	4.2	0.2	5.8	5.2
	MSS 25	24.0	11.5	25.8	93.3	6.1	0.2	6.8	6.7
	MSS 50	35.9	12.3	31.0	98.3	9.1	0.2	8.1	7.4
	FS 12.5	19.5	12.7	23.7	61.7	4.8	0.2	4.4	3.7
	FS 25	26.1	13.4	23.5	70.4	6.8	0.3	3.4	3.2
	FS 50	42.1	14.1	23.0	73.1	11.3	0.2	1.7	2.5
	PMC 12.5	28.1	10.4	22.6	98.6	8.8	0.4	4.0	13.5
	PMC 25	46.8	11.4	23.1	128.4	13.5	0.4	3.7	26.1
	PMC 50	74.3	11.9	24.4	188.3	20.1	0.6	2.6	60.7

NPK: Chemical fertilizer treatment; MSS: Municipal sewage sludge; FS: Alcohol fermentation processing sludge; PMC: Pig manure compost; a: ton/ha/yr.

For nickel, the species fraction in the soils received organic wastes for seven years were in the order of F5 > F3 > F4 > F2 > F1 in 2000 but changes in F3 and F4 were occurred during post ten years without no more input of organic wastes depending on the type of organic wastes applied (**Figure 2**). All experimental soils, except NPK treated soil, showed increase in F3 fraction from 23% - 25% to 29% - 30% in particular with PMC treated soil being increased the most. In contrast, F4 fraction with NPK and MSS treated soils decreased while it was increased with FS and PMC treated soils implied that the F4 fraction was influenced by the type of organic wastes treated.

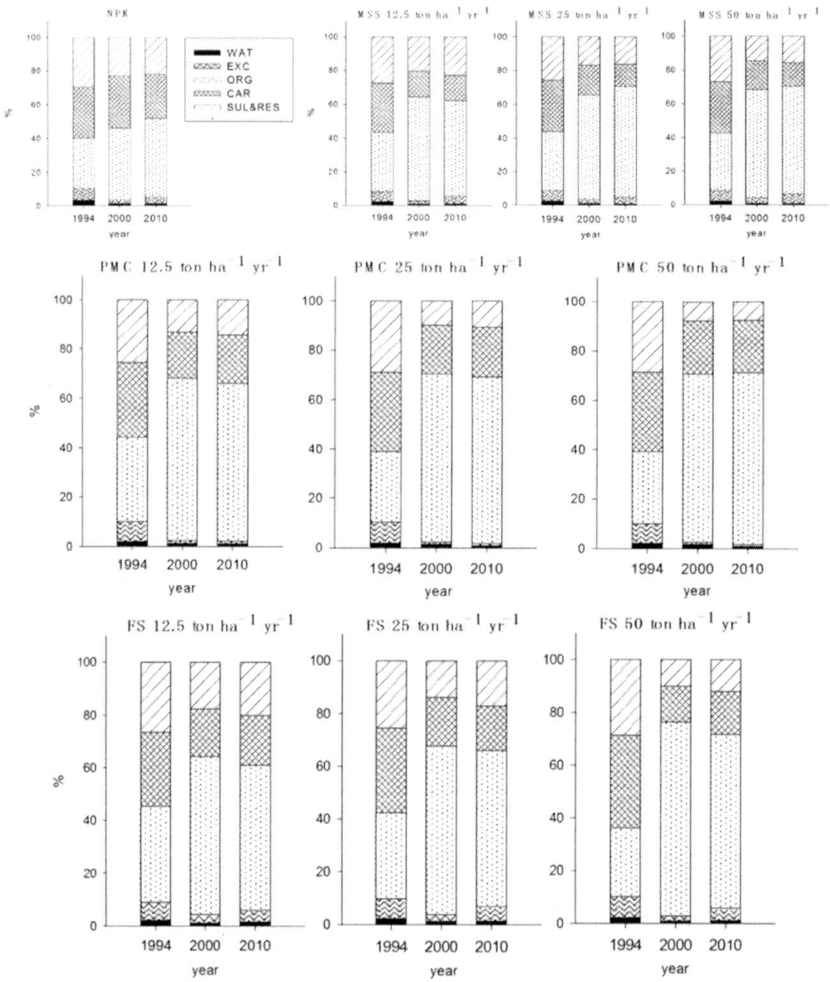

Figure 1. Fraction of Cu in the soils received long-term three different organic wastes applications. The values were determined in 1994, 2000 and 2010 (WAT, water soluble; EXC, exchangeable; ORG, organically bound; CAR, carbonate form; SUL&RES, sulfide and residual forms).

Pb speciation showed higher F4 fraction in both 2000 and 2010 than any other metal ions examined (**Figure 3**). In the previous studies conducted by Jung et al. (2000) and Yoo et al. (1995) also evidenced that EDTA extractable (F4) form was the main species of Pb in soil [21,22]. The post ten years management resulted in increase of F4 fraction of Pb in soils with all treatments while there were no significant changes in F2 and F3 fractions in soils determined in 2010 compared to 2000 soil reflecting less mobility of Pb in soils compared to other metal ions. Yoo et al. (1995) reported that metal mobility is likely retarded as the F3 and F4 fraction increased and hence the lower mobility of Pb in soil was contributed to the higher F3 fraction in soil [22].

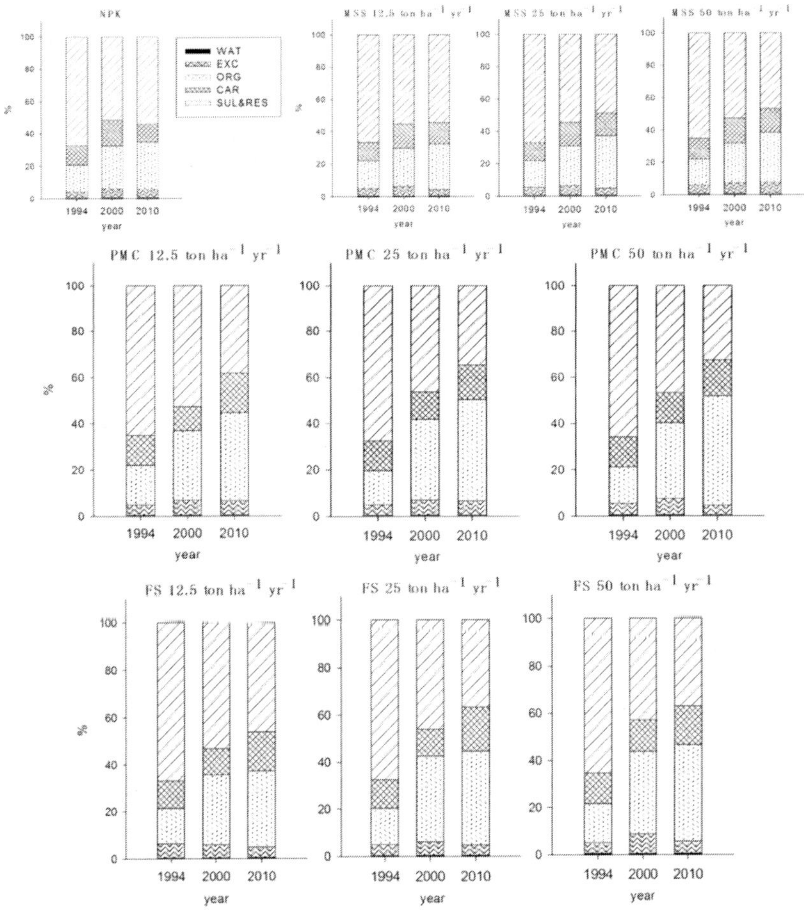

Figure 2. Fraction of Ni in the soils received long-term three different organic wastes applications. The values were determined in 1994, 2000 and 2010 (WAT, water soluble; EXC, exchangeable; ORG, organically bound; CAR, carbonate form; SUL&RES, sulfide and residual forms).

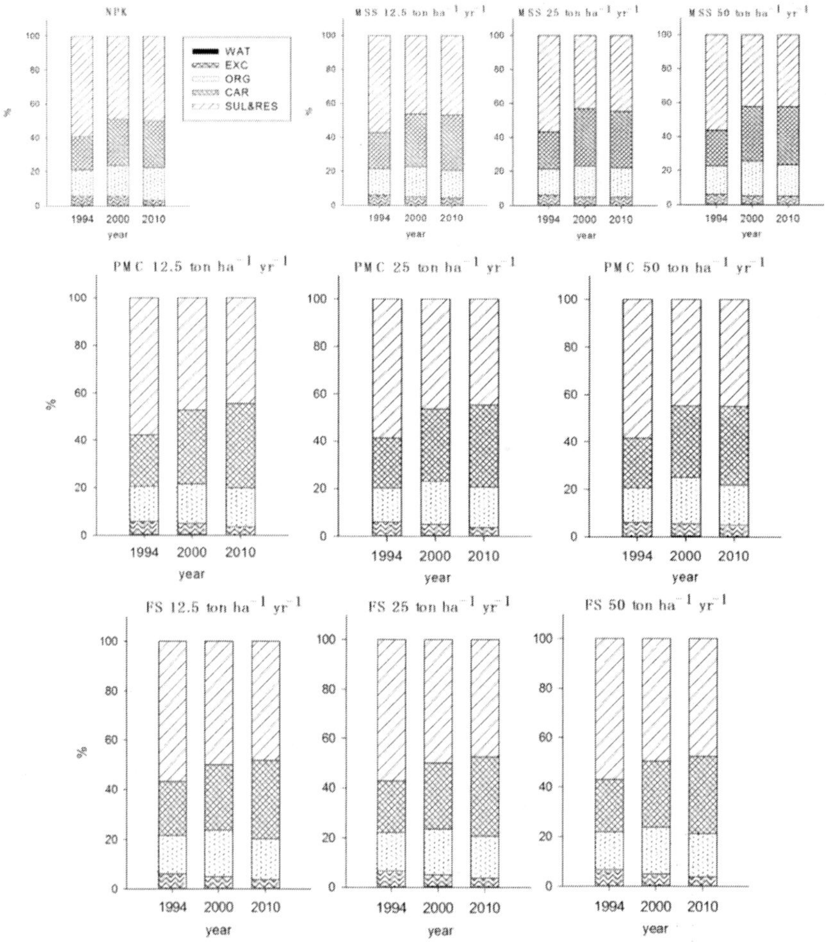

Figure 3. Fraction of Pb in the soils received long-term three different organic wastes applications. The values were determined in 1994, 2000 and 2010 (WAT, water soluble; EXC, exchangeable; ORG, organically bound; CAR, carbonate form; SUL&RES, sulfide and residual forms).

In Zn fractionation, F2 fraction was the major species existing in soils in both 2000 and 2010, which was agreed well with the report by Chang et al. (1984) in which exchangeable and carbonated forms of Zn was the most dominant species in soils received municipal solid sludge [20]. In 2000, the Zn fraction was in the order of F5 > F4 > F2 > F3 > F1 irrespective with the type and amount of organic wastes treated and the fractions of each species were changed during post ten years. While the organic wastes were applied for seven years, F3 fraction of Zn showed no significant change but this increased during post ten years in all experimental soils, in particular with FS treated soil. As a result of increased F3 fractions of Zn in FS and MSS treated soils, F2 and F4 fractions were proportionally decreased in the corre-sponding soils. In contrast, PMC treated soils appeared to be less in the magnitude of F3 fraction increment compared to other organic waste treated soils while increment of F4 fraction with PMC treatment was noticeable. Also, F2 fraction of Zn in the PMC treated soil showed higher value as it was in 2000.

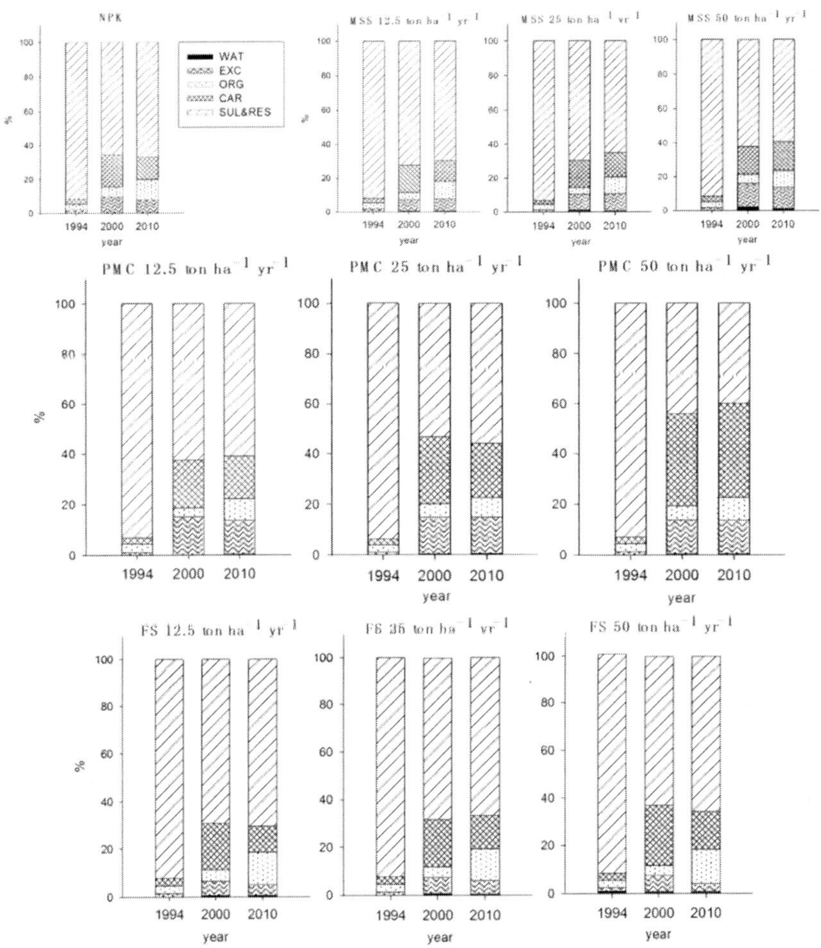

Figure 4. Fraction of Zn in the soils received long-term three different organic wastes applications. The values were determined in 1994, 2000 and 2010 (WAT, water soluble; EXC, exchangeable; ORG, organically bound; CAR, carbonate form; SUL&RES, sulfide and residual forms).

4. SUMMARY

The current study was conducted to examine the changes in soil properties and heavy metal chemistry in the soils received organic wastes (municipal sewage sludge, MSS; alcohol fermentation processing sludge, FS; pig manure compost, PMC) for seven years followed by no more input of organic wastes for another post ten years. Soil pH, EC, total-N, organic matter, available-P, exchangeable cations, total heavy metal concentrations decreased during post ten years. Heavy metal speciation showed increase in exchangeable Cu with MSS and FS treated soils which was contributed to decline in organically bound and carbonate Cu. In contrast with Cu, organically bound Ni was increased from 25% - 30% to 32% - 45% during post ten years. For Pb, carbonated Pb increased in all experimental soils during post ten years, while there were no significant changes in exchangeable and organically bound Pb reflecting low mobility of Pb in soil. Unlike other metal ions, Zn showed increase in organically bound fraction which induced decrease of exchangeable and carbonated Zn fractions in the MSS and FS treated soils.

ACKNOWLEDGEMENTS

Soon-Ik Kwon, Yeon-Ah Jang, Kye-Hoon Kim, Goo-Bok Jung, Min-Kyeong Kim, Hae Hwang, Mi-Jin Chae, Seung-Chang Hong, Kyu-Ho So, Sun-Gang Yun, Kwon-Rae Kim (Project No. PJ008566012012)", National Academy of Agricultural Science, Rural Development Administration, Republic of Korea and this work was also supported by Gyeongnam National University of Science and Technology Grant.

REFERENCES

1. NIER (2004) Resource recycling of waste management status and prospects. National Institute of Environmental Research, Chon, 129-153.
2. NIAST (2003) Agricultural utilization of organic wastes and its environmental risk assessments. National Institute of Agricultural Science and Technology, RDA, Suwon, 49-103.
3. Epstein, E., Taylor, J.M. and Chaney, R.L. (1976) Effects of sewage sludge and sludge compost applied to soils on some physical and chemical properties. Journal of Environmental Quality, 5, 422-426.
4. Smith, S.R. (1996) Agricultural recycling of sewage sludge and the environment. CAB International. Wallingford, 382.
5. Kim, K.-R., Owens, G. and Naidu, R. (2009) Heavy metal distribution, bioaccessibility and phytoavailability in longterm contaminated soils from Lake Macquarie. Australian Journal of Soil Research, 47, 166-176.

6. Nicholson, F.A., Chambers, B.J. and Alloway, B.J. (1997) Effect of soil pH on heavy metal bioavailability. Proceeding of Fourth International Conference on the Biogeochemistry of Trace Elements, Berkeley, 23-26 June 1997, 499-500.
7. Zhao, F.J., Dunham, S.J. and McGrath, S.P. (1997) Lessons to be learned about soil-plant metal transfers from the 50-year sewage sludge experiment at Woburn, UK. Fourth International Conference on the Biogeochemistry of Trace Elements, Berkeley, 23-26 June 1997, 693-694.
8. Horn, A.L., Düring, R.-A. and Gäth, S. (2003) Comparison of decision support systems for an optimised application of compost and sewage sludge on agricultural land based on heavy metal accumulation in soil. Science of the Total Environment, 311, 35-48.
9. Moolenaar, S.W., Lexmond, T.M. and van der Zee, S.E.A.T.M. (1997) Calculating heavy metal accumulation in soil: A comparison of methods illustrated by a casestudy on compost application. Agricultural Ecosystem and Environment, 66, 71-82.
10. MIFAFF (2012) Fertilizer Management Act 2012. Ministry for Food, Agriculture, Forestry and Fisheries, Seoul.
11. Nelson, D.W. and Sommers, L.E. (1996) Total carbon, organic carbon, and organic matter. In: Page, A.L., et al. (Eds.), Methods of Soil Analysis, Part 2. Chemical Analysis, 2nd Edition, Soil Science Society of America, Madison, 961-1110.
12. Bremner, J.M. (1996) Nirogen-total. In: Sparks, D.L., et al. (Eds.), Methods of soil analysis. Part 3. Chemical Methods. SSSA and ASA, Madison, 1085-1121.
13. Miller, W.P. and Miller, M. (1987) A micro pipette method for soil mechanical analysis. Communications in Soil Science and Plant Analysis, 18, 1-15.
14. Sumner, M.E. and Miller, W.P. (1996) Cation exchange capacity and exchange coefficients. In: Sparks, D.L., et al. Eds., Methods of Soil Analysis Part 3. Chemical Methods, Soil Science Society of America, Madison.
15. Kim, K.-R., Owens, G. and Naidu, R. (2010) Effect of root-induced chemical changes on dynamics and plant uptake of heavy metals in rhizosphere soils. Pedosphere, 20, 494-504.
16. MoE (2007) Soil Environment Conservation Act 2007. Ministry of Environment, Seoul.
17. Chang, A.C., Page, A.L., Warneke, J.E. and Grgurevic, E. (1984) Sequential extraction of heavy metals following a sludge application. Journal of Environmental Quality, 13, 33-38.
18. Kim, J.G., Lee, K.B., Lee, S.B., Lee, D.B. and Kim, S.J. (2000) The effect of long-term application of different organic material sources on chemical properties of upland soil. Korean Journal of Soil Science and Fertilizer, 33, 416-431.
19. Businelli, D., Massaccesi, L., Said-Pullicino, D. and Gigliotti, G. (2009) Long-term distribution, mobility and plant availability of compost-derived heavy metals in a landfill covering soil. Science of the Total Environment, 407, 1426-1435.
20. Kim, K.-R., Owens, G., Naidu, R. and Kim, K.-H. (2007) Hyperaccumulation mechanism in plants and the effects of roots on rhizosphere soil chemistry: A critical review. Korean Journal of Soil Science and Fertilizer, 40, 280- 291.
21. Jung, G.B., Kim, W.I. and Ryu, I.S. (2000) Fractionation and availability of heavy metals in paddy soils near abandoned mining areas. Korean Journal of Environmental Agriculture, 19, 319-323.
22. Yoo, S.H., Lee, J.R. and Kim, K.H. (1995) Sequential extraction of Cd, Zn, Cu, and Pb from the polluted paddy soils and their behavior. Korean Journal of Soil Science and Fertilizer, 28, 207-217.

CHAPTER 5

A First Pilot Study on the Sorption of Environmental Pollutants on Various Microplastic Materials

Caroline Goedecke[1,2#*], Ulrike Mülow-Stollin[1,2#*], Selina Hering[2], Janine Richter[1], Christian Piechotta[1], Andrea Paul[1] and Ulrike Braun[1]

[1]*Bundesanstalt für Materialforschung und –prüfung (BAM), Richard-Willstätter-Str. 11, 12489 Berlin, Germany*

[2]*Technische Universität Berlin, Department of Environmental Technology, Straße des 17. Juni 135, 10623 Berlin, Germany*

ABSTRACT

With the drastic increase in plastic production, the input of plastic particles into the environment has become a recognised problem. Xenobiotics are able to sorb to polymer materials, and this process is further enhanced where they encounter microplastics (plastic fragments <5 mm). In this work we studied the sorption of metformin, a type-2 diabetes drug, and difenoconazole, a fungicide, onto the virgin polymer materials polyamide (PA), polypropylene (PP), and polystyrene (PS). Additionally, PP was cryo-milled and PA was treated with acid to investigate the influence of an increase in surface area and chemical modification. The material properties were also studied by dynamic scanning calorimetry (DSC), gel permeation chromatography (GPC) and Fourier transform infrared spectroscopy (FTIR). Sorption experiments were performed on the basis of a full factorial design examining the effect of agitation, pH value, and salinity.

Experimental results showed that difenoconazole sorbs readily to all microplastics, whereas the more polar analyte metformin did not show any affinity to the materials used. For difenoconazole the governing factor in all cases is agitation, while both pH and salinity exhibited only a slight influence. The modification of polymers leads to enhanced sorption, indicating that an increase in surface area (cryo-milled PP) or inner volume (acid-treated PA) strongly favours adsorption. Moreover, long-term experiments demonstrated that the time until equilibrium is reached depends strongly on the particle size.

Keywords: Difenoconazole; Metformin; Plastic debris; Polymer; Dynamic scanning calorimetry

1. INTRODUCTION

Over the last decades, plastic has become one of the most important materials in a wide variety of fields. Global plastic production has risen sharply from 1.7 million tons in 1950 to 311 million in 2014 [1,2]. The polymers with the highest production volumes are polyethyelene (PE), polypropylene (PP), polystyrene (PS), poly(ethylene terephthalate) (PET) and poly(vinyl chloride) (PVC) [3].

Due to the light weight and the durability of plastics, they are used extensively for packaging and in the construction industry [2]. Through these applications the uncontrolled input of plastic litter (macroplastic, >5 mm) into the environment has increased distinctly [4]. There the plastics decompose slowly due to aging processes, often triggered by (photo-)oxidation and mechanical abrasion [5]. Thus small polymer particles, also called secondary microplastics (<5 mm), are formed. Primary microplastics, in contrast, are released directly into the environment as ingredients in cleaning products, paints and cosmetics. Although most of the content is removed in waste water plants, microplastics are still present in sewage treatment plants [6-8].

Hydrophobic chemicals like polychlorinated biphenyls [9] and polycyclic aromatic hydrocarbons [10] are known to adsorb on microplastic surfaces [11]. The surface properties of plastics, such as point zero charge, surface area, surface topography, functional groups, and acid-base character, stongly influence the adsorption of other compounds [12]. Wu et al. reported on the adsorption effects of carbamazepine, 4-methylbenzylidene camphor, triclosan, and 17α-ethinyl estradiol to PE, and evaluated the effects of different salt concentrations and the addition of dissolved organic matter [13]. They determined linear absorption relationships for all substances, which were attributed to their hydrophobicity. Wang et al. investigated the adsorption of perfluorooctanesulfonate (PFOS) and perfluorooctanesulfonamide (PFOSA) on PE, PS, and PVC. They showed that the sorption efficiency depends strongly on the molecular structure and the composition of the microplastics [14].

However, the adsorption ability of many important environmental pollutants has not been studied sufficiently. Especially polar analytes, which are found frequently in surface water and in municipal wastewater, have not been considered in publications. In this study the polar antidiabetic drug metformin and the non-polar triazole fungicide difenoconazole were selected as model compounds for interaction studies with microplastics. Relevant data on the two substances are shown in **TABLE 1**. Metformin is the drug of choice for treating type- 2 diabetes. More than half of the total amount of pharmaceuticals in the environment consist of the antidiabetic agent metformin and its major transformation product guanylurea [15]. In the year 2013, 598 million standard daily doses of metformin were prescribed in Germany [16]. First indications are that metformin could potentially act as an endocrine disruptor in male fish under chronic exposure [17]. Difenoconazole belongs to the group of triazole fungicides that are widely used in agriculture. In 2014, their market share in Germany was 19 %, making them the most commonly used organic fungicides [18]. Difenoconazole is generally perceived as persistent [19]. Furthermore, studies by Mu et al. suggest that difenoconazole acts as an endocrine disruptor [20-22].

Table 1: Structures and properties of the model substances.

	Metformin [22]	*Difenoconazole* [19]
	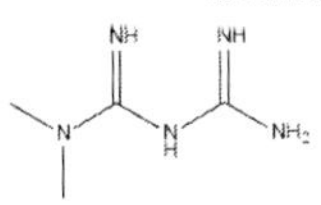	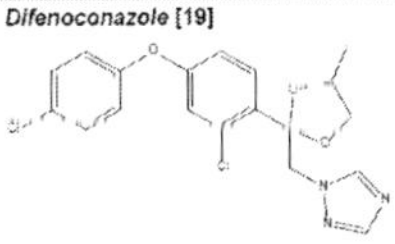
pKa	10.23, 12.33	1.07
logP	-4.30	4.36
water solubility	10^6 mg L^{-1}	15 mg L^{-1}

Due to their high occurrence and suspected endocrine activity, it is important to investigate the adsorption capacities of these compounds on microplastics in order to evaluate possible transportation routes or accumulation in the food chain.

Only a few publications systematically investigate the influence of different surface water conditions like pH value, salinity, and agitation and their influence on the adsorption capacity on microplastics [14,23].

In this study the adsorption of metformin and difenoconazole on polypropylene (PP), polystyrene (PS), and polyamide (PA) microplastics was studied. These three materials were chosen because of their relevance to the environment and the differences in their material properties. PS and PP are often used as packing materials. The densities of the materials vary around that of water. PS and PP are hydrophobic polymers with glass transition temperatures Tglass below and above those in the experimental design (**TABLE 2**). PS is an amorphous polymer, whereas PP is semi-crystalline. PA is also known as nylon, and exhibits excellent mechanical properties, especially as fibres. In this study PA6 was used, as it has a more hydrophilic character. Due to the presence of polar functional groups in PA, an affinity to metformin was expected. The experiments were performed on virgin and aged pellets. Milled PP and acid-treated PA, both of which were characterised physico-chemically, were used to simulate aged microplastics. The parameters salinity, pH, and agitation were varied on the basis of realistic natural conditions and the effects on the sorption of the substances were evaluated.

Table 2: Structures and glass transition temperatures of several polymers.

Polymer		Structure [24]	Glass transition temperature/°C [24]	Density/(g cm^{-3})
PA6	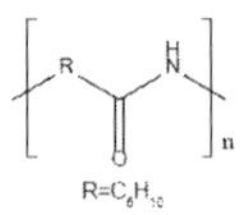	partially crystalline	~78	1.02-1.16
PP	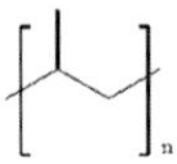	partially crystalline	0-20	0.83-0.92
PS	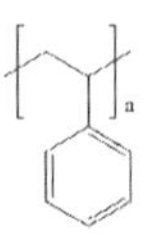	amorphous	90-110	1.04-1.10

2. EXPERIMENTAL

2.1 Chemicals and standards

Metformin hydrochloride (97 %) and formic acid (reagent grade, ≥ 95 %) were purchased from Sigma Aldrich (Taufkirchen, Germany). N-butylbiguanide hydrochloride at a purity of 95 % was received from Select Lab Chemicals GmbH (Bönen, Germany). Methanol and acetonitrile of HPLC grade were obtained from OMNILAB Laborzentrum GmbH & Co. KG

(Bremen, Germany). Difenoconazole (98.5 %) and deuterium-labelled difenoconazole (100 μg mL^{-1} in acetonitrile) were acquired from Dr. Ehrenstorfer GmbH (Augsburg, Germany). Stock solutions of metformin hydrochloride and the internal standard buformin (n-butylbiguanide hydrochloride) were prepared by dissolving an appropriate amount of the substances in ultrapure water to reach a concentration of about 10 mg mL^{-1}. The working solutions were made by diluting the stock solution to achieve a final concentration of 100 μg mL^{-1}. The stock solution of difenoconazole was prepared by dissolving 10 mg of the original substance in 10 mL of methanol. The solution was diluted to reach a final concentration of 0.02 μg mL^{-1} for the spiking solution. 100 μL of the spiking solution were added to 50 mL of ultrapure water to yield the working solution. The internal standard solution of deuterium-labelled difenoconazole was made by diluting the purchased solution with methanol to yield a concentration of 2.5 μg mL^{-1}

2.2 Polymer Materials

The polymer materials for sorption experiments were used as granulate (3-5 mm). All materials, i.e., a polystyrene (PS), a polypropylene (PP) as well as a polyamide 6 (PA6), were kindly provided by partners of PlasticsEurope (Frankfurt, Germany) or originated from previous projects at BAM. None of them contained additives for specific applications (i.e., long-term antioxidants, flame retardants, reinforcement additives, pigments, etc.). PP and PS are hydrophobic polyolefin materials with densities of ~ 0.90 g cm^{-3} and ~ 1.05 g cm^{-3}, respectively, whereas PP displays a semi-crystalline phase morphology. PS is completely amorphous, with a Tglass=95°C. The properties of PA6 (density of ~1.14 g cm^{-3}) are dominated by the polar amide bonds in the chain. PA is semi-crystalline and has a Tglass=40-60 °C, whereas in PA6 Tglass depends strongly on the water content in the sample(**TABLE 2**). For a compilation of the material properties see **TABLE 2**.

MILLED PP: Virgin PP pellets were milled at 16000 rpm in a cryogenic mill (Retsch, ZM 200, Haan, Germany) under nitrogen cooling, using a 2 mm ring sieve. Sieving analysis found that around 55 % of the particles have dimensions of 2.0 to 0.6 mm, ~40 % are between 0.2 to 0.6 mm and only 5 % are smaller than 0.2 mm.

ACID-TREATED PA: To simulate aged microplastics, virgin PA granulates were submerged for 24 h in a mixture of HCl (w=12 %) and 20 % acetone [24,25]. For most polymers the aging process is characterised by a reduction in molecular weight. This degradation process is often initiated by oxidative attack. However, this process is time-consuming, because polymers are usually equipped with antioxidants that supress this reaction mechanism. For our lab experiments we avoided this procedure by using treatment with an HCl/acetone solution. After filtering, the treated pellets were neutralised with sodium hydroxide solution and rinsed with water several times. Finally, the microplastics were dried overnight at 80°C. Afterwards the previously compact granulate pieces had changed into brittle flakes. The aging process was characterised using FTIR, DSC, and GPC analysis.

2.3 Sample preparation and analysis

One gram of the respective microplastic pellets was placed in a glass vial and 50 mL of the appropriate standard solution was added. The pH values of the samples were adjusted with sodium hydroxide solution (0.1 M) and hydrochloric acid (0.1 M). The salinity was regulated by the addition of NaCl to reach final salt concentrations of 0.1 wt % and 3.5 wt %. The samples were stored or agitated on a horizontal shaker for a defined time period.

All experiments were carried out in duplicate. Additionally, one sample without microplastics was treated in the same way, in order to determine the adsorption of the analytes onto the glass surface. This value was then subtracted from the adsorption values for polymer materials.

The analysis of metformin was performed using high-performance liquid chromatography with diode array detection (HPLC-DAD). The microplastics were separated by decantation. Afterwards, 200 µL of the samples and 200 µL of buformin standard solution were combined in a sampling vial and 600 µL of ultrapure water added.

For gas chromatography mass spectrometry (GC-MS) measurements of difenoconazole, 100 µL of deuterium-labelled difenoconazole internal standard solution were added to the sample and the solution was filtered using a folded filter (MN 615, 0.16 mm, 70 g m^{-2}, Macherey-Nagel GmbH & Co. KG, Düren, Germany) to remove the microplastics. The filtrate was extracted for 15 minutes using 10 mL of ethyl acetate. Subsequently, 10 mL of cyclohexane were added and the filtrate was extracted again for 15 minutes. The organic layer was separated and dried over sodium sulfate. The solution was then concentrated to 1 mL under a gentle stream of nitrogen prior to GC-MS measurement.

2.4 Analytical methods

DYNAMIC SCANNING CALORIMETRY: The dynamic scanning calorimetry (DSC) measurements were carried out with a DSC 7020 (Seiko, THASS, Germany) equipped with an autosampler. About 3 mg of the sample were weighed in an aluminium pan and heated up from -80 to 260°C.

Subsequently, the sample was cooled to -80°C and reheated to 260°C. For all measurements a heating rate of 10°C min^{-1} was used under a nitrogen atmosphere (50 mL min^{-1}). In DSC measurements the first heating curve reflects the "thermal history" of the material (in this case, also the exposure to an aqueous solution), whereas the first cooling and second heating can be used for material characterisation.

Gel permeation chromatography: The gel permeation chromatography (GPC) measurements were performed with a Merck- Hitachi L 7100 HPLC-pump and hexafluoroisopropanol (HFIP) as eluent, equipped with a refractive index detector (ERC-7510). The three columns (1 x HFIP gel 30 x 0.8 cm and 2 x PL-HFIP gel 25 x 0.46 cm) were heated to 35°C using a Merck-Hitachi L-5025 column thermostat. Samples dissolved in HFIP with a concentration of 5 mg mL^{-1} were prepared and filtered through a 2 µm PTFE filter. 100 µL of the solution were injected into the system. The flow rate was 0.5 mL min^{-1}. The molar mass distribution (Mn, MW) was calculated using WinGPC software (PSS GmbH, Mainz). Poly(methyl methacrylates) (PMMA) samples of different molar masses were used as calibration standards.

LC-DAD: Analysis of metformin was carried out using an Agilent Technologies HP-1100 Series (Agilent Technologies, Waldbronn, Germany) HPLC system. The HPLC was equipped with a DAD operating at a detection wavelength of 232 nm. Chromatographic separation was achieved with an Acclaim Polar Advantage C16 analytical column (4.1 x 250 mm, 5 µm particle size, Dionex) at a flow rate of 1 mL min^{-1}. The mobile phases consisted of A: H_2O with 10 mmol CH_3COONH_4 and B: MeOH/ACN, 1/1 (v/v). The gradient was 10 % A at 0 min, 90 % A at 6 min, 90 % A at 10 min, 10 % A at 15 min and 10 % A at 18 min.

GC-MS: For analysis of difenoconazole, an Agilent GC 7890A coupled to a 5975C inert XL MSD (Agilent Technologies, Waldbronn, Germany) was used in combination with an MPS2XL Multipurpose Sampler (Gerstel, Mühlheim an der Ruhr, Germany). He 5.0 with a flow rate of 1 mL min^{-1} was used as the carrier gas, and the column was a DB5-MS (30 m × 250 µm × 0.25 µm, Agilent Technologies). Measurements were done in selected ion mode with an EI source set to 70 eV. The MS ion source was maintained at 230°C. One µl of the sample was injected using a KAS4 injector (Gerstel GmbH & Co. KG, Mühlheim an der Ruhr, Germany) which was programmed to start injection at 40°C and ramp up to 275°C at a rate of 12 K s^{-1}, maintaining the target temperature for 5 min. The temperature of the column oven was programmed as follows: 60°C to 210°C at 20 K min^{-1}, hold for 5 min, at 10 K min^{-1} to 255°C, hold for 5 min, at 4 K min^{-1} to 310°C, hold for 10 min. The total runtime was 45.75 min. The signals for the native and the deuterated diastereomers were each integrated separately. For the adsorption values both signals were considered in total.

The sorption of the pollutants was determined via the remaining content of the analytes in the aqueous phase. Therefore, the difference between the ratio of analyte to internal standard at the beginning of the experiment as opposed to the ratio after the end describes the adsorption on the microplastics. The blank value (adsorption on glassware, see section 2.3) was subtracted from this value.

2.5 Experimental design

A full factorial design with two levels for the three design variables "agitation", "pH", and "salinity" was created for all microplastics (PA, PP, PS, PA-acid, and PP-milled) using the Design Expert 9 (Stat-Ease). All designs were tested with true replicates but did not contain centre samples. The designs were evaluated by analysis of variance (ANOVA), including second and third order interactions using "adsorption" as the response variable.

3. RESULTS AND DISCUSSION

3.1 Selection of experimental parameters

Apart from the polymer materials used for the studies, pH value, salinity, and agitation were identified as the main parameters influencing the sorption behaviour of analytes. As the pH value of surface waters usually varies between pH=6 and pH=8 [26], these values were chosen as the lower and upper limits. The salt contents were selected to represent freshwater (0.1 %) and sea water (Mediterranean Sea: 3.5 %) [26-28]. Not all of the salt content in bodies of water is composed exclusively of sodium chloride. Nevertheless, NaCl is by far the most abundant component and has previously been used without further additions to model salt waters [29]. Furthermore, standing bodies of water (agitation 0 min^{-1}) were compared with turbulent waters (240 min^{-1}) [30].

3.2 Adsorption onto virgin polymers

Since difenoconazole is a predominantly hydrophobic analyte (**TABLE 1**), a pronounced sorption affinity to all selected polymer materials was expected. Among the polymers tested, PS was of special interest because of possible interactions of the π-electron systems [31,32] and its amorphous structure [24].

FIGURE 1 illustrates the results of the 24-hour adsorption experiments using virgin polymer material and difenoconazole solution. The graphs depict the adsorption of the analyte on the microplastics as a function of pH, salinity, and agitation. The values for pH and salinity are denoted on the right-hand axis. The absence and presence of agitation are designated by open or filled columns. At first glance it is noticeable that for PA and PP the filled columns correspond to higher adsorption values in all cases. The values obtained for PS exhibit higher variations in the adsorption intensities than PP and PA.

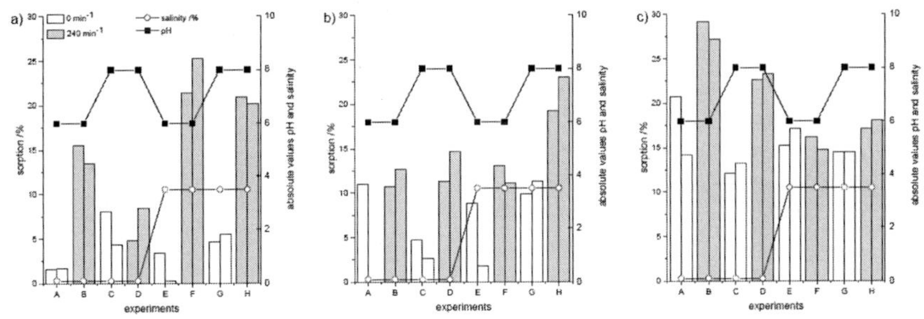

Figure 1: Replicates for adsorption experiments of difenoconazole onto a) PA, b) PP, and c) PS microplastic pellets after 24 hours.

In order to provide a sound hypothesis about the sorption process of difenoconazole on microplastics, the data were evaluated by ANOVA, which yielded significant models in all cases. The results are provided in **TABLE 3**, which summarises both non-significant and significant model terms; the latter are marked with asterisks. From **TABLE 3** it can be seen that among the tested design variables, "agitation" has the strongest effect on adsorption for all microplastics, followed by salinity and pH, both with much weaker effects. The parameter "pH" was a significant term only for PS, but not for PP and PA. In addition, some significant second-order interactions with low effects were identified, but no significant third-order interaction was found. In general, these ANOVA results are in line with the graphs.

Especially for PA and PP, higher adsorption can be observed for the agitated samples than for the non-agitated samples. With these polymers a maximum adsorption of up to 25 % is reached when samples are agitated. **FIGURE 1** shows a trend of increased adsorption at higher salinity for PA and PP. However, the effect of agitating seems to be inverse to the effect of varying pH. When samples are agitated and salinity is high, a change in pH does not lead to a change in adsorption on PA. If the samples are not agitated, increasing the pH value results in higher adsorption. Nevertheless, according to ANOVA the pH value is not a significant term for PA and PP.

Table 3: Parameters of equation in terms of coded factors (+1, -1) derived from ANOVA. Significant variables, i.e., when p-values are found lower than 0.05, are marked with an asterisk.

	PP	PA	PS	PPmilled	PAacid
A: agitation	+4.11*	+6.28*	+2.94*	+33.04*	+24.76*
B: pH	+1.71	-0.37	-1.20*	-3.30*	
C: salinity	+1.90*	+2.74*	-2.10*	+2.35*	1.12
AB		+2.29*			
AC		+2.97*	-2.33*	1.19	+2.24*
BC	+1.85*	+0.47	+1.31*	-2.05*	
ABC					

For PS in general, the adsorption levels reached in the non-agitated samples are higher than for PA and PP. Although upon agitation an increase in adsorption can be observed, the relative increase is less than for the other polymers. Moreover, the effect of salinity on PS is now inverse the effect on PA and PP (negative sign for term "salinity" in **TABLE 3**).

We suppose that the effect of agitation may be related to the inherent properties of the polymer materials, for example their floating behaviour: PA sank to the bottom of the vial, PP floated, and PS was located both on the surface of the water and the bottom of the vial (**TABLE 2**).

Adsorption of difenoconazole in non-agitated solutions was much higher for PS than for PP and PA. PS is the only polymer tested that is completely amorphous (**TABLE 2**). Therefore, adsorption occurs not only on the surface, but may also result from effects like permeation and diffusion into the free volume of the polymer. The observed influence of salinity on the adsorption could be attributed to either a salting-in or salting-out effect of difenoconazole. It can also be assumed that the salt content of the matrix, in interplay with the structure of the respective polymer, is responsible for the changes in adsorption.

Difenoconazole is a weak base (**TABLE 1**). Still, under the parameters used in this study it is mostly neutral, thereby diminishing the effect of a change in pH on adsorption. As the acetale moiety of difenoconazole is also stable under these conditions, no changes in the interaction of the model compound and the polymers are expected due to a chemical change in the molecule.

There are already several reports on the interaction of the employed polymer materials with non-polar analytes. In the case of PP, a high sorption capacity to PAH has already been described. According to Krüger et al., PP containers may reduce PAH recovery by up to 80 % [33]. In a similar study, it was found that higher salinity favours the sorption of phenanthrene (logP=4.46) [34] onto PP [35]. Rochman et al. showed that more low-molecular PAHs (logP <5) sorbed to microplastics made from PS than from PP [10], supporting the results of this study.

With regard to pH, fewer studies with definitive results have been published. Fotopoulou et al. showed through mass potentiometry that PP does not exhibit any acid-base interactions [12]. Interaction between PP and difenoconazole should therefore be more or less the same at pH=6 and pH=8. PA is influenced by pH, however, and may undergo acid-base reactions or even form complexes with metal cations [36]. Still, under the experimental conditions used in this study, no strongly enhanced interaction with difenoconazole under changed pH is to be expected, as proven by our experiments. In their studies dealing with the adsorption effects of PFOS and PFOA, Wang et al. suggested that salinity may play an important role in sorption processes because it can enhance or diminish repulsive electrostatic forces.

Concerning the hydrophilic analyte metformin, under the given parameters no significant adsorption to PA could be observed after 24 hours (data not shown here). Below pH=9, the polar analyte metformin (logP: -4.30, pka: 10.23 and 12.33, see **TABLE 1**) carries a double positive charge [15]. This leads to increased solubility in water and inhibits interaction with non-polar microplastics. Even though the chemical structure of the polymers may also be influenced to some extent by changes in pH, the combined effects do not seem to lead to increased metformin sorption.

It was expected that salinity might strongly affect the adsorption behaviour of metformin because of its charges. However, no change in adsorption could be observed at 3.5 % salt content, which is an environmentally relevant value. Experiments with this salinity were also repeated at pH=10 and pH=3, but did not lead to any detectable adsorption, either. To remain close to environmental conditions, experiments with more extreme pH values were not carried out. In order to avoid the possibility that degradation was not observed due to excessively high concentrations, experiments were carried out with more dilute metformin solutions (10 μg mL^{-1} and 1 μg mL^{-1}).

3.3 Influence of particle size

In order to simulate the adsorption on microplastics with an environmentally relevant size distribution, experiments were repeated using milled PP. The results of the adsorption studies using difenoconazole are shown in **FIGURE 2**. Agitation of samples leads to a strong increase in adsorption compared to the non-agitated samples. It can be assumed that increasing the surface area has the strongest influence on adsorption. In addition, pH, salinity, and the combined effect of both (BC) have a significant effect on the adsorption of difenoconazole (**TABLE 3**).

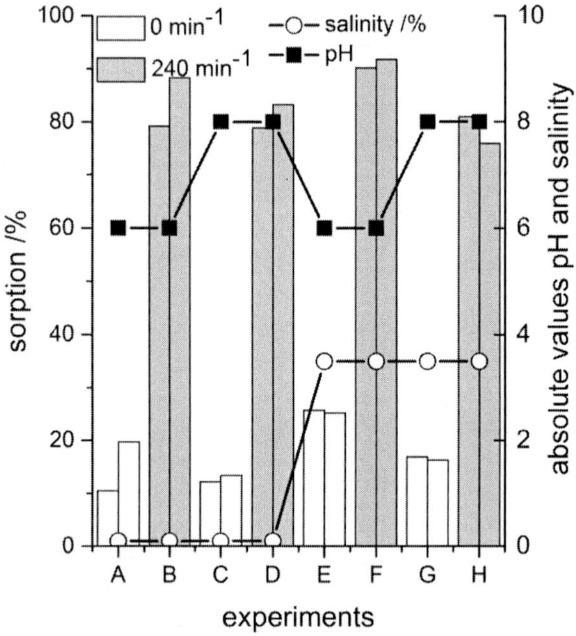

Figure 2: Replicates for adsorption experiments of difenoconazole onto milled PP after 24 hours.

Compared to the virgin PP pellets, adsorption on the milled material increased by a factor of four, to approximately 80 %. Previous studies indicated that sorption on PP is governed by a surface mechanism [35]. Milling of the microplastics leads to a strong increase in the surface area (according to the rough estimation of particle dimensions, by a factor of about 5 to 14). Thus, the sorption processes on the surface of the particles are intensified. In addition, diffusion pathways into the interior of the PP pellets could become accessible due to the milling process, just as Karapanagioti et al. proposed for eroded PP [37]. Our results are consistent with the studies by Bakir et al., who suggested that the sorption equilibrium of organic pollutants on plastics is reached much faster for smaller particles [38].

In order to assess whether the polar analyte metformin had not shown adsorption to any of the virgin microplastics in the previous experiments due to a limited surface area of the polymers, experiments were repeated with milled PP. Since no adsorption of metformin was observed after 24 h, the experiments on milled PP were performed over a period of 34 days with a pH of 8 and a salt content of 3.5 % (for the general practical implementation, see section 2.3). However, no significant adsorption was observed.

3.4 Influence of polymer treatment

FIGURE 3 shows the results of the sorption experiments on difenoconazole with the chemically treated PA. In all cases, agitated samples exhibited adsorption about seven times higher than did the non-agitated samples. This implies that agitation has the most important influence on the adsorption effect and is further in line with the results of ANOVA, suggesting that, besides agitation, only the interaction of agitation and salinity has a significant effect (**TABLE 3**). The adsorption of difenoconazole onto the treated PA is about three times higher than that of the virgin pellets.

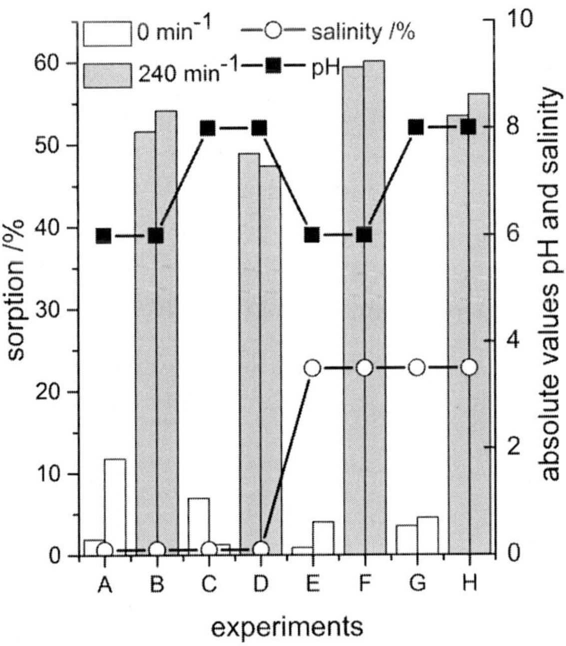

Figure 3: Replicates for adsorption experiments of difenoconazole on acid-treated PA after 24 hours.

The FTIR spectra (see supporting material) do not show any difference between the two materials, indicating that no significant change in the chemical composition (oxidation) of the material took place. In contrast, DSC and GPC analysis show clear differences (**FIGURE 4**). The comparison of the GPC measurements of the virgin and the aged PA6 reveals a clear shift of the peak towards lower molar masses (Mn=42.700 g mol^{-1} to 30.700 g mol^{-1}, MW=84.600 g mol^{-1} to 66.400 g mol^{-1}). The reduction of all relevant key parameters by about 25 % indicates the systematic chain scission of all polymer molecules.

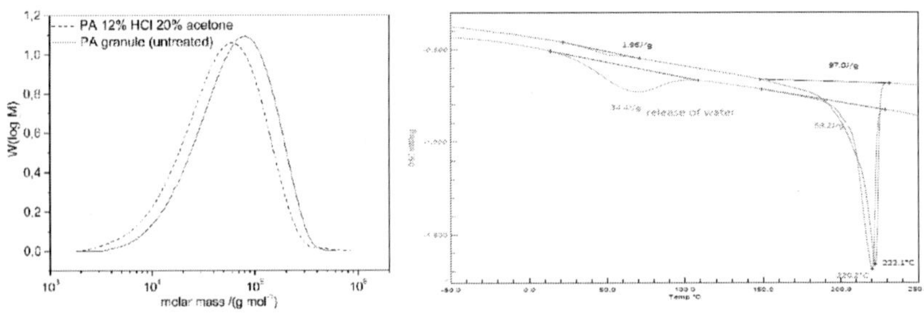

Figure 4: Molar mass distribution in GPC measurements (left) and first heating run of DSC analysis (right).

The result of GPC investigation – the reduction in molar mass – is illustrated in DSC measurements by the shift of the crystallisation peak to higher temperatures during the cooling process, as well as by the shift of the melting peak and the glass transition temperature to lower temperatures during the second heating process [39] (data not shown). However, the first heating cycle offers new results. A small endothermic peak for the virgin PA 6 and a broad peak for the aged PA6 can be observed below 100°C. These peaks are caused by the vaporisation of water. With a sample weight of 2.489 mg and an evaporation enthalpy of water of 2.257 J g^{-1}, a water content of about 1.5 % can be calculated. Around 220°C the endothermic signal of melting is observed. The melting process is related to the crystalline phase of the material. For the aged PA 6 the melting enthalpy and therefore the crystallinity is decreased by 39 % compared to the virgin PA 6.

It can be concluded that treatment with the HCl/acetone solution degrades the molecular weight, but also results in increased water uptake and reduction of crystalline content in the sample. In first approximation, this can be interpreted as a swelling of the molecular chain arrangement. The increased adsorption of difenoconazole onto the treated PA must be related to the reduction of the molecular mass and crystallinity, due to the cleavage of the polymer chains from an otherwise non-altered chemical structure. We therefore suppose that the treatment led to an increase in the inner volume of the polymer particles.

3.5 Long-term experiments

Adsorption processes in the environment have a lot of time to unfold. After a certain period, a dynamic equilibrium can be reached, in which desorption and adsorption take place at the same time with the same velocity. To provide a statement about the time until the equilibrium was reached for difenoconazole in interaction with the microplastics used, samples were evaluated over the course of 28 days. In **FIGURE 5** the adsorption of difenoconazole on several microplastics is shown as a function of time. When an adsorption curve reaches a plateau, equilibrium has been achieved during the sampling period.

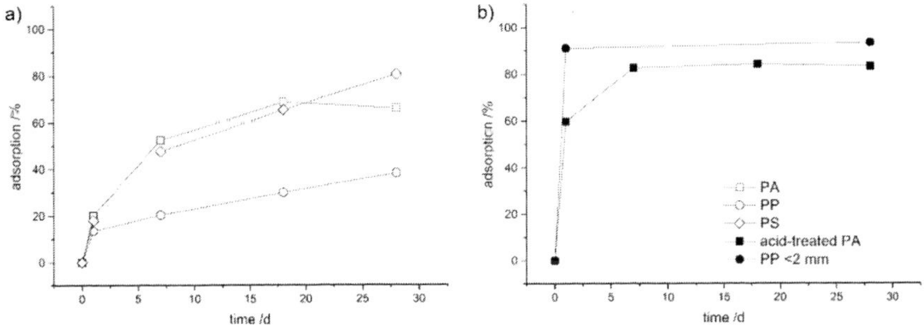

Figure 5: Sorption of difenoconazole onto different polymeric materials over the course of 28 days (pH=6, salinity=3.5 %, agitation=240 min^{-1}).

For the virgin pellets, similar equilibrium rates can be observed for PA and PS (**FIGURE 5A**). This indicates that the physico-chemical properties of these different polymers (amorphous morphology of PS, chemical structure of PA) are of minor importance, or that the different properties compensate for each other. PP adsorbs difenoconazole more slowly.

In contrast to this, the influence of the particle size is obvious for PP: With the milled PP, 91 % of difenoconazole had sorbed to the microplastics after 24 hours (**FIGURE 5B**). Further fluctuations in concentration can be attributed to measurement uncertainties; therefore the equilibrium is approximated. Similarly, for acid-treated PA the equilibrium is already in place after seven days, whereas it is not reached until the 18-day mark by untreated PA. This surprisingly fast adsorption of PP powder and treated

PA indicate a complex mechanism of adsorption and desorption of the permeating species into the polymer particles, as well as changed diffusion, especially through the treated polymer. Because no particle size or particle form distribution is documented for these materials, no further conclusions can be drawn; however, this result indicates the urgent need for exact particle descriptions when microplastic powders are used.

For their studies on the adsorption of different organic compounds on microplastics, Hüffer et al. determined the equilibrium of the sorption process using naphthalene (logP=3.29) [23]. They found that it took longer to reach equilibrium on PS than on PA (17 d vs. 5 d). This verifies the results of the presented study. As for adsorption of phenanthrene (logP=4.53) on PP pellets, it has been reported that it took 20-40 days to reach equilibrium [35]. It is unclear whether a similar value is obtained for difenoconazole, as the duration of the experiment was not sufficiently long.

4. CONCLUSIONS

For the first time the adsorption behaviour of difenoconazole and metformin on microplastics was evaluated systematically. The parameters agitation, salinity, and pH were investigated concerning their influence on the adsorption of metformin and difenoconazole on microplastics. Experiments were based on a 23 full factorial design with the above parameters adjusted to environmentally relevant values. PA, PP, and PS were used as virgin pellets. In addition, PP was cryomilled to increase its surface area and PA was chemically pre-treated to simulate a loss of molecular weight.

Difenoconazole shows sorption onto all hydrophobic polymers used in the study. The adsorption reached its maximum values when particle size (PP) or molecular weights were reduced (PA). Among the untreated polymers, the adsorption of difenoconazole was highest for PS, followed by PP and PA. This is in line with previous studies on the adsorption of PAHs to different synthetic polymers. Our experimental results showed that adsorption onto all microplastics studied here is strongly governed by agitation. The effect of salinity is comparably small, and pH seems to be of least relevance.

Long-term experiments showed an increase in the adsorption of difenoconazole over the entire experimental period of 30 days. Among the untreated microplastics, PS and PA reached adsorption levels of about 80 % and 60 %, respectively, whereas adsorption to PP amounted to 30 % at the end of the long-term study. As no saturation was reached, further adsorption over longer periods must be assumed.

No adsorption of metformin onto untreated polymer pellets was found under the experimental conditions. It can be assumed that because of the high polarity of metformin, sorption is not energetically favoured. Hence, the accumulation on polymer particles and thus long-range transport in water is not of concern for this analyte.

ACKNOWLEDGEMENTS

The authors thank J. Falkenhagen and E. Dümichen for GPC and DSC investigations, as well as A. Müller for providing the milled polypropylene.

REFERENCES

1. Duis K, Coors A (2016) Microplastics in the aquatic and terrestrial environment: sources (with a specific focus on personal care products), fate and effects. Environmental Sciences Europe 28: 2.
2. PlasticsEurope (2015) Plastics - the facts 2014/15, PlasticsEurope, Brussels.
3. Andrady AL (2011) Microplastics in the marine environment. Mar Pollut Bull 62: 1596-1605.
4. Thompson RC, Moore CJ, Vom Saal FS, Swan SH (2009) Plastics, the environment and human health: current consensus and future trends. Philos Trans R SocLond B Biol Sci 364: 2153-2166.
5. Ryan PG, Moore CJ, van Franeker JA, Moloney CL (2009) Monitoring the abundance of plastic debris in the marine environment. Philos Trans R SocLond B Biol Sci 364: 1999-2012.
6. Lassen C, Hansen SF, Magnusson K, Hartmann NB, Rehne Jensen P, et al. (2015) Microplastics- Occurrence, effects and sources of releases to the environment in Denmark, Ministry of Environment and Food of Denmark.
7. Van Cauwenberghe L, Vanreusel A, Mees J, Janssen CR (2013) Microplastic pollution in deep-sea sediments. Environ Pollut 182: 495-499.
8. Browne MA, Galloway T, Thompson R (2007) Microplastic - an emerging contaminant of potential concern? Integrated Environmental Assessment and Management 3: 559-566.
9. Mato Y, Isobe T, Takada H, Kanehiro H, Ohtake C, et al. (2001) Plastic Resin Pellets as a Transport Medium for Toxic Chemicals in the Marine Environment. Environmental Science & Technology 35: 318-324.
10. Rochman CM, Manzano C, Hentschel BT, Simonich SL, Hoh E (2013) Polystyrene plastic: a source and sink for polycyclic aromatic hydrocarbons in the marine environment. Environ Sci Technol 47: 13976-13984.
11. Teuten EL, Saquing JM, Knappe DR, Barlaz MA, Jonsson S, et al. (2009) Transport and release of chemicals from plastics to the environment and to wildlife. Philos Trans R SocLond B Biol Sci 364: 2027-2045.
12. Fotopoulou KN, Karapanagioti HK (2012) Surface properties of beached plastic pellets. Mar Environ Res 81: 70-77.
13. Wu C, Zhang K, Huang X, et al. (2016) Sorption of pharmaceuticals and personal care products to polyethylene debris. Environ Sci Pollut Res Int 23: 8819-8826.
14. Wang F, Shih KM, Li XY (2015) The partition behavior of perfluorooctanesulfonate (PFOS) and perfluorooctanesulfonamide (FOSA) on microplastics. Chemosphere 119: 841-847.

15. Ter Laak TL, Kooij PJ, Tolkamp H, Hofman J (2014) Different compositions of pharmaceuticals in Dutch and Belgian rivers explained by consumption patterns and treatment efficiency. Environmental science and pollution research international 21: 12843-12855.
16. Schwabe U, Paffrath D (2014) Arzneiverordnungs-Report 2014, Berlin, Heidelberg.
17. Niemuth NJ, Jordan R, Crago J, Blanksma C, Johnson R, et al. (2015) Metformin exposure at environmentally relevant concentrations causes potential endocrine disruption in adult male fish. Environmental toxicology and chemistry 34: 291-296.
18. BundesamtfürVerbraucherschutz, Lebensmittelsicherheit(2014) Absatz an Pflanzenschutzmitteln in der Bundesrepublik Deutschland.
19. Lewis KA, Tzilivakis J, Warner DJ, Green A (2016) An international database for pesticide risk assessments and management. Human and Ecological Risk Assessment: An International Journal 22:1050-1064.
20. Mu X, Chai T, Wang K, Zhang J, Zhu L, et al. (2015) Occurrence and origin of sensitivity toward difenoconazole in zebrafish (Danio reio) during different life stages. AquatToxicol 160: 57-68.
21. Mu X, Chai T, Wang K, Zhu L, Huang Y, et al. (2016) The developmental effect of difenoconazole on zebrafish embryos: A mechanism research. Environmental Pollution 212: 18-26.
22. Baken TL(2014) The occurrence, fate and ecological and human health risks of metformin and guanylurea in the water cycle - A literature review, In: KWR Watercycle Research Institute, Nieuwegeln.
23. Hüffer T, Hofmann T (2016) Sorption of non-polar organic compounds by micro-sized plastic particles in aqueous solution. Environmental Pollution 214: 194-201.
24. Frick A, Stern C (2013) DSC-Prüfung in der Anwendung, Carl Hanser Verlag, München.
25. Thöner M (1981) VerfahrenzurVorbehandlung von Polyamid-Formteilenfür das Aufbringenhaftfester, stromlosabgeschiedenerMetalle, Schering AG, Germany.
26. Grohmann AN, Jekel M, Grohmann A, Szewzyk U, Szewzyk R (2011) Wasser. Chemie, Mikrobiologie und nachhaltigeNutzung, De Gruyter, Berlin.
27. Eilers JM, Sullivan TJ, Hurley KC (1990) The most dilute lake in the world? Hydrobiologia 199: 1-6.
28. Dietrich G (1950) Continental influences on temperature and salinity of the ocean water. Deutsche HydrografischeZeitschrift 3: 33-39.
29. Yang G, Ying L, Haichao L (2001) Experimental studies on the local corrosion of low alloy steels in 3.5% NaCl. Corrosion Science 43: 397-411.
30. Bakir A, Rowland SJ, Thompson RC (2012) Competitive sorption of persistent organic pollutants onto microplastics in the marine environment. Marine Pollution Bulletin 64: 2782-2789.
31. Sun S, Bernstein ER (1996) Aromatic van der Waals Clusters:? Structure and Nonrigidity. The Journal of Physical Chemistry 100: 13348-13366.
32. Nakano T (2010) Synthesis, structure and function of [pi]-stacked polymers. Polym J 42: 103-123.
33. Krüger O, Kalbe U, Meißner K, Sobottka S (2014) Sorption effects interfering with the analysis of polycyclic aromatic hydrocarbons (PAH) in aqueous samples. Talanta 122: 151-156.
34. Willighagen EL, Denissen HM, Wehrens R, Buydens LM (2006) On the use of 1H and 13C 1D NMR spectra as QSPR descriptors. J ChemInf Model 46: 487-494.
35. Karapanagioti HK, Klontza I (2008) Testing phenanthrene distribution properties of virgin plastic pellets and plastic eroded pellets found on Lesvos island beaches (Greece). Mar Environ Res 65: 283-290.
36. Roberts MF, Jenekhe SA (1991) Site-specific reversible scission of hydrogen bonds in polymers: an investigation of polyamides and their Lewis acid-base complexes by infrared spectroscopy. Macromolecules 24: 3142-3146.
37. Karapanagioti HK, Ogata Y, Takada H (2010) Eroded plastic pellets as monitoring tools for polycyclic aromatic hydrocarbons (PAH): Laboratory and field studies. Global NEST Journal 12: 327-334.

38. Bakir A, Rowland SJ, Thompson RC (2014) Enhanced desorption of persistent organic pollutants from microplastics under simulated physiological conditions. Environ Pollut 185: 16-23.
39. Fornes TD, Paul DR (2003) Crystallization behavior of nylon 6 nanocomposites. Polymer 44: 3945-3961.

CHAPTER 6

The emergence of stable isotopes in environmental and forensic geochemistry studies: A review

R. Paul Philp

School of Geology and Geophysics, University of Oklahoma, Norman, OK 73019, USA

ABSTRACT

In the past decade, environmental forensics has emerged as a discipline directed toward determining parties liable for causing spills of contaminants into the environment. Such investigations, while geared toward determining the guilty parties in order to recover costs of the cleanup and remediation, require various questions to be addressed. These include determination of the nature of the product; Where did it come from?; Extent of weathering, if any; How long has it been there?; and Is it degrading naturally? Traditionally, these studies have been addressed through utilization of techniques such as gas chromatography (GC) and gas chromatography–mass spectrometry (GCMS). However, in recent years, stable isotopes, primarily determined through the use of combined gas chromatography–isotope ratio mass spectrometry (GCIRMS), have emerged as an equally important tool in environmental forensics. For relatively low molecular, volatile compounds such as MTBE, BTEX, or chlorinated solvents, the isotopes, primarily carbon and hydrogen, have been used extensively for evaluating the onset of natural attenuation. For larger molecules such as PCBs or PAHs, in which the effects of biodegradation on the isotope composition of these molecules is minimal, the isotopic fingerprints of the individual compounds can be used for correlation purposes. In this paper, a brief introduction to isotope geochemistry will be given, followed by a review of applications of stable isotopes to a variety of environmental problems. While the review may not necessarily be exhaustive, it will provide a comprehensive overview of areas where isotopes have been used and potential applications for the future. Most of the review is concerned with carbon and hydrogen isotopes, although a brief overview of the emerging area of chlorine isotopes will also be discussed.

Keywords: Stable isotopes, Environmental Forensic, GCIRMS, Carbon Chlorine, MTBE, PCE, PCBs, PAHs

1. INTRODUCTION

The past decade has seen an exponential increase in the number of forensic geochemical applications in the literature and utilized in litigation. The reasons for this are numerous, but two, in particular, should be noted. First, many of the techniques developed for correlation of crude oils and suspected source rocks, or other oils, based on the biomarker concept are used routinely in forensic applications for correlation of spilled products, both crude oils and refined products. Second, the past decade has seen a tremendous increase in the number of isotope applications to forensic problems. Determination of stable isotopes is not new, having been around for over 50 years. The ability to determine the isotopic composition of individual compounds in complex mixtures is relatively new and came about with the development and commercial availability of combined gas chromatograph–isotope ratio mass spectrometers (GCIRMS) in the late 1980s and early 1990s. Initially, only carbon isotopes could be determined with this approach, but now it is possible to determine hydrogen and nitrogen isotopes in the same manner.

The developments in isotope geochemistry have lead to significant advances in the ability to undertake source determinations at contaminated sites; unravel commingled plumes of contaminants; and monitor natural attenuation at contaminated sites. Any forensic geochemistry study is basically directed toward answering three or four major questions:

1) What is the product(s)?
2) Where did it come from?
3) How long has it been there?
4) Is it going away or degrading naturally?

These four questions, to some degree or other, are geared to answering the ultimate question: Who is going to pay for the cleanup?

This paper will address how the use of stable isotopes has evolved as a tool that can assist in addressing the first four questions. It is important to emphasize that wherever possible, the isotopic data must be combined with complimentary fingerprint data available from techniques such as gas chromatography–isotope ratio mass spectrometry (GCMS).

Sources and types of contaminants in the environment and resulting problems are many and varied and include natural seepage of crude oils, ship traffic, supertankers spilling crude oils, leaking storage tanks, pipelines, hydrocarbons, polychlorinated biphenyls (PCBs), pesticides, dioxins, and many other types of chemical spills from a wide variety of sources (Aislabie et al. 2004; Alzaga et al. 2004; Ohe et al. 2004; Chapman and Riddle 2005; Franco et al. 2006; Hong et al. 2005; Lee et al. 2005; Chen et al. 2006; Samara et al. 2006). In certain cases, characterization of these spills by GC and GCMS data may be ambiguous, and inconclusive, as result of weathering processes such as evaporation, photooxidation, water-washing, and biodegradation changing the distribution of components in the spilled material versus the original sample (Milner et al. 1977; Lafargue and Barker 1988; Palmer 1993). Depending on the nature of the compounds involved and environmental conditions, evaporation may occur in the first few hours after a spill and remove the more volatile components. If the incident occurs in an aquatic environment, water-washing will remove the more water-soluble components from crude oils or refined products, such as hydrocarbons below C_{15}, and some of the C_{15+} aromatic compounds, which are more water-soluble than paraffins (Lafargue and Barker 1988; Palmer 1993; Fuentes et al. 1996). At the same time, biodegradation may start to affect the distribution of individual compounds, although the rate will depend on the nature of the spill and environment into which it spilled (Thornton et al. 2001; Brenner et al. 2002; Gray et al. 2002; Mancini et al. 2002; Meckenstock et al. 2002; Schüth et al. 2003; Griebler et al. 2004; Schirmer and Butler 2004; Kuder et al. 2005; Haws et al. 2006).

Correlation of spilled material with suspected sources, whether hydrocarbon-based, or of some other chemical origin, requires discriminative parameters that are relatively insensitive to weathering processes. In recent years, stable isotope compositions, typically carbon or hydrogen isotopic values, have been used to provide an additional and complimentary tool for correlation purposes (Hartman and Hammond 1981; Macko et al. 1981; Macko and Parker 1983; Farran et al. 1987; Wassenaar and Hobson 2000; Pond et al. 2002; Fang et al. 2002; Wang et al. 2004; Glaser et al. 2005; Gürgey et al. 2005; Horii et al. 2005; Wen et al. 2005). In selected cases, chlorine, nitrogen, oxygen, or sulfur isotope compositions have also been used (Warmerdarm et al. 1995; Sturchio et al. 1998; Beneteau et al. 1999; Reddy et al. 2000; Coffin et al. 2001; Jendrzejewski et al. 2001; Drenzek et al. 2002; Numata et al. 2002; Shouakar-Stash et al. 2003, 2005; Philp et al. 2002; Pond et al. 2002; Griebler et al. 2004; Kuder et al. 2005; Zwank et al. 2005). Potential areas of environmental chemistry that may also benefit from incorporation of stable isotopes as an additional analytical tool can be found in a recent monograph (Lichtfouse et al. 2005). It should also be noted that utilization of isotopes is not restricted to environmental problems and applications to other areas were summarized in a review article by Lichtfouse (2000).

2. ISOTOPE GEOCHEMISTRY

2.1 Principles

For a comprehensive overview of stable isotope chemistry, the reader is advised to consult some of the excellent textbooks recently published in this field (Melander and Saunders 1980; Galimov 1985; Hoeffs 2004; Faure and Mensing 2005). The elements that comprise molecules of common organic contaminants (such as C, H, O, N, S, and Cl) each have at least two stable isotope species. In the recent decades, isotope data published in earth and environmental sciences are reported relative to the same international standard using the delta notation. Delta units are referred to as ‰, or permil, and represent deviation of the measured ratio versus the international standard. In case of carbon, the ratio of ^{13}C to ^{12}C is represented in the delta notation as follows:

$$\delta^{13}C = \left(\frac{R_{\text{sample}}}{R_{\text{standard}} - 1}\right) \times 1000$$

where R is $^{13}C/^{12}C$, and the standard is VPDB, or Vienna Pee Dee Belemnite.

For bulk isotope values, the sample to be characterized is placed in a sealed glass tube with a small amount of CuO. The tube is evacuated, sealed, and combusted at 600°C for several hours to ensure complete combustion of all the organic material in the sample. The tube is attached to the vacuum line and opened under vacuum, water is removed, and CO_2 is pulsed into the isotope ratio mass spectrometer along with the standard gas and the relative proportions of the $^{12}CO_2$ and $^{13}CO_2$ measured and ultimately converted into a $\delta^{13}C$ value using the previous equation.

Isotope values of organic compounds will vary for a variety of reasons. Carbon exists as a mixture of two stable isotopes, ^{12}C and ^{13}C, with the approximate natural abundance of $^{12}C/^{13}C$ ratio being 99:1. The carbon isotopic composition of living organic matter, in part, not only depends on the species but is also determined by a number of environmental properties. Atmospheric carbon dioxide is assimilated by living plants during photosynthesis. The nature of the plants and whether they assimilate CO_2 via a C_3 or C_4 photosynthetic cycle determines the extent of preferential assimilation of the lighter ^{12}C isotope, resulting in a 5–25‰ depletion in ^{13}C. The C_4 plants are typically associated with warmer and more arid climates and, in general, have isotopic values in the −10 to −18‰ range. The C_3 plants, representing 85% of plant species and dominating most terrestrial ecosystems, are generally associated with cooler and wetter climates and have lighter isotopic values in the −22 to −30‰ range. The natural vegetation in temperate and high-latitude regions, as well as tropical forests, is almost exclusively C_3 along with major crops such as wheat, rye, barley, legumes, cotton, tobacco, tubers (including sugar beets), and fallow grasses. The C_4 plants represent less than 5% of floral species but dominate in hot open ecosystems such as tropical and temperate grasslands and include common crops such as sugar cane, corn, and sorghum. Desert plants such as cacti operate under yet another mechanism of photosynthesis referred to as the CAM photosynthesis cycle and can operate under both the C_3 and C_4 cycles.

In summary, the stable isotopic composition of any compound or material reflects of the relative proportions of the two stable isotopes and a measure of how depleted the compound or material is in the heavier isotope relative to a standard. These isotopic source differences are ultimately responsible for the isotopic differences in organic contaminants, most of which are derived from crude oils. The crude oils, in turn, have different isotope signatures as a result of the varying source inputs.

2.2 Isotope fractionation

Changes of isotope ratios caused by various chemical or physical processes are referred to as isotope fractionation or isotope effects. Kinetic isotope effects from degradation result from the different reaction rates for ^{13}C and ^{12}C isotopes. The presence of the heavier isotope in the bond targeted by the degradation increases the bond cleavage activation energy and, as a consequence, reduces the rate of degradation. During the progress of degradation, the residual substrate becomes progressively enriched in the "heavier" isotope species (i.e., $^{13}C/^{12}C$ increases). Degradation-related isotope effects are strong for the atoms included in the chemical bond being broken (primary isotope effects), while the remaining atoms experience weak secondary isotope effects. Isotope effects resulting from biodegradation reflect the first rate-limiting step of the reaction sequence. Biodegradation isotope effects typically are limited to small molecules, readily permeating cell membranes of degrading organisms. Biodegradation of chlorinated ethanes and ethanes, methyl *tert*-butyl ether (MTBE), and monoaromatic compounds have been most widely studied, and isotope fractionation has been observed. Typically, biodegradation of semivolatiles such as long chain *n*-alkanes, multi-ring PAHs (polyaromatic hydrocarbons), and PCBs does not result in measurable isotope fractionation. This, in turn, has advantages, since with the larger molecules, this means that the isotope fingerprints from those compounds can be used for correlation purposes.

2.3 Bulk isotope values

The bulk isotopic composition of a contaminant, such as a crude oil, represents a weighted average of the isotopic composition of all components in a mixture. For crude oils, correlations can be made using the bulk isotopic compositions of the saturate and aromatic fractions rather than the whole oil itself. A cross plot of the isotopic values for the saturate and aromatic fractions will show that samples related to each other will plot very close to each other, while those that are not related will plot in totally different areas. The use of bulk isotopes in this manner is a preliminary screening tool and simply because two samples plot close to each other does not necessarily mean that they are unequivocally related to each other. Once a preliminary relationship between samples has been established, more definitive correlations should be undertaken using techniques such as GC and GCMS.

A classic example of the use of bulk carbon isotopes in a forensic type study would be from the work of Kvenvolden et al. (1995). Tar ball residues from the beaches of Prince William Sound were collected several years after the Exxon Valdez accident and characterized on the basis of their bulk carbon isotope ratios. Based on the bulk isotope values for several residues, two distinct families of tar ball residues could be identified. One family corresponded closely to the Exxon Valdez oil with carbon isotope values around −29‰. The second group was isotopically heavier with values around −24‰, and based on these values plus the biomarker data, it was concluded that this group was derived from Californian crude oils derived from the Monterey Formation. The California crude was imported into Alaska prior to 1964 when a massive earthquake struck the region of Valdez, the tanks ruptured, and the oil spilled into the Sound, where it was left to degrade naturally. Bulk isotope values have been used in many other areas of forensic investigations including determination of the geographic origin of illicit drugs, adulterated foodstuffs, diluted liquor, synthetic drugs, and many others (Hillaire-Marcel 1986).

However, bulk isotope values have limited application, since they reflect a weighted average of the isotopic compositions of all the compounds in a sample. Furthermore, with volatile spills, there is the possibility that evaporation will alter the spilled mixture to such a degree that it will no longer be possible to make correlations based on the bulk isotope values.

Several years ago, the combination of the gas chromatograph and isotope ratio mass spectrometer was finally achieved, and from that point on, it was possible to obtain the isotopic compositions of individual compounds in complex mixtures (Hayes et al. 1990). This provided the opportunity to obtain an isotopic fingerprint which, in the case of complex mixtures, could be used as an additional correlation tool. In the case of single components, it might be the only correlation tool available. The following sections will provide some examples of where this approach has been used in an ever expanding number of applications.

2.4 Isotopic compositions of individual compounds determined by GCIRMS

2.4.1 Solvent extractable fractions and free products

In a typical GCIRMS analysis to determine carbon isotope compositions, chromatographic separation is followed by online high-temperature conversion of individual compounds to CO_2which passes into a continuous-flow IRMS for isotope ratio determination (Merritt et al. 1995). Similar methods are followed for hydrogen (Burgoyne and Hayes 1998). Determination and utilization of chlorine isotopes have been reviewed and discussed by Slater (2003), Schmidt et al. (2004), Meckenstock et al. (2004a,b), and Scow and Hicks (2005). Compound-specific analysis of Cl (Holt et al. 1997) and Br (Shoukar-Stash et al. 2005) is only possible after offline chromatographic separation of the target analytes and offline chemical processing to IRMS-amenable gaseous products (CH_3Cl and CH_3Br, respectively). Two different concepts of interfacing LC (liquid chromatography) and IRMS have been proposed

(Caimi and Brenna 1993; Krummen et al. 2004) but no applications of LC-IRMS to contaminant studies are available at present.

For aqueous, volatile range organic contaminants, optimal results are obtained with a purge-and-trap interfaced to the GCIRMS instrument (Zwank et al. 2003; Kuder et al. 2005). Excellent quantitation limits, in low microgram per liter range or even nanogram per liter for certain species, are possible with this approach. $\delta^{13}C$ values for benzene, toluene, etc. have been determined at concentrations of approximately 700 ng/l and MTBE at 1.5 μg/l using purge-and-trap. Limits of detection are slightly higher for polar contaminants–e.g., 200 μg/l for ethanol and 20 μg/l for *tert*-butyl alcohol (TBA). Hydrogen isotope ratios can also be determined by purge–and-trap but concentrations need to be approximately 10 times higher than for carbon determination. By comparison, carbon isotope analysis of volatile species via headspace analyses requires concentrations in the range of 500 μg/l or more (detection limits from Hunkeler and Aravena 2000). Solid-phase microextraction (SPME) is an improvement over headspace analysis and permits better detection limits but is still not a reproducible as purge-and-trap (e.g., Hunkeler and Aravena 2000).

Chemical fingerprinting of the *n*-alkane fraction in crude oils and refined products in combination with isotopic characterization of carbon in the individual homologues has been used successfully to allocate sources of sediment contamination (Mansuy et al. 1997; Rogers et al. 1999; Abrajano and Sherwood Lollar, 1999; Smallwood et al. 2002) and oiling of bird feathers (Mansuy et al. 1997; Mazeas and Budzinski 2002). Hough et al. (2006) used the $\delta^{13}C$ values of individual *n*-alkanes to identify the source of hydrocarbons in soils contaminated with heavily weathered petroleum wastes.

Pond et al. (2002) proposed the use of hydrogen isotopic composition of longer chain *n*-alkanes (*n*-C_{19} to *n*-C_{27}) for source discrimination, since the hydrogen isotopic signature of crude oil components vary more significantly than the carbon values and do not show discernable changes during weathering or degradation of crude oils. There have been a number of applications in the petroleum geochemistry literature describing the use of both the C and H isotope values of crude oils and saturate hydrocarbon fractions for undertaking oil/oil and oil/source rock correlation studies (Peters and Fowler 2002; Philp et al. 2002; Milkov 2005). However, the number of applications of H/D determinations in the environmental field has been fairly limited and most of them have been simply designed to show applicability of the technique (Hilkert et al. 1999; Sessions et al. 1999; Wiesenberg et al. 2004).

GCIRMS is particularly important in the case of refined products such as gasolines, which may be derived from different feedstocks but may be very similar chromatographically and difficult to discriminate using conventional techniques such as GC and GCMS. In early environmental applications of GCIRMS, Kelley et al. (1995, 1997) measured the concentrations and isotopic compositions of BTEX (benzene, toluene, ethylbenzene, and xylene) compounds collected from a gasoline contaminated site in Southern California. From the results, it was evident that there were two isotopically distinct families of BTEX compounds, permitting one to draw the conclusion that the leaded and unleaded gasoline at this site were derived from different sources. Since that initial report, Smallwood et al. (2002) published a study demonstrating that for gasolines derived from different crude oil feedstocks, an isotopic fingerprint of individual compounds in the gasolines could be used to discriminate the gasolines. It is important to emphasize that this approach is not necessarily going to work every time but, at the same time, the more conventional GC and GCMS will probably not discriminate the samples either if GCIRMS is unsuccessful.

Polycyclic aromatic hydrocarbons in the environment may be derived from sources such as coal tar, crude oils, or creosotes for example. In many cases, the chromatographic distributions of these compounds from different sources may be very similar and of little use in terms of distinguishing one source from another. In addition, weathering can often result in additional ambiguities. However, with the development of GCIRMS, the possibility of determining the isotopic composition of the individual components provides a potential method for differentiating these sources of the PAHs. On a regional scale, source apportionment of PAHs both in the atmosphere and in sediment records has been studied intensely using $\delta^{13}C$ analysis. A combination of concentration measurements and $\delta^{13}C$ isotopic analysis of individual PAHs in sediments from Lake Erie permitted three areas with different contamination histories to be distinguished with the main emission pathway for PAHs being fluvial input (Smirnov et al. 1998). A comprehensive study to evaluate the use of stable isotopes to differentiate PAHs derived from urban backgrounds versus those derived from coal gasification plants has been recently completed (Saber et al. 2003, 2005; Craig et al. 2005; Mauro et al. 2005). PAH samples that appear similar chromatographically may be differentiated on the basis of their isotopic signatures, and similarly those that are isotopically different may be chromatographically similar. Furthermore, the isotopic signatures of samples derived from the manufactured gas plants are distinct from those typically derived from PAHs from the urban background signature (Figure 4.13). Hammer et al. (1998) studied the PAHs from two creosote-contaminated sites and showed that at those sites, there was a suite of compounds that had isotopic values within 1‰ of each other and could be useful in the identification of PAHs of creosote origin. Various sources of PAHs were distinguished in sediments along the St. Lawrence River with relatively heavy $\delta^{13}C$ values found for the three-ring PAHs originating from an aluminum smelter in one area (Stark et al. 2003). McRae et al. (1999, 2000) showed it was possible to relate coal-derived PAHs released during different thermal conversion processes (combustion, pyrolysis, and gasification) with their isotopic compositions conserved in soil PAHs. Extremely isotopically light PAHs ($\delta^{13}C$ from −31 to −62‰) in lagoon sediments near Ravenna (Italy) led to the conclusion that emissions were dominated by a former plant that used biogenic methane ($\delta^{13}C$ from −69 to −73‰) as feedstock rather than by operating plants using petrogenic feedstocks that typically are isotopically heavier (McRae et al. 2000). Wilcke et al. (2002) used $\delta^{13}C$ analysis of perylene to substantiate their earlier hypothesis that recent biological sources of PAHs related to termites are important in tropical environments, whereas pyrolytic sources dominate in temperate climates. PAHs in atmospheric particles resulting from natural burning processes could be distinguished from those stemming from various anthropogenic combustion processes by using fingerprinting and $\delta^{13}C$ analysis of individual compounds (Norman et al. 1999; Okuda et al. 2002a). In Chinese urban areas, PAH $\delta^{13}C$ analysis was successfully used to identify either vehicle exhaust or coal combustion as major PAH source (Okuda et al. 2002b). In contrast, PAH fingerprinting by GC did not yield equivalent information. Many of these studies show the necessity to combine chemical fingerprinting techniques and compound-specific isotope analysis.

2.4.2 Characterization of chlorinated compounds

The widespread occurrence and stability of chlorinated contaminants in the environment, PCBs, dioxins, dichloro-diphenyl-trichloroethane (DDT), and a wide variety of other compounds has lead to an increased interest in the possible use of the chlorine isotope ratios as a means of studying the origin, transport, and fate of these compounds in the environment. The topic of chlorinated solvents can be divided into two areas–nonvolatiles such as PCBs, etc. and volatiles such as perchloroethylene (PCE) and trichloroethylene (TCE), which are common groundwater contaminants. Reddy et al. (2000) studied a number of PCBs from a variety of sources and noted that the $\delta^{37}Cl$ values fell within a narrow range from −3.4 to 2.1‰ but found no correlation between the mass percent of chlorine and $\delta^{37}Cl$ values. It should be noted that the samples used in this particular study were obtained from environmental standard suppliers who, in turn, had obtained the materials from Bayer (Germany) and Caffaro (Italy). The variability observed in the standards was suggested to arise from differences in the starting materials, synthetic methods, purification, or postproduction storage and handling of these products or some combination of these factors. PCBs extracted from sediments ranged from −4.54 to −2.25‰. These isotopic variations suggest that $\delta^{37}Cl$ could be a useful tool for tracing the sources and fate of PCBs. In a laboratory study that monitored the aerobic degradation of dichloromethane (DCM), Heraty et al. (1999) noted that the $\delta^{37}Cl$ values of the residual DCM increased as a function of biodegradation.

In another slightly different approach with PCBs, Jarman et al. (1998) examined the carbon isotope variations for individual congeners in a variety of commercial PCBs. The bulk carbon isotope values for individual PCBs varied from −22.34 to −26.95‰, whereas individual congeners showed values ranging from −18.65 to −27.98‰. Large differences were observed for individual congeners between mixtures. Individual congeners showed a depletion of ^{13}C with increasing chlorine content, probably as result of a kinetic isotope effect caused by the position of the chlorine atom on the biphenyl molecule. Such a wide diversity of carbon isotope values could prove to be very powerful in determining the source for PCBs in the environment.

In the past few years, attention has turned to the more volatile chlorinated compounds such as PCE, TCE, and related degradation products. Since the 1940s, these chlorinated hydrocarbons have been used in a variety of industries such as electronic, instruments manufacturing, aerospace, machinery, printing, and dry cleaning (Beneteau et al. 1999; Hunkeler et al. 1999; Jendrzejewski et al. 2001).

Hunkeler et al. (1999) presented $\delta^{13}C$ values of PCE and its dechlorination products from microcosms and field samples. The pattern of isotope fractionation matched complete dechlorination of PCE in the microcosm and a similar pattern was observed in the field data set. Sherwood Lollar et al. (2001) studied stable carbon isotope fractionation of TCE and PCE at Dover AFB (USA). Isotope fractionation of TCE, in particular, was indicative of extensive dechlorination. Song et al. (2002) used Compound-specific isotope analyses (CSIA) to monitor TCE dechlorination in a pilot study of in-situ lactate amendment (at a test site of Idaho National Engineering and Environmental Laboratory, USA). CSIA results confirmed quantitative conversion of TCE to ethene. Hunkeler et al. (2005) presented data on a chlorinated hydrocarbon plume study (at an undisclosed former solvent disposal site). The typical reductive dechlorination of chlorinated ethenes was found to be of minor importance, and the predominant reaction pathways were dehydrochlorination of 1,1,2,2-PCE to TCE, reductive dechlorination of chloroform to dichloromethane, dichloroelimination of 1,1,2-TCA (trichloroethane) to vinyl chloride (VC), and dichloroelimination of 1,2-DCA (dichloroethane) to ethene. Morrill et al. (2005) showed that enrichment of ^{13}C in *cis*-DCE validated in-situ dechlorination of the compound in a bioaugmentation project (Kelly AFB, USA). Recently, Chartrand et al. (2005) used CSIA to monitor reductive dechlorination process in a fractured bedrock plume during a pilot-scale bioaugmentation project. Chlorinated ethene concentration profiles were not convincing evidence for the progress of dechlorination due to variable hydraulic gradients in the fractured bedrock and ongoing flux of contaminants from the NAPL source. However, enrichments of ^{13}C in *cis*-DCE and VC were indicative of dechlorination, and a quantitative estimate of the CSIA data permitted to evaluate the efficiency of bioaugmentation treatment.

3. MTBE

Gasoline oxygenates, particularly MTBE, represent a significant groundwater contamination problem but one well suited for isotope studies. A number of studies have shown MTBE degradation in anaerobic microcosms and in the field, under a range of electron acceptor conditions (Bradley et al. 1999; Finneran et al. 2001). Degradation under anaerobic conditions (Somsamak et al. 2001) is slower and, in most cases, may be restricted to biotransformation of MTBE to TBA. As a consequence, natural attenuation of anaerobic plumes tends to be relatively slow, resulting in the accumulation of TBA as MTBE disappears.

Kolhatkar et al. (2002) showed enrichment of ^{13}C in an anaerobic MTBE plume at a retail gasoline station in New Jersey, USA. Kuder et al. (2005) presented combined carbon and hydrogen CSIA data for nine MTBE sites (gasoline retail stations in California and New Jersey, USA) and from attendant microcosm experiments. The isotopic enrichments were indicative of biodegradation, and the two-dimensional isotope fractionation pattern was indicative of anaerobic biodegradation pathway for all of the sites studied. The pattern of isotope fractionation was indicative of demethylation for MTBE to TBA. Assessment of the progress of in-situ degradation based on the Rayleigh model demonstrated significant removal of MTBE, exceeded 90% for some of the samples. Zwank et al. (2005) presented carbon and hydrogen data set from a chemical waste storage site in Brasilia and reached similar conclusions on the MTBE degradation pathway. Day et al. (2002) showed an increase in the $\delta^{13}C$ values for TBA corresponding to plume downgradient concentration reduction at an undisclosed chemical plant in the United States. This appears to be the only available example of CSIA confirmation of in-situ TBA degradation that has been published to date.

The most important application for the isotope values, in addition to source evaluation or correlation, would be monitoring the rate and/or extent of biodegradation of individual compounds in the environment. The potential of this approach is enormous, particularly in the case of groundwater samples and specifically for compounds that are very soluble in water such as MTBE (Gray et al. 2002). One of the problems with compounds such as MTBE is that they are mobile and concentration changes at the monitoring wells do not necessary reflect changes in concentrations resulting from biodegradation. The changes in concentration at a specific water well may simply reflect movement of the compounds of interest within the plume. However, the potential of this approach has been demonstrated in a number of field and laboratory studies, both published and unpublished. For example, in a number of studies from our laboratory, and others, it has been shown that MTBE becomes progressively heavier isotopically with increasing biodegradation (Gray et al. 2002; Mancini et al. 2003; Chartrand et al. 2005; Kuder et al. 2005). This effect has now been observed at many sites and appears to be reproducible. In their papers, Meckenstock et al. (1999, 2002) used toluene as a substrate to demonstrate that isotopic enrichments of up to 10‰ could be observed, depending on the particular strain of bacterium being utilized.

4. SUMMARY

For many years, it was only possible to determine the bulk isotopic composition of a mixture of compounds. Bulk isotope values were used extensively in the petroleum industry and for a variety of other applications. However, for complex mixtures, bulk isotope values have been of limited use, since they represent weighted averages of all the compounds in the mixture. Hence, it was fortuitous for environmental geochemistry, and other areas, when the combination of the gas chromatograph-isotope ratio mass spectrometer was developed. Now it is possible to determine isotopic composition of individual compounds in complex mixtures at relatively low concentrations. These values can be used for correlation purposes of monitoring natural attenuation in groundwater studies. This paper has attempted to summarize many of the important papers that have appeared as a result of these developments. There are many others but, in general, it is felt the information here will provide a good overview of the developments and applications that have taken place in the past few years.

In environmental forensic studies, stable isotopes are being used to determine sources or monitor natural attenutation. One has to determine whether or not any degradation has occurred prior to using the isotope fingerprints for correlation purposes, since the only significant process to alter the isotope ratio of the contaminant is degradation–biodegradation or abiotic chemical degradation. If no degradation is occurring, isotope fingerprinting can be considered as a reliable source-tracking tool. Qualitative verification of in-situ biodegradation only requires that isotope fractionation is confirmed relative to the initial source isotope ratio (enrichment of the heavier isotope, e.g., increase of $\delta^{13}C$ value).

In brief, the exponential increase in applications and utilization of stable isotopes in environmental studies bodes well for the future of this technique. With time, procedures for determination of other isotopes will become routine. Carbon and hydrogen are already routine and before long it can be assumed that a method will also be developed for online determination of chlorine. It is not proposed that the isotope data should be used on its own but rather as a complimentary tool for use in conjunction with the established techniques of GC and GCMS.

REFERENCES

1. Abrajano J, Sherwood Lollar B (1999) Compound-specific isotope analysis: tracing organic contaminant sources and processes in geochemical systems. Org Geochem 30(8B):v–vii
2. Aislabie JM, Balks MR, Foght JM, Waterhouse EJ (2004) Hydrocarbon spills on Antarctic soils: effects and management. Environ Sci Technol 38(5):1265–1274
3. Alzaga R, Montuori P, Ortiz L, Bayona JM, Albaigés J (2004) Fast solid-phase extraction–gas chromatography–mass spectrometry procedure for oil fingerprinting: application to the Prestige oil spill. J Chromatogr A 1025(1):133–138
4. Beneteau KM, Aravena R, Frape SK (1999) Isotopic characterization of chlorinated solvents–laboratory and field results. Org Geochem 30(8):739–753
5. Bradley PM, Landmeyer JE, Chapelle FH (1999) Aerobic mineralization of MTBE and tert-butyl alcohol by stream-bed sediment micro-organisms. Environ Sci Technol 33(11):1877–1879
6. Brenner RC, Magar VS, Ickes JA, Abbott JE, Stout SA, Crecelius EA, Bingler LS (2002) Characterization and fate of PAH-contaminated sediments at the Wyckoff/Eagle Harbor superfund site. Environ Sci Technol 36(12):2605–2613
7. Burgoyne TW, Hayes JM (1998) Quantitative production of H2 by pyrolysis of gas chromatographic effluents. Anal Chem 70:5136–5141
8. Caimi RJ, Brenna JT (1993) High-precision liquid chromatography–combustion isotope ratio mass spectrometry. Anal Chem 65:3497–500
9. Chapman PM, Riddle MJ (2005) Toxic effects of contaminants in polar marine environments. Environ Sci Technol 39(9):200A–206A
10. Chartrand MMG, Morrill PL, Lacrampe-Couloume G, Sherwood Lollar B (2005) Stable isotope evidence for biodegradation of chlorinated ethenes at a fractured bedrock site. Environ Sci Technol 39(13):4848–4856
11. Chen S-J, Luo X-J, Mai B-X, Sheng G-Y, Fu J-M, Zeng EY (2006) Distribution and mass inventories of polycyclic aromatic hydrocarbons and organochlorine pesticides in sediments of the Pearl River estuary and the Northern South China Sea. Environ Sci Technol 40(3):709–714

12. Coffin RB, Miyares PH, Kelley CA, Cifuentes LA, Reynolds CM (2001) δ13C and δ15N isotope analysis of TNT: two dimensional source identification. Environ Toxicol Chem 20(12):2676–2680
13. Craig DR, Mauro D, Saber D, Sirivedhin T, Philp RP, Allen JA (2005). Carbon isotope ratios of PAHs in urban background soil. In: Proceedings of the 3rd international conference on remediation of contaminated sediments. 57th annual meeting American Academy of Forensic Sciences, February 21–26, New Orleans
14. Day M, Aravena R, Hunkeler D, Gulliver T (2002) Application of carbon isotopes to document biodegradation of tert-butyl alcohol under field conditions. Contaminated Soil Sediment & Water, August, pp 88–92
15. Drenzek NJ, Tarr CH, Eglinton TI, Heraty LJ, Sturchio NC, Shiner VJ, Reddy CM (2002) Stable chlorine and carbon isotopic compositions of selected semi-volatile organochlorine compounds. Org Geochem 33(4):437–444
16. Fang J, Kawamura K, Ishimura Y, Matsumoto K (2002) Carbon isotopic composition of fatty acids in the marine aerosols from the western North Pacific: implication for the source and atmospheric transport. Environ Sci Technol 36(12):2598–2604
17. Farran A, Grimalt J, Albaiges J, Botello AV, Macko SE (1987) Assessment of petroleum pollution in a Mexican river by molecular markers and carbon isotope ratios. Marine Pollut Bull 18(6):284–288
18. Faure G, Mensing TM (2005) Isotopes: principles and applications. Wiley, Hoboken, NJ, pp 897
19. Finneran KT, Lovley DR (2001) Anaerobic degradation of methyl tert-butyl ether (MTBE) and tert-butyl alcohol (TBA). Environ Sci Technol 35(9):1785–1790
20. Franco MA, Viñas L, Soriano JA, de Armas D, González JJ, Beiras R, Salas N, Bayona JM, Albaigés J (2006) Spatial distribution and ecotoxicity of petroleum hydrocarbons in sediments from the Galicia continental shelf (NW Spain) after the Prestige oil spill. Marine Pollut Bull 53(5–7):260–271
21. Fuentes HR, Jaffe R, Shrotriya RV (1996) Weathering of crude oil spills in seawater under simulated laboratory conditions. In: Proceedings of the 5th Latin American Congress on organic geochemistry, Cancun, Mexico, Abstracts. Mexican Petroleum Institute, Mexico City, Mexico, pp 310–312
22. Galimov EM (1985) The biological fractionation of isotopes. Academic Press, Orlando, pp 261
23. Glaser B, Dreyer A, Bock M, Fiedler S, Mehring M, Heitmann T (2005) Source apportionment of organic pollutants of a highway-traffic influenced urban area in Bayreuth (Germany) using biomarker and stable carbon isotope signatures. Environ Sci Technol 39(11):3911–3917
24. Gray R, Lacrampe-Couloume G, Gandhi D, Scow KM, Wilson RD, Mackay DM, Sherwood Lollar B (2002) Carbon and hydrogen isotopic fractionation during biodegradation of methyl tert-butyl ether. Environ Sci Technol 36(9):1931–1938
25. Griebler C, Safinowski M, Vieth A, Richnow HH, Meckenstock RU (2004) Combined application of stable carbon isotope analysis and specific metabolites determination for assessing in situ degradation of aromatic hydrocarbons in a tar oil-contaminated aquifer. Environ Sci Technol 38(2):617–631
26. Gürgey K, Philp RP, Clayton C, Emiroğlu H, Siyako M (2005) Geochemical and isotopic approach to maturity/source/mixing estimations for natural gas and associated condensates in the Thrace Basin, NW Turkey. Appl Geochem 20(11):2017–2037
27. Hammer BT, Kelley CA, Coffin RB, Cifuentes LA, Mueller JG (1998) 13C values of polycyclic aromatic hydrocarbons collected from two creosote-contaminated sites. Org Geochem 152(1/2):43–58
28. Hartman B, Hammond D (1981) The use of carbon and sulfur isotopes as a correlation parameter for source identification of beach tars in southern California. Geochim Cosmochim Acta 45:309–319
29. Haws NW, Ball WP, Bouwer EJ (2006) Modeling and interpreting bioavailability of organic contaminant mixtures in subsurface environments. J Contam Hydrol 82(3/4):255–292

30. Hayes JM, Freeman KH, Popp BN, Hoham CH (1990) Compound specific isotope analysis: a novel tool for reconstruction of ancient biogeochemical processes. Org Geochem 16(4-6):1115–1128
31. Heraty LJ, Fuller ME, Huang L, Abrajano T, Sturchio NC (1999) Isotopic fractionation of carbon and chlorine by microbial degradation of dichloromethane. Org Geochem 30(8):739–753
32. Hilkert AW, Douthitt CB, Schluter HJ, Brand WA (1999) Isotope ratio monitoring gas chromatography/mass spectrometry of D/H by high temperature conversion isotope ratio mass spectrometry. Rapid Commun Mass Spectrom 13:1226–1230
33. Hillaire-Marcel C (1986) Isotopes and food. In: Fritz P, Fontes JCh (eds) Handbook of environmental isotope geochemistry. vol 2: the terrestrial environment, B, Chapter 12. Elsevier, Amsterdam, pp 507–548
34. Hoeffs J (2004) Stable isotope geochemistry. Springer, Berlin, New York, pp 244
35. Holt BD, Sturchio NC, Abrajano TA, Heraty LJ (1997) Conversion of chlorinated volatile organic compounds to carbon dioxide and methyl chloride for isotopic analysis of carbon and chlorine. Anal Chem 69:2727–2733
36. Hong SH, Yim UH, Shim WJ, Oh JR (2005) Congener-specific survey for polychlorinated biphenlys in sediments of industrialized bays in Korea: regional characteristics and pollution sources. Environ Sci Technol 39(19):7380–7388
37. Horii Y, Kannan K, Petrick G, Gamo T, Falandysz J, Yamashita N (2005) Congener-specific carbon isotopic analysis of technical PCB and PCN mixtures using two-dimensional gas chromatography-isotope ratio mass spectrometry. Environ Sci Technol 39(11):4206–4212
38. Hough RL, Whittaker M, Fallick AE, Preston T, Farmer JG, Pollard SJT (2006) Identifying source correlation parameters for hydrocarbon wastes using compound-specific isotope analysis. Environ Pollut, in press
39. Hunkeler D, Aravena R (2000) Determination of compound-specific carbon isotope ratios of chlorinated methanes, ethanes, ethenes in aqueous samples. Environ Sci Technol 34:2839–2844
40. Hunkeler D, Aravena R, Butler BJ (1999) Monitoring microbial dechlorination of tetrachloroethene (PCE) in groundwater using compound-specific stable carbon isotope ratios: microcosm and field studies. Environ Sci Technol 33:2733–2738
41. Jarman WM, Hilkert A, Bacon CE, Collister JW, Ballschmitter P, Risebrough RW (1998) Compound specific carbon isotope analysis of Aroclors, Clophens, Kaneclors, Phenoclors. Environ Sci Technol 32(6):833–836
42. Jendrzejewski N, Eggenkamp HGM, Coleman ML (2001) Characterization of chlorinated hydrocarbons from chlorine and carbon isotopic compositions: scope of application to environmental problems. Appl Geochem 16(9/10):1021–1031
43. Kelley CA, Coffin RB, Cifuentes LA, Lantz SE, Mueller JG (1995) The use of GC/C/IRMS coupled with GC/MS to monitor biodegradation of petroleum hydrocarbons. Platform Abstracts of the Third International Symposium in Situ and On Situ Bioreclaimation, San Diego, CA
44. Kelley CA, Hammar BT, Coffin RB (1997) Concentrations and stable isotope values of BTEX in gasoline contaminated groundwater. Environ Sci Technol 31(9):2469–2472
45. Kolhatkar R, Kuder T, Philp P, Allen J, Wilson JT (2002) Use of compound-specific stable carbon isotope analyses to demonstrate anaerobic biodegradation of MTBE in groundwater at a gasoline release site. Environ Sci Technol 36:5139–5146
46. Krummen ML, Hilkert AW, Juchelka D, Duhr A, Schlueter H, Pesch R (2004) A new concept for isotope ratio monitoring liquid chromatography/mass spectrometry. Rapid Commun Mass Spectrom 18:2260–2266
47. Kuder T, Wilson JT, Kaiser P, Kolhatkar R, Philp P, Allen J (2005) Enrichment of stable carbon and hydrogen isotopes during anaerobic biodegradation of MTBE: microcosm and field evidence. Environ Sci Technol 39:213–220
48. Kvenvolden KA, Hostettler FD, Carlson PR, Bapp JB, Threlkeld CN, warden A (1995) Ubiquitous tar balls with a California-source signature on shorelines of Prince William Sound, Alaska. Environ Sci Technol 29(10):2684–2694

49. Lafargue E, Barker C (1988) Effect of water-washing on crude oil compositions. Am Assoc Petroleum Geolo Bull 72:263–276
50. Lee B-C, Shimizu Y, Matsuda T, Matsui S (2005) Characterization of polycyclic aromatic hydrocarbons (PAHs) in different size fractions in deposited road particles (DRPs) from Lake Biwa area, Japan. Environ Sci Technol 39(19):7402–7409
51. Lichtfouse E (2000) Compound-specific isotope analysis (CSIA). Application to archaeology, biomedical sciences, biosynthesis, environment, extraterrestrial chemistry, food science, forensic science, humic substances, microbiology, organic geochemistry, soil science and sport. Rapid Commun Mass Spectrom 14:1337–1344
52. Lichtfouse E, Schwarzbauer J, Robert D (eds) (2005). Environmental chemistry, green chemistry and pollutants in ecosystems. Sections: 1. Analytical chemistry. 2. Toxic metals. 3. Organic pollutants. 4. Polycyclic aromatic compounds. 5. Pesticides. 6. Green chemistry. 7. Ecotoxicology. Springer, Berlin Heidelberg New York, 69 chapters, 780 p
53. Macko SA, Parker PL, Botello AV (1981) Persistence of spilled oil in a Texas salt marsh. Environ Pollut Bull 2:119–128
54. Macko SA, Parker PL (1983) Stable nitrogen and carbon isotope ratios of beach tars on South Texas barrier islands. Marine Environ Res 10:93–103
55. Mancini SA, Lacrampe-Couloume G, Jonker H, van Breukelen BN, Groen J, Volkering F, Sherwood Lollar B (2002) Hydrogen isotopic enrichment: an indicator of biodegradation at a petroleum hydrocarbon contaminated field site. Environ Sci Technol 36(11):2464–2470
56. Mancini SA, Ulrich AC, Lacrampe-Couloume G, Sleep B, Edwards EA, Sherwood Lollar B (2003) Carbon and hydrogen isotopic fractionation during anaerobic biodegradation of benzene. Appl Environ Microbiol 69:191–198
57. Mansuy L, Philp RP, Allen J (1997) Source identification of oil spills based on the isotopic composition of individual components in weathered oil samples. Environ Sci Technol 31(12):3417–3425
58. Mauro D, Saber D, Sirivedhin T, Philp RP, Allen JA (2005). Carbon isotope ratios of PAHs in urban background soil. In: Proceedings of the 3rd international conference on remediation of contaminated sediments, New Orleans, January 2005, Abstracts
59. Mazeas L, Budzinski H (2002) Molecular and stable carbon isotopic source identification of oil residues and oiled bird feathers sampled along the Atlantic Coast of France after the Erika oil spill. Environ Sci Technol 36(2):130–137
60. McRae C, Sun CG, Snape CE, Fallick AE, Taylor D (1999) δ13C values of coal-derived PAHs from different processes and their application to source apportionment. Org Geochem 30(8):881–889
61. McRae C, Snape CE, Sun C-G, Fabbri D, Tartari D, Trombini C, Fallick AE (2000) Use of compound-specific stable isotope analysis to source anthropogenic natural gas-derived polycyclic aromatic hydrocarbons in a lagoon sediment. Environ Sci Technol 34(22):4684–4686
62. Meckenstock RU, Morasch B, Warthmann RW, Schink B, Annweller E, Michaelis W, Richnow HH (1999) 13C/12C Isotope fractionation of aromatic hydrocarbons during microbial degradation. Environ Microbiol 1(5):409–414
63. Meckenstock RU, Morasch B, Kastner M, Vieth A, Richnow HH (2002) Assessment of bacterial degradation of aromatic hydrocarbons in the environment by analysis of stable carbon isotope fractionation. Water Air Soil Pollut Focus 2:141–152
64. Meckenstock RU, Morasch B, Griebler C, Richnow HH (2004a) Stable isotope fractionation analysis as a tool to monitor biodegradation in contaminated aquifers. J Contam Hydrol 75:215–255
65. Meckenstock RU, Safinowski M, Griebler C (2004b) Anaerobic degradation of polycyclic aromatic hydrocarbons. FEMS Microbiol Ecol 49:27–36
66. Melander L, Saunders WH (1980) Reaction rates of isotopic molecules. Wiley, New York, pp 331
67. Merritt DA, Freeman KH, Ricci MP, Studley SA, Hayes JM (1995) Performance and optimization of a combustion interface for isotope ratio monitoring gas chromatography/mass spectrometry. Anal Chem 67:2461–73

68. Milkov AV (2005) Molecular and stable isotope compositions of natural gas hydrates: a revised global dataset and basic interpretations in the context of geological settings. Org Geochem 36(5):681–702
69. Milner CWD, Rogers MA, Evans CR (1977) Petroleum transformations in reservoirs. J Geochem Expl 7:101–153
70. Morrill PL, LaCrampe-Couloume G, Slater GF, Sleep BE, Edwards EA,McMaster ML, Major DW, Sherwood Lollar B (2005) Quantifying chlorinated ethene degradation during reductive dechlorination at Kelly AFB using stable carbon isotopes. J Contam Hydrol 76:279–293
71. Norman AL, Hopper JF, Blanchard P, Ernst D, Brice K, Alexandroi N, Klouda G (1999) The stable carbon isotope composition of atmospheric PAHs. Atmos Environ 33(17):2807–2814
72. Numata M, Nakamura N, Koshikawa H, Terashima Y (2002) Chlorine isotope fractionation during reductive dechlorination of chlorinated ethenes by anaerobic bacteria. Environ Sci Technol 36(20):4389–4394
73. Ohe T, Watanabe T, Wakabayashi K (2004) Mutagens in surface waters: a review. Mutat Res Rev Mutat Res 567(2/3):109–149
74. Okuda T, Kumata H, Zakaria MP, Naraoka H, Ishiwatari R, Takada H (2002a) Source identification of Malaysian atmospheric polycyclic aromatic hydrocarbons nearby forest fires using molecular and isotopic compositions. Atmos Environ 36:611–618
75. Okuda T, Kumata H, Naraoka H, Ishiwatari R, Takada H (2002b) Vertical distributions and δ13C isotopic compositions of PAHs in Chidorigafuchi Moat sediment, Japan. Org Geochem 33(7):843–848
76. Palmer SE (1993) Effect of biodegradation and water-washing on crude oil composition. In: Engel MH, Macko SE (eds) Organic geochemistry. Plenum Press, New York, pp 511–534
77. Peters KE, Fowler MG (2002) Applications of petroleum geochemistry to exploration and reservoir management. Org Geochem 332(1):5–36
78. Philp RP, Allen J, Kuder T (2002) The use of the isotopic composition of individual compounds for correlating spilled oils and refined products in the environment with suspected sources. Environ Forensics 3(3/4):341–348
79. Pond KL, Huang Y, Wang Y, Kulpa CF (2002) Hydrogen isotopic composition of individual n-alkanes as an intrinsic tracer for bioremediation and source identification of petroleum contamination. Environ Sci Technol 36(4):724–728
80. Reddy CM, Heraty LJ, Holt BD, Sturchio NC, Eglinton TI, Drenzek NJ, Xu L, Lake JL, Maruya KA (2000) Stable carbon isotope compositions of Aroclors and Aroclor contaminated sediments. Environ Sci Technol 34(13):2866–2870
81. Rogers KMMM (1999) Detection of petroleum contamination in river sediments from Quebec City region using GC–IRMS. Org Geochem 30(12):1559–1569
82. Saber D, Mauro D, Philp RP, Allen J, Kuder T (2003) Carbon isotope ratios of PAHs in urban background soil. In: Proceedings of the Battelle meeting, Orlando
83. Saber D, Mauro D, Philp RP, Allen JA (2005) Chemical fingerprinting of sediments and soils near a former MPG site. In: Proceedings of the 3rd international conference on remediation of contaminated sediments, New Orleans. Abstracts
84. Samara F, Tsai CW, Aga DS (2006) Determination of potential sources of PCBs and PBDEs in sediments of the Niagara River. Environ Pollut 139(3):489–497
85. Schirmer M, Butler BJ (2004) Transport behavior and natural attenuation of organic contaminants at spill sites. Toxicology 205(3):173–179
86. Schmidt TC, Zwank L, Elsner M, Berg M, Meckenstock RU, Haderlein SB (2004) Compound-specific stable isotope analysis of organic contaminants in natural environments: a critical review of the state of the art, prospects, and future challenges. Anal Bioanal Chem 378:283–300
87. Schüth C, Taubald H, Bolaño N, Maciejczyk K (2003) Carbon and hydrogen isotope effects during sorption of organic contaminants on carbonaceous materials. J Contam Hydrol 64(3/4):269–281

88. Scow KM, Hicks KA (2005) Natural attenuation and enhanced bioremediation of organic contaminants in groundwater. Curr Opin Biotechnol 16:246–253
89. Sessions AL, Burgoyne TW, Schimmelmann A, Hayes JM (1999) Fractionation of hydrogen isotopes in lipid biosynthesis. Org Geochem 30(9):1193–1200
90. Sherwood Lollar B, Slater GF, Sleep B, Witt M, Klecka GM, Harkness M, Spivack J (2001) Stable carbon isotope evidence for intrinsic bioremediation of tetrachloroethene and trichloroethene at area 6, Dover Air Force Base. Environ Sci Technol 35:261–269
91. Shouakar-Stash O, Frape SK, Drimmie RJ (2003) Stable hydrogen, carbon and chlorine isotope measurements of selected chlorinated organic solvents. J Contam Hydrol 60(3/4):211–228
92. Shouakar-Stash O, Drimmie RJ, Frape SK (2005) Determination of inorganic chlorine stable isotopes by conditions flow isotope ratio mass spectrometry. Rapid Commun Mass Spectrom 19:121–127
93. Slater GF (2003) Stable isotope forensics–when isotopes work. Environ Forensics 4:13–23
94. Smallwood BJ, Philp RP, Allen JD (2002) Stable carbon isotopic composition of gasolines determined by isotope ratio monitoring gas chromatography mass spectrometry. Org Geochem 33:149–159
95. Smirnov A, Abrajano TA, Smirnov A, Stark A (1998) Distribution and sources of polycyclic aromatic hydrocarbons in the sediments of Lake Erie, Part 1. Spatial distribution, transport, and deposition. Org Geochem 29(5–7):1813–1828
96. Somsamak P, Cowan RM, Haggblom MM (2001) Anaerobic biotransformation of fuel oxygenates under different anoxic conditions. FEMS Microbiol Ecol 37:259–264
97. Song DL, Conrad ME, Sorenson KS, Alvarez-Cohen L (2002) Stable carbon isotope fractionation during enhanced in situ bioremediation of trichloroethene. Environ Sci Technol 36:2262–2268
98. Stark A, Abrajano T, Hellou J, Metcalf-Smith JL (2003) Molecular and isotopic characterization of polycyclic aromatic hydrocarbon distribution and sources at the international segment of the St. Lawrence River. Org Geochem 34(2):225–237
99. Sturchio NC, Clausen JL, Heraty L, Huang L, Holt BD, Abrajano T (1998) Chlorine isotope investigation of natural attenuation of trichloroethene in an aerobic aquifer. Environ Sci Technol 32(20):3037–3042
100. Thornton SF, Lerner DN, Banwart SA (2001) Assessing the natural attenuation of organic contaminants in aquifers using plume-scale electron and carbon balances: model development with analysis of uncertainty and parameter sensitivity. J Contam Hydrol 53(3/4):199–232
101. Wang Y, Huang Y, Huckins JN, Petty JD (2004) Compound-specific carbon and hydrogen. Environ Sci Technol 38(13):3689–3697
102. Warmerdam EM, Frape SK, Aravena R, Drimmie RJ, Flatt H, Cherry JA (1995) Stable chlorine and carbon isotope measurements of selected chlorinated organic solvents. Appl Geochem 10(5):547–552
103. Wassenaar LI, Hobson KA (2000) Improved method for determining the stable-hydrogen isotopic composition (δD) of complex organic materials of environmental interest. Environ Sci Technol 34(11):2354–2360
104. Wen S, Feng Y, Yu Y, Bi X, Wang X, Sheng G, Fu J, Peng P (2005) Development of a compound-specific isotope analysis method for atmospheric formaldehyde and acetaldehyde. Environ Sci Technol 39(16):6202–6207
105. Wiesenberg GLB, Schwarzbauer J, Schmidt MWI, Schwark L (2004) Source and turnover of organic matter in agricultural soils derived from n-alkane/n-carboxylic acid compositions and C-isotope signatures. Org Geochem 35(11/12):1371–1393
106. Wilcke W, Krauss M, Amelung W (2002) Carbon isotope signature of polycyclic aromatic hydrocarbons (PAHs): evidence for different sources in tropical and temperate environments? Environ Sci Technol 36(16):3530–3535
107. Zwank L, Berg M, Elsner M, Schmidt TC, Schwarzenbach RP, Haderlein SB (2005) New evaluation scheme for two-dimensional isotope analysis to decipher biodegradation

processes: application to groundwater contamination by MTBE. Environ Sci Technol 39(18):7344–7344

108. Zwank L, Berg M, Schmidt TC, Haderlein SB (2003) Compound-specific carbon isotope analysis of volatile organic compounds in the low-microgram per liter range. Anal Chem 75:5575–5583

CHAPTER 7

Antiaflatoxigenic activity of *Carum copticum* essential oil

Mohammad Reza Rezaei Kahkha1, Saeed Amanloo1, Massoud Kaykhaii2

1School of Health, Zabol University of Medical Sciences, 98616- 15881 Zabol, Iran

2 Department of Chemistry, Faculty of Sciences, University of Sistan and Baluchestan, Zahedan, Iran

ABSTRACT

Plants are unique sources of useful metabolites. Plant essential oils display a wide range of antimicrobial effects against various pathogens. Here, we studied the essential oil from the seeds of *Carum copticum*. We monitored aflatoxin by high-performance liquid chromatography. Results show that *Carum copticum* essential oil inhibits *Asergillus parasiticus* growth and prevents aflatoxin production. The half-maximal inhibitory concentration (IC_{50}) is 127.5 μg mL^{-1} for aflatoxin B_1 and 23.22 μg mL^{-1} for aflatoxin G_1. Our findings show that *Carum copticum* essential oil is a potential candidate for the protection of foodstuff and feeds from toxigenic fungus growth and their subsequent aflatoxin contamination.

Keywords: Aflatoxin, Essential oil, Carum, copticum

1. INTRODUCTION

Aflatoxins are closely related toxic secondary metabolites produced by *Aspergillus flavus, Aspergillus parasiticus, Aspergillus nomius* and *Aspergillus tamarii* (Fooladi and Farahnaky 2003). There are four principal types of aflatoxins, namely B1, B2, G1 and G2. They are extremely potent, naturally occurring hepatocarcinogens (Henry et al. 1999). Considerable attention is being focused on aflatoxin control in agricultural produce because of health risks involved in the consumption of contaminated agricultural commodities. Prevention of mold growth on agricultural produce offers the best means of aflatoxin control because decontamination of foods and feeds containing preformed toxin is a challenging proposition. According to Rustom (1997), occurrence of mold growth is often unavoidable and still remains a serious problem in food commodities, emphasizing the need for a suitable detoxification protocol. Numerous attempts have been made to detoxify aflatoxin in feedstuff while ensuring the safety and nutritional adequacy of the produce (Keyl and Booth 1971). However, no single method of detoxification has completely solved the problem. Hence, an efficient method to remove aflatoxin either completely or to acceptable low levels while retaining the nutritional value of food commodity will be an attractive alternative.

In recent years, the essential oils and extracts of several plant species have become popular as fungi toxicants. Plants are considered as unique sources of useful metabolites. The plant essential oils received major considerations with regard to possessing a wide range of antimicrobial effects against different groups of pathogenic organisms (Bakkali et al. 2008). So, essential oils with antimicrobial activity are potential candidates, as natural antimicrobial preservatives, that can be used in controlling microbial food contaminations. Different plant preparations including the essential oils and extracts prepared from leaves, roots and seeds have successfully been used as potential inhibitors of growth and/or aflatoxin production by *A. flavus* and *A. parasiticus* (Mahmoud 1994; Mahoney et al. 2000; Rasooli and Razzaghi-Abyaneh 2004; Razzaghi-Abyaneh et al. 2005). These plants may be useful for controlling aflatoxin contamination of crops in field conditions.

Carum copticum seeds are used as a common household remedy for cholera, colic, diarrhea, dyspepsia, hypertension, asthma and hepato-biliary complications (Avesina 1985). The plant is known to possess antiallergic, antibacterial, anthelmintic, antifungal (Dubey and Mishra 1990), hypocholesterolemic, bronchodilator (Boskabady et al. 1998) and cholinergic activities. The phytochemical studies on Carum seeds have revealed the presence of multiple constituents including steroptin, ecumene, thymene, amino acids such as lysine and threonine, calcium, iron, starch, tannins and dietary fiber. The seeds also contain essential oil (2–3 %), which contain thymol (40–50 %), γ-terpinene, *p*-cymene, α-pinene, β-pinene and carvacrol (Ballba et al. 1973). Thymol is a strong germicide, antispasmodic agent and a known antifungal agent (Mahmoud 1994). Antimicrobial (Kavoosi et al. 2013) and antifungal activity of essential oil and alcoholic extract of *Carum copticum* (Ghahfarkhi et al. 2008) was known already. Also, it showed activities against food-borne and nosocomial pathogens (Zomorodian et al. 2011). So, it was interesting to study its activity against fungal growth *A. parasiticus* NRRL 2999. In this study, EO of *Carum copticum* was evaluated with regard to its ability to inhibit growth and aflatoxins (AFB1 and AFG1) production by *A. parasiticus* species.

2. EXPERIMENTAL

2.1 Materials and methods

The *Carum copticum* seeds were purchased from a local market. All chemicals used were of analytical grade and were obtained from Merck Chemicals Co., Germany. Aflatoxigenic Aspergillus parasiticus NRRL 2999 was used as a source of aflatoxins. Aflatoxigenic activity of fungal strain of Aspergillus parasiticus was known for many years (Meier and Marth 1977). It produces all 4 major aflatoxins, i.e., B1, B2, G1 and G2. The strain was maintained on potato dextrose broth slants by regular subculturing.

2.2 Preparation of *Carum copticum* extract

Twenty-five grams of *Carum copticum* was steam distilled for 90 min at a Clevenger-type apparatus. The extraction was carried out for 120 min in 500 mL of water. The extracted essential oils were kept in dark at 4 °C before use.

2.3 Microbioassay

Aflatoxigenic *A. parasiticus* NRRL 2999 was cultured on potato dextrose broth in six-well flat-bottomed microplates in the presence of *c. copticum* EOs. The culture medium was added to the microplates in amount of 5 mL well^{-1} and then inoculated with fungal spore suspension (5 × 10^6 spore well^{-1}) prepared in distilled water. Different concentrations of *C. copticum* essential oils (1, 10, 50, 100, 500 and 1,000 μg mL^{-1}) prepared in methanol were added to the test wells. The control wells were treated in the same manner with the test wells except that they did not contained *C. copticum* essential oils. The microplates were incubated for 7 days on a rotary shaker at temperature of 28 °C.

All contents of each well including the culture media and fungal biomass were filtered through a thin layer of cheesecloth and thoroughly washed with distilled water. A known weight of mycelia was placed in a container and allowed to dry at 80 °C till a constant weight was obtained.

2.4 Aflatoxin assay

The toxin was analyzed qualitatively for the presence of aflatoxins by thin-layer chromatography (TLC) using silica gel 60 F_{254} plates under ultraviolet light (maximum intensity at wavelength of 365 nm). The amount of toxin detected by TLC was 10 ng g^{-1}.

2.5 Quantitative analysis

Aflatoxin content of cultures was measured using an HPLC (Cecil UV–VIS system, England) equipped with a UV–Vis detector according to Razzaghi-Abyaneh et al. (2007). Fifty microliters of sample (culture filtrate) was injected into the HPLC column (ACE 5-C_{18}; 4.6 × 150 mm) and eluted at a flow rate of 1 mL min^{-1} by water/acetonitrile/methanol (60:25:15, v/v/v) solvent mixture. The amount of aflatoxin B1 (AFB1) and aflatoxin G1 (AFG1) was detected at a wavelength of 365 nm by comparison of under-curved area of unknown samples with authentic standards treated in the same manner. The retention times of AFB1 and AFG1 were 5.0 and 6.3 min, respectively.

3. RESULTS AND DISCUSSION

3.1 Effect of plant essential oils on *A. parasiticus* growth

As illustrated in Figure 1, the inhibitory *C. copticums* essential oils show strong capabilities to suppress *A. parasiticus* grown in microplates. Generally, the inhibitory effect of essential oils increased in proportion to their concentrations and reached to a maximum in the final concentration of 1,000 μg mL^{-1}. In this concentration, *C. copticum* showed the highest growth inhibition (100 %).

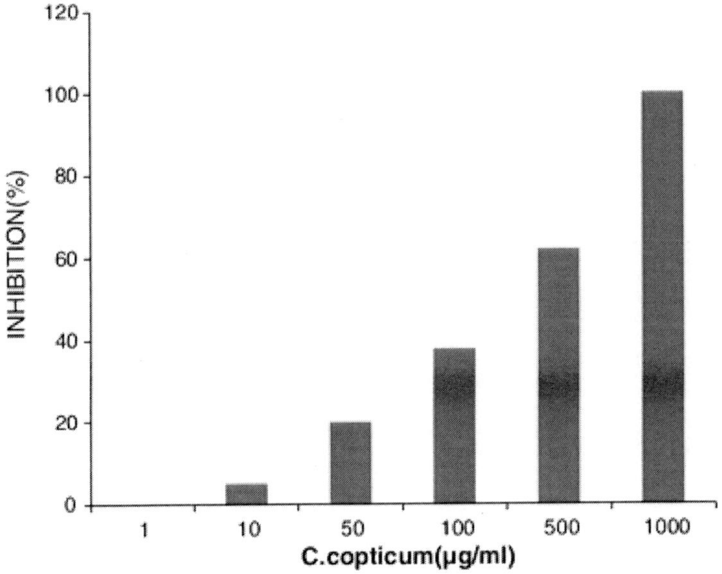

Figure 1. Effect of *C. copticums* essential oils on *A. parasiticus* NRRL 2999 growth in microbioassay (All data points are the average of 3 experiments performed in triplicate.)

3.2 Effect of plant EOs on aflatoxin production

As indicated in Figure 2, *C. copticums* essential oils also were able to suppress aflatoxin production by *A. parasiticus*. In the final concentration of 1,000 μg mL^{-1}, *C. copticum* showed the highest growth inhibition (100 %). Minimum inhibitory concentration (MIC) was found to be 0.87 μg mL^{-1}. As indicated in Table 1, the IC50 values of *C. copticums* for aflatoxin B1 were calculated as 127.5 μg mL^{-1}, where as for aflatoxin G1, it is 23.2 μg mL^{-1} (Table 1).

These results clearly show that the essential oil of *Carum copticum* may have the potential to be used as natural preservatives in controlling aflatoxin contamination of foods, feeds and agricultural commodities in practice. A comprehensive ecological study is needed for evaluating the effectiveness of these inhibitory plants and their effective constituents in field condition. In addition, antimicrobial and antioxidant activities of *Carum copticum* should be considered by other scientists, because the combination of these properties may act in a synergistic manner by reducing lipid oxidation and *Carum copticum* could be a good candidate for extending the shelf-life of perishable foods.

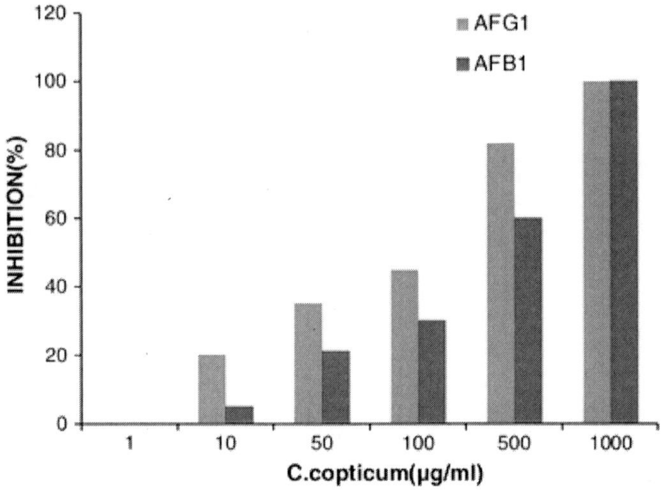

Figure 2. Inhibitory effects of *C. copticums* essential oils on production of aflatoxins B1 (*graycolumns*) and G1 (*bluecolumns*) by *A. parasiticus* NRRL 2999 (All data points are the average of 3 experiments performed in triplicate.)

Table 1. Inhibition percentage and IC_{50} values of essential oils of *C. copticum* for aflatoxins produced by *A. parasiticus* NRRL 2999

EOs concentration ($\mu g\ mL^{-1}$)	Aflatoxin B1		Aflatoxin G1	
	IC_{50} ($\mu g\ mL^{-1}$)	Inhibition (‰)	IC_{50} ($\mu g\ mL^{-1}$)	Inhibition (‰)
10	90	127.5	85	23.22
50	80		75	
100	75		60	
500	50		15	
1,000	ND		ND	

4. CONCLUSION

A novel method of toxin inactivation using aqueous *Carum copticum* extract has been described in the present work. The essential oil from *C. copticum* has effectively suppressed fungal growth. *C. copticum* was used as inhibitor of aflatoxin production by *A. parasiticus*. This plant suppressed aflatoxin B_1 and aflatoxin G_1 production dose dependently, consistent with its inhibitory effects on fungal growth. From data on antifungal activity of thymol and carvacrol as effective constituents of *Satureja hortensis* essential oil (Razzaghi-Abyaneh et al. 2008), it can be concluded that thymol, the main essential oil constituent of *C. copticum*, is responsible for inhibiting growth and aflatoxin production by the fungus. Our results clearly show that the essential oil of *C. copticum* may have the potential to be used as natural preservatives in controlling aflatoxin contamination of foods, feeds and agricultural commodities in practice.

REFERENCES

1. Avesina A (1985) Law in Medicine, Vol. 2, 2nd edn. Soroush Press, Tehran, p 187
2. Bakkali F, Averbeck S, Averbeck D, Idaomar M (2008) Biological effects of essential oils—a review. Food Chem Toxicol 46:446–475
3. Ballba SI, Hilal SH, Haggag MY (1973) The volatile oil from the herb and fruits of Carum copticum at different stages of growth. Planta Med 23:312–320
4. Boskabady MH, Rakhshanda H, Shariati VM (1998) Bronchodilatory and anticholinergic effects of Carum copticum on isolated guinea-pig tracheal chains. Med J Islam Repub Iran 11:329–334
5. Dubey NK, Mishra AK (1990) Evaluation of some essential oils against dermatophytes. Indian Drugs 27:529–531
6. Fooladi MH, Farahnaky A (2003) Aflatoxin removal from pistachio nuts by natural natrolite. J Food Sci 68:1225–1228
7. Ghahfarkhi MN, Sattari M, Yadegari MH, Goudarzi GR, Saharkhiz MJ (2008) Antifungal Activity of Essential Oil and Alcoholic Extract of Carum Copticum Against Fluconazole-Resistant and Susceptible Candida Albicans Isolated. Modares J Med Sci: Pathobiol 11(1–2):91–97
8. Henry SH, Bosch FX, Troxell TC, Bolger PM (1999) Reducing liver cancer—global control of aflatoxin. Science 286:2453–2454
9. Kavoosi G, Tafsiry A, Ebdam AA, Rowshan V (2013) Evaluation of antioxidant and antimicrobial activities of essential oils from Carum copticum seed and Ferula assafoetida latex. J Food Sci 78(2):T356–T361
10. Keyl AC, Booth AN (1971) Aflatoxin effects in livestock. J Am Oil Soc 48:599–604
11. Mahmoud AL (1994) Antifungal action and antiaflatoxigenic properties of some essential oil constituents. Lett Appl Microbiol 2:110–113
12. Mahoney N, Molyneux RJ, Campbell BC (2000) Regulation of aflatoxin production by naphthoquinones of walnut (Juglans regia). J Agric Food Chem 48:4418–4421
13. Meier KR, Marth EH (1977) Production of aflatoxin by Aspergillus parasiticus NRRL-2999 during growth in the presence of curing salts. Mycopathologia 61:77–83
14. Rasooli I, Razzaghi-Abyaneh M (2004) Inhibitory effects of Thyme oils on growth and aflatoxin production by Aspergillus parasiticus. Food Control 15:479–483
15. Razzaghi-Abyaneh M, Allameh A, Tiraihi T, Shams-Ghahfarokhi M, Ghorbanian M (2005) Morphological alterations in toxigenic Aspergillus parasiticus exposed to neem (Azadirachta indica) leaf and seed aqueous extracts. Mycopathologia 159:565–570
16. Razzaghi-Abyaneh M, Yoshinari T, Shams-Ghahfarokhi M, Rezaee MB, Nagasawa H, Sakuda S (2007) Dillapiol and apiol as specific inhibitors for the biosynthesis of aflatoxin G1 in Aspergillus parasiticus. Biosci Biotechnol Biochem 71:2329–2332
17. Razzaghi-Abyaneh M, Shams-Ghahfarokhi M, Yoshinari T, Rezaee MB, Jaimand K, Nagasawa H (2008) Inhibitory effects of Satureja hortensis L. essential oil on growth and aflatoxin production by Aspergillus parasiticus. Int J Food Microbiol 123:228–233

18. Rustom IYS (1997) Aflatoxin in food and feed: occurrence, legislation and inactivation by physical methods. Food Chem 59:57–67
19. Zomorodian K, Moein MN, Rahimi MJ, Pakshir K, Ghasemi Y, Sharbatfar S (2011) Possible application and chemical compositions of Carum copticum essential oils against food borne and nosocomial pathogens. Middle-East J Sci Res 9(2):239–245

CHAPTER 8

Sulphur and heavy metals contents in soils and Grewia bicolor leaves around the Selibe Pikwe Cu-Ni mine (BCL), Botswana

Moagi Letshwenyo

Department of Biological Sciences, University of Botswana, P/Bag 0022, Gaborone, Botswana.

ABSTRACT

This study investigated environmental pollution due to mining activities at the BCL mine in Selibe-Phikwe town, Botswana and assessed the level of heavy metals and sulphur in soil and Grewia bicolor leaves adjacent to the mine. Soil and G. bicolor were contaminated with heavy metals on the downwind side, where also concentrations of sulphur in the soil and G. bicolor were found to be elevated when compared to the upwind side. The leaves samples, were significantly difference at P = 0.00 and soil were significantly different at P = 0.03 for both west and east side. A mean soil pH of 4.12 in the prevailing wind direction suggested that atmospheric deposition from the mine has caused acidification. With further distance from the mine concentration in soil and leaves appeared to reflect a decreasing deposition of heavy metals and sulphur. Elevated concentration of heavy metals in leaves represent a high risk of bioaccumulation and biomagnifications in the food chain because G. bicolor leaves are grazed on by farmed animals and are common ingredients of traditional medicines.

Keywords: BCL, heavy metals, Grewia bicolor, sulphur, pH, environmental pollution

1. INTRODUCTION

Pollution due to mining activities is a great concern throughout the world since it has been recognized as a major contributor to environmental pollution over the past years (Zhou et al., 2007). Mining and smelting affect the air, terrestrial and aquatic ecosystems through discharge of solid and liquid waste, as well as particulate matter (Nagajyoti et al., 2010). Developing countries which are rich in minerals like Botswana are at a greater risk from mining pollution. Mining of minerals is associated with benefits such as economic development (Rashed, 2010), increased employment and foreign exchange reserves. Notwithstanding the considerable economic benefits of mining, it should also be noted that mines have the potential to leave a long lasting legacy for the environment as evidenced by case studies in Sudbury, Canada and Kola Peninsula in Russia (Whitby et al., 1976; Poikolainen, 1997).

The presence of heavy metals in soils may pose risk and hazards to humans and the ecosystem through: direct ingestion or contact with contaminated soil, the food chain, drinking of contaminated water (Candeias et al., 2011), reduction in food quality via phytotoxicity, reduction in land usability for agricultural production causing food insecurity and land tenure problems (Wuana and Okieimen, 2011).

Although the extraction of metals such as copper-nickel is necessary for the economic or industrial development of societies (Rashed, 2010), their significant risk to the environment comes from the disposal of mining waste (Figure 1). If the mine waste is not properly disposed, it can contribute immensely to heavy metal pollution in the surrounding environment, which can result in the destruction of ecological landscape, ground water pollution and decrease in biological diversity (Sheoran et al., 2011; Rashed, 2010; Kiikkila, 2003).

Copper-Nickel mines releases not only potential toxic metals but also other particulate matter and gaseous materials such as dust, SO_2, CO, CO_2, N_2O, NO_2 and NO in the atmosphere (Nagajyoti et al., 2010). This can have grave consequences for the ecosystem as evidenced by the forests devastation and reduced diversity of vascular plants in Sudbury, Canada (Nkongolo et al., 2008).

The threat that heavy metals pose to humans and animal health is aggravated by their long term persistence in the environment (Yoon et al., 2006). Heavy metals such as Lead and Cadmium are non-biodegradable and have no known biological function. Copper, Zinc and Nickel are source of nutrients to plants at low concentrations but they become toxic at elevated concentrations.

Heavy metals tend to bioaccumulate and biomagnify in animal tissues. Since domestic animals such as cattle and goats do graze in the surrounding area of the BCL mine, they are likely to ingest soil and plants material contaminated with heavy metals which in turn can pose a risk to people consuming animal meat. Cadmium and Lead tend to accumulate in the liver and kidneys of animals (Kirkham, 2006; Wilkinson et al., 2003). These tissues are regarded as delicacies in Southern Africa (Ikenaka et al., 2010; Yabe et al., 2011) and their consumption may pose a risk in contaminated areas.

The Bamangwato Concessions Limited mine (BCL) at Selibe Phikwe is rich in copper and nickel deposits. The main concern with this particular mining and smelting operations is the release of sulphur dioxide emissions into the atmosphere which can damage the environment through acidification. Another concern is the significant volumes of wastes that are deposited on the soil which eventually affect plants and animals through exposure to elements contained in the residue (Figure 1).

Given the environmental impact caused by sulphur dioxide in the adjacent area of the BCL mine, the objective of the present study is to asses sulphur and heavy metals content in soil and *Grewia bicolor* leaves along a 4 km west and east transect in BCL mine. The soil pH of the area around the latter will also be investigated to further demonstrate the impact of mining on the environment.

2. MATERIALS AND METHODS

2.1 Study site

The study site is located in Selibe Phikwe which is about 400km north from the Capital city Gaborone in eastern Botswana (Figure 2). The area is known for mining and smelting of copper and nickel since 1960's but full operation started in 1973. The BCL mine is the most important anthropogenic polluter in the town and the country. A lot of sulphur dioxide is pumped directly into the atmosphere through a 153 m stack.

Figure 1. Mining pollution and its environmental impact from the BCL mine in Selibe-Phikwe (Western Side).

The soils are classified as Eutric Regosols and Haplic Luvisols (Dept of Surveys and Mapping, Botswana). The major vegetation is of hardveld dominated by Colophospermum mopane. The mean annual rainfall is 415 mm and maximum daily temperature is 31.6°C. The prevailing wind comes from south east and east of the town according to the 2000-2006 wind rose data, Dept of Meteorology (Figure 3).

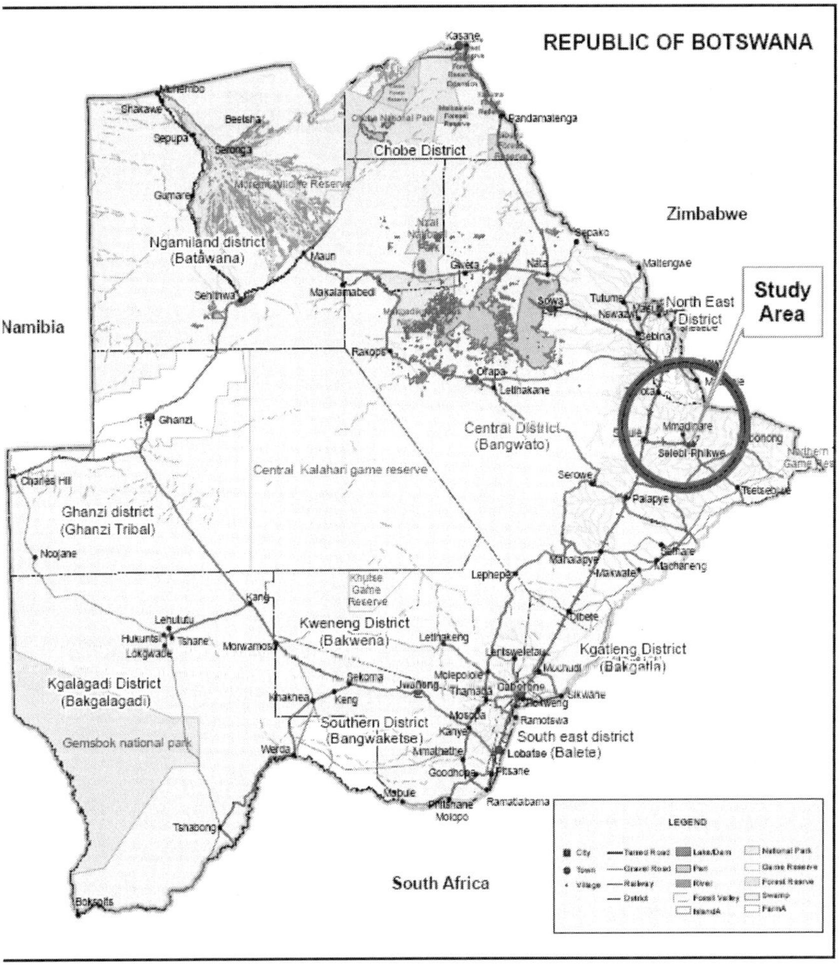

Figure 2. Map showing the study area in Selibe Phikwe (BCL mine), Botswana (Department of Surveys and Mapping).

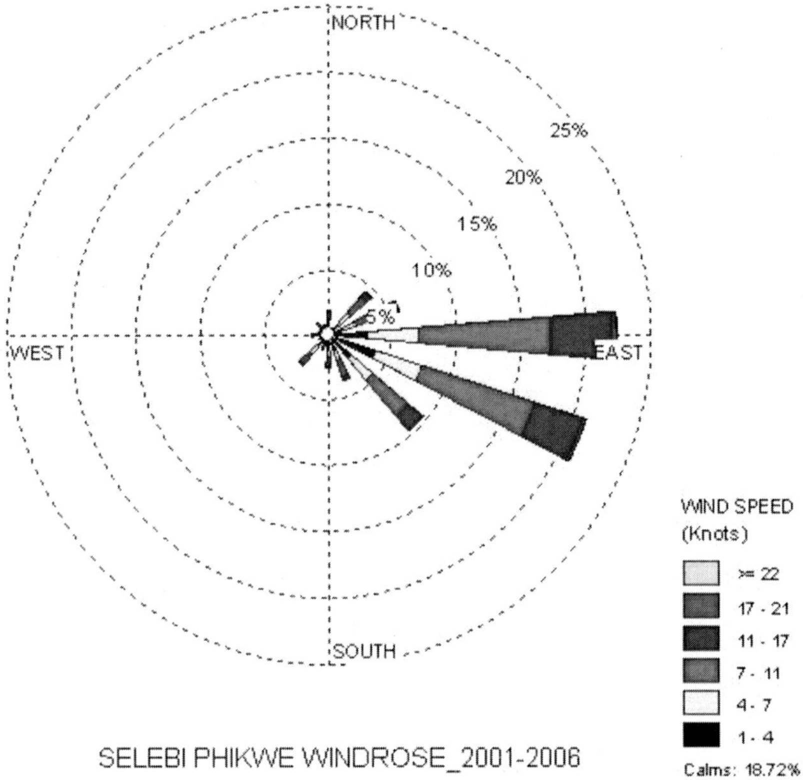

Figure 3. Prevailing wind direction in Selibe Phikwe from 2001-2006 (Department of Meteorology, Botswana).

2.2 Sampling

The soil and plants were sampled in mid February 2012. Samples were collected along a 4km transect running from the western (downwind of the mine) and eastern part of the BCL mine in Selibe Phikwe. The geographical coordinates for all sampled plots were taken by a Global Position System (Garmin GPS 76). The sampling plots (10 m x 10 m) were made every 0.5 km along each transect. Three soil samples were collected by a soil auger at a depth of 0-10 cm. They were then combined into one composite soil sample. Two composite soil samples were taken per plot and a total of thirty two soil samples were collected in the study area. Mature and healthy leaves of G. bicolor were randomly picked around each plant at breast height (1.4 m).

2.3 Sample preparation

Soil samples were air dried in the laboratory before chemical analysis was performed. After two weeks of air drying, the samples were sieved in a <2 mm stainless steel sieve. G. bicolor leaves (unwashed) were dried in an oven at a temperature of 70°C for 24 h. After the leaves were dried, they were ground to powder with Fritsch pulverisette five machines. The pulverised leaves were stored in air tight plastic bottles until they were analysed.

2.4 Chemical analysis

For the determination of the total soil heavy metal concentration, the homogenized sample, about 0.500 g±0.05 g was digested in a closed microwave vessel system (Milestone Ethos EZ, equipped with Mutliprep-41 rotor) according to USEPA 3051 Method. Standard soil reference materials from Institute of Geological Sciences Japan (JSO-1) were used for the quality control and to verify the accuracy of the metal analysis.

To determine heavy metals in plants, 0.500 g±0.05 g plant materials was weighed into porcelain crucibles and incinerated in a muffle furnace (Thermolyne, Model F6000, USA) at a temperature of 550°C for 6 h until a white ash was obtained. Before loading with plant samples the furnace was conditioned for 2 h at a temperature of 800°C. After incineration, the ash was dissolved in 10 ml diluted hydrochloric acid (6N) and made up to a volume of 50 ml before being analysed for heavy metals with AAS. Reference material and blanks were included with the samples as a quality control measure. While analysing samples, a standard was analysed after every ten samples to monitor the instrument drift from the initial calibration. AAS analysis was performed in triplicates and reslope standard was performed to update the calibration during analysis.

An Atomic Absorption Spectrometer (Perkin Elmer 3110 with S10 auto sampler) was calibrated using quality standards from Merck Chemicals, German. Samples were performed under an air and acetylene flame gas mixture. Lead and Cadmium were determined by a Varian 220FS AAS.

Sulphur in soil and plant samples was analysed using a Leco Sulphur analyzer (S-144DR). A sample of 0.350 g was subjected to a temperature of 1350°C in an oxygen rich environment. The instrument was calibrated with appropriate Leco standards and samples were run in duplicate inclusive of reference material STD-2 and Tobacco leaves from Leco Corporation.

The soil pH was determined according to Soil Sampling and Methods of Analysis (Carter, 1993). The samples were run in duplicate.

2.5 Data analysis

The data was calculated using SPSS Statistical software version 19 and graphs were performed with Ms Excel 2010. Wilcoxon sign rank test was used to find out if there is any significant different between samples from the west and east (Moreno-Jimenez et al., 2009). Spearman's rank correlation was used to determine the strength of correlation between different parameters in G. bicolor leaves and soil in the west and east transect respectively (Gauthier, 2001; Ogunkunle and Fatoba, 2014).

3. RESULTS AND DISCUSSION

3.1 Heavy metals and sulphur concentration in soil

Previous research has shown that soils around Copper-Nickel mines tend to have high concentrations of heavy metals (Nkongolo et al., 2008; Meadows and Watmough, 2012) especially Cu, Ni and Pb. In most cases, the downwind side is the most affected as depicted by the mean concentrations of heavy metals in Tables 3 and 5. Concentrations of heavy metals in soil and plants from the upwind and downwind sides were significantly different at P=0.00 and P=0.03, respectively.

Figures 4 and 5 further depicts an elevation of heavy metals concentration in the western side of the mine whereas it is opposite with the upwind side. This increase could be a result of dust particles emanating from mine waste, crushed stones and fumes from the smelter which with concentration above three fold in the downwind side. Both heavy metals do not bio-magnify up the food chain; although at higher concentrations they can be detrimental to animals and people. On the other hand, 62.5% of samples had Pb content which exceeded one fold in the downwind side. Although Cd and Pb are within the world soil range, the concern is that they are highly toxic metals and can accumulate in body organs and further biomagnify through the food chain. In Figure 6A, most of the elements are elevated in the inner 0.5 km from the mine except Sulphur which peaked at 1.5-2.0 km. Cu and S peaked from 0.5-2 km, then flattened thereafter (Figure 6C). Overall, the western side has higher concentration of heavy metals in both soils and plants. The concentration of heavy metals further shows a decrease as you go away from the mine and this has been observed by other authors.

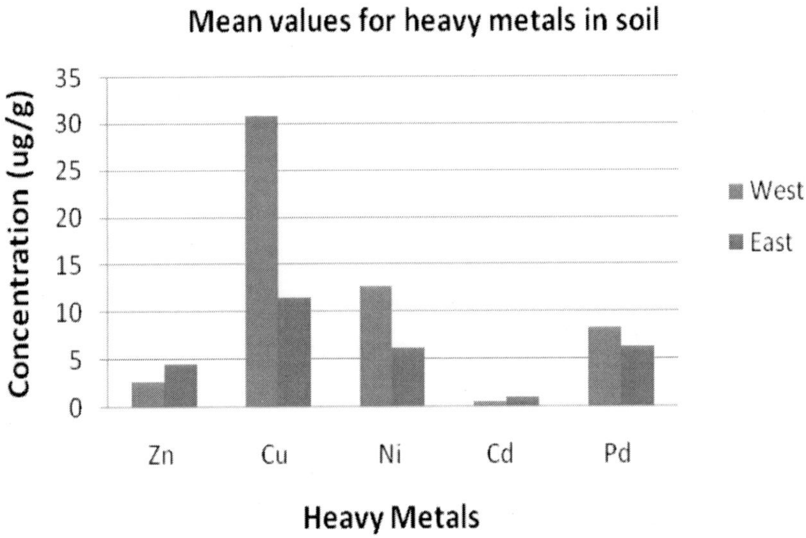

Figure 4. Comparison of heavy metals concentration in soil samples.

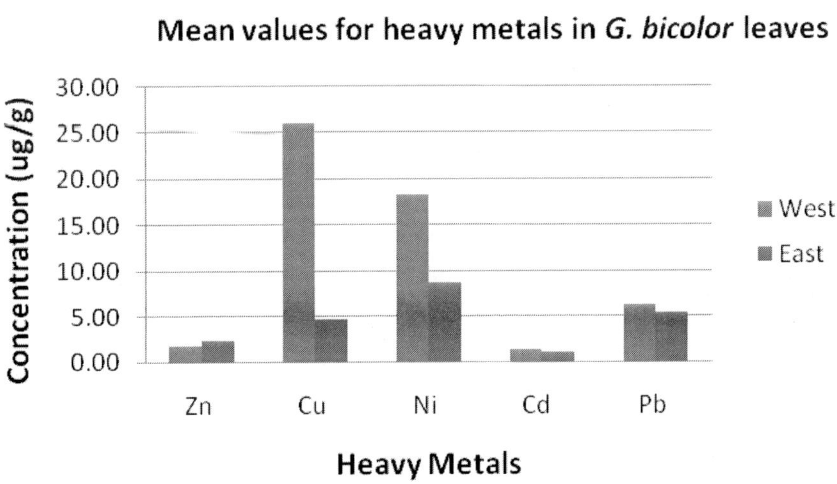

Figure 5. Comparison of heavy metal concentration in *Grewia bicolor* leaves.

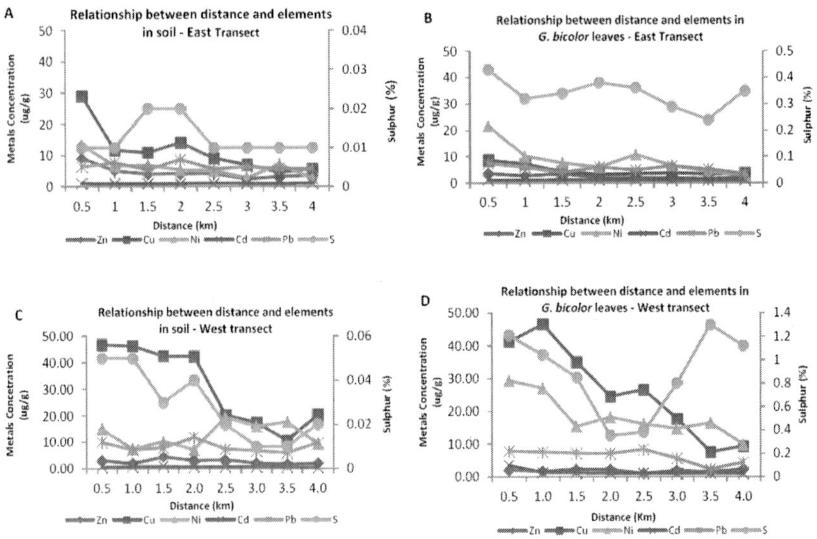

Figure 6. Changes in heavy metals and sulphur concentrations in relation to distance from the BCL mine.

In Tables 5 and 6, the ranges for heavy metals and sulphur were Zn (1.64-9.14 µg/g) Cu (4.55-46.67 µg/g), Ni (2.63-19.33 µg/g), Cd (0.39-1.06 µg/g), Pb (5.40-9.91 µg/g) whereas Sulphur and pH ranged from 0.01-0.05(%) and 3.85-5.91 respectively. In this study, the soil heavy metals contents were within the world soil ranges (Rojo et al., 2004) except for Cd which is elevated on both sides. Cd could be emanating from the parent rock. The concentrations of Zn were very low, maximum of 14 ug/g, which could be the result of Sulphur as it is known to decreases soil pH and increases the solubility, availability and mobility of heavy metals, which can eventually leaches into streams (Skwierawska et al., 2012; Fijalkowski et al., 2012).

In Table 5, Sulphur content was higher in the first inner 2 km; depicting the impact of the mining pollution in the soil adjacent to it. The upwind side had lower mean concentration of 0.01% S while downwind side had 0.03% S. According to Linzon et al. (1979), average background Sulphur concentration for soil sampled in Southern Ontario, Canada ranged from 0.05-0.08% for rural and urban areas respectively. The Sulphur content in this study is below the background concentration compared to the latter region.

The mean soil pH in the downwind side is 4.12 whereas in the upwind side is 5.12. These acidic environment could be the consequences of sulphur dioxide emission from the smelter and as such it has negatively affected the vegetation therefore there is no grass and seedlings (Figure 1). The pH of 5.12 in the upwind side shows that the soil in the study area is acidic in nature; however 75% of samples from the same sampling site had pH values above 5 (Table 6). It should be noted that soil pH is considered as one of the most important factors that determine the concentration of metals in the soil solution, their mobility and availability to plants (Fijalkowski et al., 2012). In highly acid soils, the mobility of heavy metals is much higher than in soils with neutral and alkaline reactions or pH (Fijalkowski et al., 2012).

The correlation analysis was employed to determine the common source of pollution (Table 2). Positive correlations were found between Cu-Pb, Cu-Ni, Pb-S, Pb-Zn and Cd-pH. This positive correlation maybe showing that this elements emanate from the same source (Yoon et al., 2006), which is the mine. Since cadmium does not correlate well with other heavy metals, it may be coming from the geological parent material.

Table 1. Correlation coefficients (R^2) of each element in *G. bicolor* leaves.

	Zn	Cu	Ni	Cd	Pb	S
Zn	1					
Cu	-0.060	1				
Ni	-0.026	0.805**	1			
Cd	0.176	0.321	0.461**	1		
Pb	0.021	0.549**	0.631**	0.345	1	
S	-0.078	0.579**	0.569**	0.229	0.017	1

Correlation is significant at **0.01 and *0.05 level (2-tailored).

Table 2. Correlation coefficients (R^2) of each element in soil.

	Zn	Cu	Ni	Cd	Pb	S
Zn	1					
Cu	-0.060	1				
Ni	-0.026	0.805**	1			
Cd	0.176	0.321	0.461**	1		
Pb	0.021	0.549**	0.631**	0.345	1	
S	-0.078	0.579**	0.569**	0.229	0.017	1

Correlation is significant at **0.01 and *0.05 level (2-tailored).

Table 3. Mean (±S.D) concentration of heavy metals & sulphur in *G. bicolor* leaves – West.

Distance (km)	Zn (µg/g)	Cu (µg/g)	Ni (µg/g)	Cd (µg/g)	Pb (µg/g)	S (%)
0.5	1.97±0.01	41.18±1.24	29.41±1.91	3.46±0.06	7.82±0.00	1.21±0.00
1.0	1.68±0.06	46.59±0.83	27.10±2.38	1.32±0.00	7.54±0.06	1.04±0.08
1.5	2.22±0.03	35.01±0.95	15.28±2.92	1.14±0.00	7.18±0.51	0.85±0.51
2.0	2.17±0.04	24.58±0.82	18.20±9.05	1.05±0.01	7.08±1.19	0.35±0.00
2.5	0.95±0.18	26.60±0.05	16.00±1.41	0.92±0.11	8.33±0.86	0.38±0.00
3.0	1.83±0.07	17.72±0.40	14.64±2.40	0.91±0.01	5.58±1.27	0.80±0.16
3.5	1.46±0.40	7.51±0.07	16.36±5.94	0.98±0.11	2.29±2.05	1.30±0.00
4.0	2.20±0.23	9.23±0.44	9.83±2.53	0.95±0.07	4.38±1.07	1.12±0.10

Table 4. Mean (±S.D) concentration of heavy metals & sulphur in *G. bicolor* leaves – East.

Distance (km)	Zn (μg/g)	Cu (μg/g)	Ni (μg/g)	Cd (μg/g)	Pb (μg/g)	S (%)
0.5	3.60±0.68	8.60±1.73	21.68±0.25	1.06±0.28	7.46±0.31	0.43±0.16
1.0	2.68±0.37	7.17±1.15	10.15±0.21	1.01±0.14	5.94±0.20	0.32±0.00
1.5	3.50±1.58	4.25±1.12	7.58±2.01	1.08±0.17	3.94±4.62	0.34±0.07
2.0	1.93±0.04	3.16±0.40	5.97±1.51	1.06±0.11	5.94±1.67	0.38±0.01
2.5	1.85±0.01	3.44±0.17	10.63±0.27	1.02±0.11	4.89±0.33	0.36±0.05
3.0	1.95±0.21	3.77±0.64	6.50±0.42	1.00±0.08	6.35±1.29	0.29±0.003
3.5	1.44±0.08	3.32±0.42	4.27±0.47	0.97±0.042	5.13±0.96	0.24±0.03
4.0	2.21±0.69	3.60±0.11	2.90±0.59	0.91±0.07	3.44±0.71	0.35±0.06

Table 5. Mean (±S.D) value of heavy metals, Sulphur and pH in soil – West.

Distance (km)	Zn (μg/g)	Cu (μg/g)	Ni (μg/g)	Cd (μg/g)	Pb (μg/g)	S (%)	pH
0.5	3.10±0.11	46.67±0.12	15.05±0.16	0.51±0.13	9.91±1.40	0.05±0.01	3.95
1.0	1.95±0.01	46.19±2.28	7.30±1.16	0.73±0.04	7.11±1.26	0.05±0.01	4.12
1.5	4.46±0.03	42.43±3.01	10.31±0.01	0.78±0.03	7.90±0.28	0.03±0.001	4.22
2.0	3.17±1.48	42.33±0.69	7.12±0.54	0.63±0.07	11.75±1.46	0.04±0.00	3.85
2.5	3.33±0.01	20.24±0.57	19.33±2.42	0.59±0.07	7.35±0.21	0.02±0.00	4.30
3.0	2.03±0.10	17.36±0.42	15.97±7.62	0.74±0.03	6.71±0.86	0.01±0.002	4.10
3.5	1.64±0.00	10.46±0.59	17.64±0.28	0.53±0.01	5.93±0.04	0.01±0.001	4.34
4.0	1.90±0.08	20.39±0.78	9.22±0.11	0.39±0.13	9.70±0.62	0.02±0.002	4.06

Table 6. Mean (±S.D) value of heavy metals, Sulphur and pH in soil – East.

Distance (km)	Zn (µg/g)	Cu (µg/g)	Ni (µg/g)	Cd (µg/g)	Pb (µg/g)	S (%)	pH
0.5	9.14±1.22	29.01±5.16	13.38±1.87	1.03±0.04	6.29±1.06	0.09±0.001	4.28
1.0	4.95±0.07	11.72±3.42	6.65±7.99	0.94±0.11	7.61±1.23	0.01±0.003	4.53
1.5	4.03±0.35	10.89±3.42	6.83±0.61	0.96±0.00	5.40±1.24	0.01±0.008	5.24
2.0	4.11±1.34	14.15±4.88	5.32±1.24	1.01±0.01	8.65±0.24	0.02±0.01	5.91
2.5	4.13±1.23	9.00±2.18	5.02±0.05	1.05±0.07	5.52±1.10	0.01±0.00	5.60
3.0	2.37±0.98	6.88±1.66	2.72±0.03	0.99±0.10	6.35±0.15	0.01±0.004	5.46
3.5	3.16±1.53	4.55±1.29	7.06±0.57	0.97±0.16	5.68±0.37	0.01±0.03	5.68
4.0	3.55±0.07	5.67±0.64	2.63±0.10	1.06±0.03	5.64±0.42	0.01±0.001	5.14

3.2 Heavy metals and sulphur in *g. Bicolor* leaves

The results of *G. bicolor* leaves show that heavy metals and sulphur concentration followed a similar trend which occurred in soil samples from west transect. The samples from the western transect had elevated concentration of Cu, Ni, Cd and Pb compared to the eastern transect whereas Zn was the only one elevated in the eastern side Figure 5. Within the sampling sites, 87.5% of plant samples in the downwind side have Cu and Ni concentrations above 2 fold compared to the upwind side. Cu and Ni are essential micronutrients for micro organisms, plants and animals; however at higher concentrations they can have toxic effects (Sheoran et al., 2011). Lead is usually concentrated in the roots of plants; however the observed elevated concentration in G. *bicolor* leaves could be due to aerial deposition. Generally, plants do not absorb or accumulate lead in their leaves except in the roots. The ranges of heavy metals and Sulphur in *G. bicolor* leaves on both transects were Zn (0.95-3.60 μg/g), Cu (3.16-46.59 μg/g), Ni (2.90-29.41 μg/g), Cd (0.91-3.46 μg/g), Pb (2.29-8.33 μg/g) and S (0.24-1.30%) (Tables 3 and 4). Since the plant samples were unwashed, the observed elevated concentrations could be due to aerial deposition. Looking at Figure 6B, the concentration of heavy metals is elevated at 0.5 km then it flattens. This is understandable because it is against the prevailing wind (Figure 3). In Figure 6D, Cu, Ni and S are elevated at 0.5-2 km and the rest of other elements does not shown any trend with regard to distance from the mine. The eastern side depicted a much lower heavy metal concentration in soils and plant samples.

However looking at the heavy metals concentrations, there is a potential risk of toxicity to animals grazing in the vicinity of the mining area. Yabe et al. (2011) has conducted a study in animals grazing in a Pb-Zn mining area (Kwabe, Zambia) and he found that the animals accumulated higher concentrations of Pb and Zinc.

Sulphur is an essential element in plants because it serves as a macronutrient; however it becomes detrimental above threshold values. Long term exposure to elevated concentration of Sulphur in plant foliage could cause chronic injuries. A study carried by Linzon et al. (1979) showed that average background Sulphur concentrations for 33 different plant species in Southern Ontario-Canada were within the range of 0.13-1.72% thus higher than the concentrations in my study area. Elevated Sulphur concentration of above 0.80% was found in the downwind side whereas 0.43% (max) was detected in upwind side (Tables 3 and 4). The concentration of Sulphur in the downwind side was more than 1.5-3 fold (min and max respectively) in the upwind wide (Tables 3 and 4). This shows that the western side is more polluted with sulphur as compared to the eastern side. In plants samples, positive correlations were found between Cu-Ni, Cu-Pb, Cu-S, Ni-Cd, Ni-Pb and Ni-S (Table 1). Cu, Ni and S correlated well with most of the elements and this trend is also showing in soil samples. Cu, Ni, S clearly demonstrates that they originate from the same source, however Zn did not correlate with this three elements and it also depicts that it doesn't originate from the mining pollution (Tables 1 and 2).

4. CONCLUSION

This research study shows that significant amount of heavy metal and sulphur has been deposited in the western side of the mine (downwind) on both soil and *G. bicolor* leaves. The soil and plants samples from the western side of the mine had high concentration of Cu, Ni and Pb. The statistical test further show that the western side is significantly more polluted than the eastern side, at P=0.00 and P=0.03 (2-tailed test) on both soil and *G. bicolor* respectively. The study also revealed that the soil in the western side of the mine is acidic, with mean pH of 4.12. Furthermore, for the first time, the mean background pH has been found to be 5.12 which demonstrate that the soils in the study area are acidic in nature. Concentration of soil parameters in the downwind side were in the following order Cu>Ni>Pb>Cd>S.

Assessment of the results of the study area against the world range depicts that heavy metals were below the maximum allowable limits. However persistence of the latter in *Grewia bicolor* and soil continues to be a treat to animals and people of Selibe-Phikwe and the surrounding areas due to potential biomagnifications and bioaccumulation, especially with regard to Pb and Cd.

The recommendation of this study therefore is that more research needs to be done on heavy metals in edible and medicinal plants and also in animals grazing in the mining area.

CONFLICT OF INTERESTS

The authors have not declared any conflict of interest.

REFERENCES

1. Candeias C, Da Silva EF, Salgueiro AR, Periera HG, Reis AP, Patinha C, Matos JX, Avila PH (2011). Assessment of soil contamination by potentially toxic elements in the Aljustrel

mining area in order to implement soil reclamation strategies. Land Degrad. Dev. 22:565-585.
2. Carter MR (1993). Soil Sampling Methods of Analysis. Canadian Society of Soil Science. Lewis Publisher. pp. 141-144.
3. Gauthier TD (2001). Detecting Trends Using Spearman's Rank Correlation Coefficient. Environ. Forensics 2:259-362.
4. Ikenaka Y, Nakayama SMM, Muzandu K, Choongo K, Teraoka H, Mizuno N, Ishizuka M (2010). Heavy metal contamination of soil and sediment in Zambia. Afr. J. Environ. Sci. Technol. 4(11):729-739.
5. Kiikkila O (2003). Heavy-metal Pollution and Remediation of Forest Soil around the Harjavalta Cu-Ni Smelter in SW Finland. Silva Fennica 37(3):399-415.
6. Kirkham MB (2006). Cadmium in plants on polluted soils: Effects of soil factors, hyperaccumulation, and amendments. Geoderma 137:19-32.
7. Linzon SN, Temple PJ, Pearson RG (1979). Sulfur Concentrations in Plant Foliage and Related Effects. J. Air Pollut. Control. Assoc. 29(5):520-525.
8. Meadows M, Watmough SA (2012). An Assessment of Long-term Risk of Metals in Sudbury: A Critical Loads Approach. Water Air Soil Pollut. 223:4343-4354.
9. Moreno-Jimenez E, Penalosa JM, Manzano R, Carpena-Ruiz RO, Gamarra R, Esteban E (2009). Heavy metals distribution in soils surrounding an abandoned mine in NW Madrid (Spain) and their transference to wild flora. J. Hazard. Mater. 162:854-859.
10. Nagajyoti PC, Lee KD, Sreekanth TVM (2010). Heavy metals, occurrence and toxicity for plants: a review. Environ. Chem. Lett. 8:199-216.
11. Nkongolo KK, Vaillancourt A, Dobrzeniecka S, Mehes M, Beckett P (2008). Metal Content in Soil and Black Spruce (Picea mariana) Trees in the Sudbury Region (Ontario, Canada): Low Concentration of Arsenic, Cadmium, and Nickel Detected near Smelter Sources. Bull. Environ. Contamin. Toxicol. 80:107-111.
12. Ogunkunle CO, Fatoba PO (2014). Contamination and spatial distribution of heavy metals in topsoil surrounding a mega cement factory. Atmos. Pollut. Res. 5:270-282.
13. Poikolainen J (1997). Sulphur and Heavy metal concentrations in Scots pine Bark in Northern Finland and Kola Peninsula. J. Water Air Soil Pollut. 93:395-408.
14. Rashed MN (2010). Monitoring of contaminated toxic and heavy metals, from mine tailings through age accumulation, in soil and some wild plants at Southeast Egypt. J. Hazard. Mater. 178:739-746.
15. Sheoran V, Poonia P, Trivedi SK (2011). Metal pollution in soil and plants near copper mining site. Int. J. Geol. Earth Environ. Sci. 1(1):2277-2081.
16. Skwierawska M, Zawartka L, Skwierawski A, Nogalska A (2012). The effect of different sulfur doses and forms in changes of soil heavy metals. Plant Soil Environ. 58(3): 135-140.
17. Whitby LM, Stokes PM, Hutchinson TC, Myslik G (1976). Ecological Consequence of acidic and heavy metal discharges from the Sudbury smelters. Can. Mineralogist. 14:47-57.
18. Wilkinson JM, Hill J, Phillips CJC (2003). The accumulation of potentially-toxic metals by grazing ruminants. Proc. Nutr. Soc. 62:267-277.
19. Wuana RA, Okieimen FE (2011). Heavy Metals in Contaminated Soils: A review of Sources, Chemistry, Risks and Best Available Strategies for Remediation. International Scholarly Research Network, ISRN Ecology. pp. 1-20.
20. Yabe J, Nakayama SMM, Ikenaka Y, Mazandu K, Ishizuka M, Umemura T (2011). Uptake of lead, Cadmium, and other metals in the liver and kidney of cattle near a Lead-Zinc Mine in Kabwe, Zambia. Environ. Toxicol. Chem. 30(8):1892-1897.
21. Yoon J, Cao X, Zhou Q, Ma LQ (2006). Accumulation of Pb, Cu, and Zn in native plants growing on a contaminated Florida site. Sci. Total Environ. 368:456-464.
22. Zhou JM, Dang Z, Cai MF, Liu CQ (2007). Soil Heavy Metal Pollution Around the Dabaoshan Mine, Guangdong Province, China. Pedosphere 17(5):588-594.

CHAPTER 9

Accumulation of heavy metals on bed sediments of Nairobi dam, Kenya

Ndeda, L. A.*, Manohar, S. and Macharia, G. M.

Department of Environmental Sciences, Kenyatta University, P. O Box 43844-00100, Nairobi, Kenya

ABSTRACT

Globally heavy metals pollutants in aquatic systems are increasing and creating major ecological disturbance(s) and direct health problems. The study on accumulation of Copper (Cu), Cadmium (Cd), Lead (Pb) and Nickel (Ni) on the bottom surface sediments of Nairobi dam was conducted between November 2012 to April 2013. Seven sampling sites were selected within the dam and sediment samples were collected once in a month during the wet and dry season. Standard methods for sampling, storage and analysing according to APHA were used; statistical analysis in form of means, independent T-test and ANOVA were used to summarize the findings. (Mean±SE) (ppm) concentrations of Pb, Cd, Cu and Ni in the dam during the dry season are: 38.01±0.26, 30.86±0.21, 21.7±0.22, 8.44± 0.15, and in wet season are: 33.73±0.22, 26.60±0.19, 19.11±0.14 and 5.94±0.12, respectively. The levels of heavy metals in the sediments at the outlet are lower compared to the other sites within the dam. These metals accumulated in the order: Pb> Cd> Cu> Ni with higher concentration during the dry than wet season. It is concluded Nairobi dam sediments are contaminated with heavy metals that are significantly higher during the dry compared to wet season ($P<0.0001$). Therefore it is recommended that sediments of this dam are not suitable for cultivation and or any agricultural activities.

Keywords: Accumulation, heavy metals, sediments, Nairobi dam.

1. INTRODUCTION

Heavy metals such as copper, nickel, cadmium and lead are not found in abundant quantities in the earth's crust and are considered to pose environmental hazards. Sediments are an important sink for a variety of pollutants, particularly heavy metals and may serve as an enriched source of such pollutants, especially in estuarine ecosystems. In some cases, sediments may hold over 99% of the total amount of heavy metals present in an ecosystem (Renfro, 1983). Sediments have high adsorptive capacities of metal pollutants. It has been noted from several studies that most metals entering the aquatic systems ultimately find their way into the sediments and inevitably the whole food chain.

Direct discharge or wet and dry depositions of contaminants increase the concentrations of trace elements of aquatic systems, thus resulting in their accumulation in sediments (Dunbabin and Bowmer, 1992; Sinicrope et al., 1992). Obodai et al. (2011) assessed the level of heavy metal pollution in water, sediments, and two species of fish in Benya and Nakwa lagoons, Ghana and it was concluded that the contamination was the highest in the sediments. Ogoyi et al. (2011) studied heavy metal in water, sediment and microalgae in Lake Victoria, Kenya and concluded that highest levels of trace metals were in the sediments of the lake

Nairobi dam is contaminated with chemical pollutants. This results to bioaccumulations and biomagnifications which causes potential ecological disturbance(s) and direct health problems to local people who depend on the water and products from the resource; and particularly residents of Kibera informal settlement adjacent to the dam who are cultivating crops for domestic and commercial purposes within Nairobi dam.

The objective of the present study is to analyse, compare and contrast the accumulations of heavy metals on bottom surface sediments of Nairobi dam during the wet and dry seasons.

There is great deal of sedimentation taking place within Nairobi dam (UNEP, 2008). Very little work has been conducted on accumulation of heavy metals in sediments of the dam. Therefore, we took the initiative to study the contamination of heavy metals in sediments of Nairobi Dam. The findings of the present study will be useful for monitoring, auditing and assessing the status of any aquatic ecosystem(s) because heavy metals can pass to human beings through food chain.

2. MATERIALS AND METHODS

2.1 Study area

The study was carried out at Nairobi dam in Nairobi, Kenya. This dam is a source of water both for domestic and commercial purposes to the surrounding community. The dam is situated at a latitude of 1° 19' (1.3167°) south; longitude of 36° 48' (36.8°) east and average elevation of 1,686 meters. Its accumulating capacity was 98,422 cu. metres of water (3,477,800 cu ft) and surface area of 356,179 m² (approximately 86 acres) when it was designed (Figure 1).

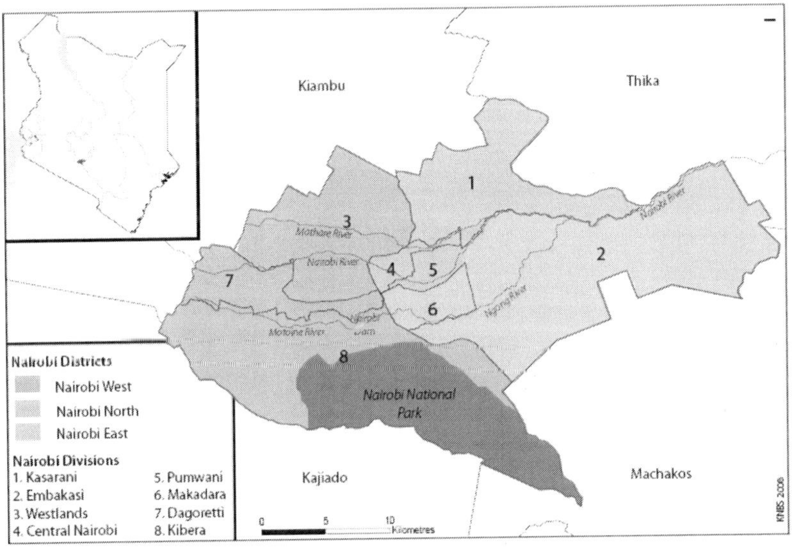

Figure 1. Map of Kenya showing study area, that is, Nairobi Dam.

2.2 Data collection

Riverbed sediment samples were collected 10 cm below the Nairobi dam bottom surface sediments using an auger on a monthly interval between

November 2012 and April 2013, from each of the seven sampling stations identified (Figure 2): the 1st sample station was along the stream entering Nairobi dam 10 m before the discharge into the dam; 2nd station was at a distance of 5 m after the discharge into the dam; 3rd station was 5 m inwards of the west shore; 4th station was 5 m inwards of the outlet, 5th station was 5 m inwards of the east shore; 6th station was 15 m inwards from the 3rd station, and the 7th station was 1 m from the outlet of Nairobi dam. The sediment samples were transferred into well labelled 250 g polypropylene jars, pre-cleaned with a metal-free non-ionic detergent solution, rinsed with tap water, soaked in 1 + 1 HNO_3 acid for 24 h at 70°C and then rinsed with metal-free water (APHA, 2005). These samples were then transported to Kenyatta University laboratory and stored in the fridge freezer. About 20 g of each sediment sample was put into labelled 250 ml beakers and dried in an oven at 105°C overnight. The dried samples were ground using a motor and pestle after which they were stored in clean well labelled polypropylene bottles awaiting digestion (APHA, 2005).

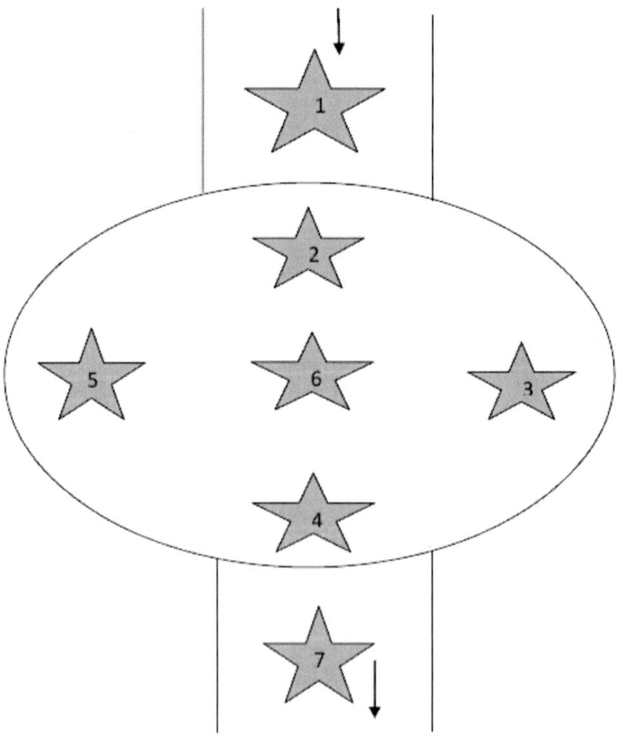

Figure 2. Schematic figure showing seven sampling sites within the study area (Nairobi Dam) for data collection.

2.3 Sample digestion and metal detection

1 g of each of the ground sediment samples was digested in Conc. HNO_3: HCl in the ratio of 2:1 and three drops of 30% Hydrogen peroxide on a hot plate in a fume hood. After total digestion and subsequent cooling, the solution was diluted with distilled water to 100 ml volumetric flask and analyzed for Cu, Cd, Pb and Ni, using an Atomic Absorption spectrophotometer (AAS) (APHA, 2005), Shimadzu flame AAS model (AA-630)

2.4 Statistical analysis

From the data, mean heavy metals (Cu, Cd, Pb, Ni) values and standard errors were calculated; Independent T-test, analysis of variance (ANOVA) and Student-Newman-Keuls test at $\alpha=0.05$ were performed on the metals mean values using SPSS version 18.

3. RESULTS AND DISCUSSION

3.1 Concentration and relationships of heavy metals on nairobi dam sediments

3.1.1 Copper

The mean copper concentrations of sediments in the different sampling sites (1 to 7) of Nairobi dam during the wet and dry season showed highly significant differences indicated by the P-value <0.001 on the columns. It can be seen that site 1, 2, 3, 5, and 6 had the same superscript letter (b) meaning that the values of copper are not significantly varied within these stations but site 4 and 7 denoted by alphabet (a) indicates that these two sites with concentrations of copper that are significantly different from those with letter (b) (Table 1). This can be explained by the fact that the purification mechanism within the wetland (Nairobi dam), resulted in removal of heavy metals from polluted sediments within the dam: Therefore the sediments at the outlet (site 4 and 7) of the dam had lower levels of copper. Copper levels are significantly higher in the dry than wet season (P<0.001) in all the sampling sites; this is attributed to concentration of metals on sediments due to higher temperatures in dry season.

Table 1. Assessment of heavy metals concentrations in sediments from different sampling sites in Nairobi dam, Kenya.

Parameter	Sampling sites	Wet season (Mean ±SE) (ppm)	Dry season (Mean ±SE) (ppm)	P-value
Cu	1	19.38 ± 0.08^{b}	22.29 ± 0.14^{b}	<0.0001
	2	19.50 ± 0.07^{b}	22.39 ± 0.13^{b}	<0.0001
	3	19.59 ± 0.14^{b}	22.49 ± 0.13^{b}	<0.0001
	4	18.23 ± 0.13^{a}	20.40 ± 0.09^{a}	<0.0001
	5	19.50 ± 0.14^{b}	22.45 ± 0.17^{b}	<0.0001
	6	19.44 ± 0.12^{b}	22.31 ± 0.11^{b}	<0.0001
	7	18.09 ± 0.07^{a}	20.18 ± 0.10^{a}	<0.0001
	P-value	<0.0001	<0.0001	
	Mean	19.11±0.14	21.79±0.22	<0.0001
Cd	1	26.58 ± 0.09^{c}	31.15 ± 0.10^{b}	<0.0001
	2	26.72 ± 0.13^{c}	31.22 ± 0.10^{b}	<0.0001
	3	27.38 ± 0.08^{de}	31.63 ± 0.09^{c}	<0.0001
	4	25.55 ± 0.06^{b}	29.59 ± 0.13^{a}	<0.0001
	5	27.62 ± 0.11^{de}	31.72 ± 0.10^{c}	<0.0001
	6	27.15 ± 0.11^{d}	31.41 ± 0.10^{bc}	<0.0001
	7	25.21 ± 0.15^{a}	29.29 ± 0.12^{a}	<0.0001
	P-value	<0.0001	<0.0001	
	Mean	26.60±0.19	30.86±0.21	<0.0001
Pb	1	33.99 ± 0.13^{b}	38.40 ± 0.14^{b}	<0.0001
	2	34.08 ± 0.13^{b}	38.52 ± 0.13^{b}	<0.0001
	3	34.39 ± 0.14^{b}	38.93 ± 0.04^{c}	<0.0001
	4	33.16 ± 0.10^{ab}	36.31 ± 0.09^{a}	<0.0001
	5	34.20 ± 0.13^{b}	38.89 ± 0.05^{c}	<0.0001
	6	34.19 ± 0.13^{b}	38.84 ± 0.06^{c}	<0.0001
	7	32.09 ± 1.05^{a}	36.18 ± 0.09^{a}	0.018
	P-value	0.016	<0.0001	
	Mean	33.73±0.22	38.01±0.25	<0.0001
Ni	1	6.16 ± 0.11^{b}	8.74 ± 0.09^{b}	<0.0001
	2	6.23 ± 0.10^{b}	8.83 ± 0.09^{b}	<0.0001
	3	6.30 ± 0.08^{b}	8.89 ± 0.10^{b}	<0.0001
	4	5.19 ± 0.08^{a}	7.52 ± 0.10^{a}	<0.0001
	5	6.44 ± 0.05^{b}	9.06 ± 0.10^{b}	<0.0001
	6	6.19 ± 0.10^{b}	8.69 ± 0.12^{b}	<0.0001
	7	5.05 ± 0.04^{a}	7.34 ± 0.11^{a}	<0.0001
	P-value	<0.0001	<0.0001	
	Mean	5.94±0.12	8.44±0.15	<0.0001

(Mean ±SE) ppm value with $p<0.05$ are significant different ($\alpha = 0.05$, Independent t-test for analysis of heavy metal concentration between the wet versus dry seasons; One way ANOVA and Student-Newman-Keuls test for analysis of heavy metal concentration within the seven sampling sites) Superscript a, b and c represents the ranking of variances in levels of heavy metal in the sediment samples across the seven sampling sites. The higher the alphabet the higher the variance in heavy metal concentration.

Cooper et al. (1996) studied reed beds and constructed wetlands for waste water treatment in the U.K. and stated that the following pollution removal mechanisms exist in wetlands: Metal adsorption and cation exchange, complexation, precipitation in sediments, plant uptake and microbial oxidation /reduction processes, therefore resulting to pollutants purification.

3.1.2 Cadmium

The P-value ($P<0.001$) on the column for cadmium metal during the wet and dry season justifies that there is significant difference on cadmium concentrations in the sampling sites 1 to 7 (Table 1). In the dry season, the lower case superscript (c) on site 3 indicates there is a significantly higher level of the contaminant compared to site 4 which is denoted by superscript (a). This can be attributed to the fact that site 3 (adjacent to Kibera slums) received high levels of Cd contaminants from the area as well as from the dams inlet but site 4 which is closer to outlet of the dam contains purer sediments due to metals clean up within the wetland. Cadmium levels on the dams sediments are significantly higher in the dry compared to wet season ($P<0.0001$) in all the seven sampling sites. This can be explained by low water flow rate in the dam during the dry season resulting in sediment settlement and also higher temperatures in the dry season resulting in concentration of heavy metals.

3.1.3 Lead

There is a significant difference in lead concentration on Nairobi dam sediments from the seven sampling points during both the wet and dry season ($P\leq0.016$). During the dry season it is shown that sediments at site 4 and 7 (Outlet) denoted by superscript (a) have significantly lower levels of Pb in comparison to site 1 and 2 having superscript (b); and even much lower Pb levels compared to site 3, 5 and 6 having (c). This can be justified by the presence of hydrophytes within the wetland that adsorbed the metals within the dam therefore lower levels of the metal in the bottom sediments at the outlet compared to inlet and the other sites.

Lead concentrations are significantly higher in the dry than wet season ($P\leq0.018$) at all sampling sites. This is due to settlement of metals concentrated sediments and higher temperatures in the dry season resulting to increased concentration of the heavy metals.

3.1.4 Nickel

There is significant difference in nickel concentration within the seven sampling sites of the dam during the dry and wet season ($P<0.0001$). During the wet season at sampling sites 1, 2, 3, 5 and 6, Ni levels are significantly higher compared to sites 4 and 7 denoted by superscript (a). In all the seven sampling sites, nickel levels are significantly higher in dry compared to wet season ($P<0.0001$). This is majorly due to higher temperatures in dry season resulting in concentration of heavy metals in the sediments.

4. DISCUSSION

A study by Schiffer (1989) on the effects of highway runoff on the quality of water and bed sediments of two wetlands in central Florida revealed that generally the levels of pH, nutrients, aluminum, lead and copper in water and sediments decreased with distance from the inlet. For example at the Island Lake wetland , Lead was highest at site 5 (2,000 μg/g) and site 7 (also 2,000 μg/L), and decreased at far sites in the wetland, the lead concentration in the sediments at the outlet was only 40 μg/g. however for the sediments Zinc and chromium concentrations did not decrease as much with distance from the inlet as water. It was also noted that concentrations in sediments among the sites are highly variable because of the relation of constituents to particle size and organic carbon content, the uneven distribution of sediment sizes in many bottom samples, and varying conditions of deposition (variation in loads and hydraulic conditions with each storm).

The order of metals accumulation on Nairobi dam sediments is: Pb> Cd> Cu> Ni. This is because during both the dry versus wet season, Pb had the highest average concentration at 38.01±0.25 versus 33.73±0.22, followed by Cd at 30.86±0.21 versus 26.60±0.19, then Cu at 21.79±0.22 versus 19.11±0.14 and the lowest was Ni at 8.44±0.15 versus 5.94±0.12 ppm respectively (Table 1). This indicates that the external environment to the sediment of the dam are polluted with heavy metals in this sequence due to socio economic, demographic and waste dumping activities including human waste, agriculture, car washing and industrial effluent observed. It has been reported that the differences in geography and geology, as well as activities in various rivers, affects the amount of metals present (Berner and Berner, 1987; Bricker and Jones, 1995).

A study on Deûle canal heavy metals sequential extraction by Nada et al. (2011) in France revealed that sediments were slightly contaminated with Cu but highly in Pb, Zn and Cd. Zerbe et al. (1999) studies of the bottom deposits from Goreckie Lake, Poland, showed accumulation of Fe to be highest at 9,150 mg/kg than Mn >Zn >Ni >Cu >Pb >Cd >Cr.

The mean Cu, Cd, Pb and Ni concentrations in the whole dam show significantly higher levels in the dry compared to wet season ($P < 0.0001$) (Table 1). This is attributed to major precipitation of decomposed organic matter (aquatic biota) on bed sediments due to elevated temperatures during the dry season thereby higher metals deposition on dam's bottom sediments. Higher flow rates observed during the wet season caused a great deal of erosion of sediments containing the heavy metals from the dam resulting to low metals on dams bottom sediments; however, during the dry season the low flow rate enhanced settlement of sediments

Similar findings were shown by Kikuchi et al. (2009) on a study of characterization of heavy metal pollution in river sediment of Hanoi City and its downstream area. That revealed concentrations of heavy metals (As, Cr, Cu, Mn, Ni and Zn) in the sediments of the Nhue River were higher in the dry season than in the rainy season. A study by Mondol et al. (2011) on seasonal variation of heavy metals concentrations in water and plant samples around Tejgaon industrial area of Bangladesh showed that in the rainy season the pollution was lower because heavy rainfall were flushed out through the canal into the adjoining vast flood zone. As the rainy season receded the soils and water were enriched with the pollution load (Chamon et al., 2009; Ullah et al., 1999).

5. CONCLUSION

Nairobi dam sediments are contaminated with Pb, Cd, Cu and Ni metals during both the wet and dry season within a range of 38.01±0.26 to 5.94±0.12 ppm. The sequence of metals accumulation on the sediments of Nairobi dam is: Pb > Cd > Cu > Ni. These results reaffirm that sediments act as a reservoir for contaminants in an ecosystem, therefore bed sediments from Nairobi dam are polluted with these heavy metals.

Heavy metals concentrations in the sediments significantly vary in the sampling sites ($P < 0.0001$), with lower metals levels in the outlet sediments compared to those from other sites within the dam. This signifies the presence of pollution removal mechanism in the wetland probably due to uptake by hydrophytes, precipitation in sediments, reduction/oxidation processes, complexation and the slower flow rate of water within the dam.

The concentration of Pb, Cd, Cu and Ni on the bed sediments of Nairobi dam are significantly higher during the dry in comparison to wet season ($P<0.0001$). This justifies that dry season gives the worst case scenario of pollution. The analysis suggests the frequent monitoring and assessment of the Nairobi dam sediments to check the activities responsible for metal contamination and the mechanisms involved into metal purification in the wetland.

6. CONFLICT OF INTERESTS

The authors have not declared any conflict of interests.

REFERENCES

1. APHA (2005). Standard methods for the Examination of water and waste water. 20th Ed. America Public Health Association, Washington DC, USA.
2. Berner EK, Berner RA (1987). The global water cycle: Geochemistry and Environment: Englewood Cliffs, NJ: Prentice-Hall, Inc. pp. 142-155.
3. Bricker OP, Jones BF (1995). Main factors affecting the composition of natural waters, In: Trace Metals in Natural Waters: Ch. 1, B Salbu, E Steines. [Eds.], CRC Press. pp. 1-19.
4. Chamon AS, Mondol MN, Faiz B, Rahman MH, Elahi SF (2009). Speciation analysis of Ni in the soils of Tejgaon industrial area of Bangladesh. Bangladesh J. Sci. Ind. Res. 44(1):87-108.
5. Cooper PF, Job GD, Green MB, Sholes RBE (1996). Reed Beds and Constructed Wetlands for Wastewater Treatment. Medmentam, Marlow U.K: WRC publications.
6. Dunbabin JS, Bowmer KH (1992). Potential use of constructed wetlands for treatment of industrial waste waters containing metals. Sci Total Environ. 111(2.3):151-168
7. Kikuchi T, Hai HT, Tanaka S (2009). Characterization of heavy metal pollution in river sediment of Hanoi City and its downstream area by multivariate analyses. Desalination Publications. 4(2009):240-247.
8. Mondol N, Chamon AS, Faiz B, Elahi DSF (2011). Seasonal variation of heavy metal concentrations in water and plant samples around tejgaon industrial area of bangladesh. J. Bangladesh Acad. Sci. 35(1):19-41.
9. Nada S, Dubourguier HC, Hamieh T (2011). Sequential Extraction and Particle Size Analysis of Heavy Metals in Sediments Dredged from the Deûle Canal France. The Open Environ. Eng. J. 4:11-17.
10. Obodai EA, Boamponsem LK, Adokoh CK, Essumang DK, Villawoe BO, Aheto DW, Debrah JS (2011). Concentrations of heavy metals in two Ghanaian Lagoons. Arch. Appl. Sci. Res. 3(3):177-187.
11. Ogoyi DO, Mwita CJ, Nguu EK, Shiundu PM (2011). Determination of Heavy Metal Content in Water, Sediment and Microalgae from Lake Victoria, East Africa, The Open Environ. Eng. J. 4:156-161.
12. Renfro WC (1983). Transfer of 65Zn from Sediments by Ma-rine Polycheate Worm. Mar. Biol. 21:305-316.
13. Schiffer DM (1989). Effects of Highway Runoff on The Quality of Water
14. Sinicrope TL, Langis R, Gersberg RM, Busnardo MJ, Zedler JB (1992). Metal removal by wetland mesocosms subjected to different hydroperiods. Ecol. Eng. 1:309-322.
15. Ullah SM, Gerzabek MH, Mondol MN, Rashid MM, Islam M (1999). Heavy metal pollution of soils and water and their transfer into plants in Bangladesh, Proc. of extended Abstracts: 5th International Conference on the Biogeochemistry of Trace Elements. WW Wenzel, DC Adriano, B Alloway, HE Doner, C Keller, NW Lepp, M Mench, R Naidu, GM Pierzynski, [Eds.], Vienna, Austria. 1:260-261.
16. UNEP (2008). NRBP Phase III: Survey and Situation Analysis of the Biological Characteristics of the Main Tributaries of the Nairobi Rivers, Reservoirs and Wetlands. Nairobi.
17. Zerbe J, Sobczyński T, Elbanowska H, Siepak J (1999). Speciation of Heavy Metals in Bottom Sediments of Lakes. J. Environ. Stud. 8(5):331-339

CHAPTER 10

Hydro-Chemistry of Gokyo Valley, Sagarmatha (Everest) National Park, Nepal

Rajendra Bhandari[1*], Subodh Sharma[1], Juerg Merz[2], Robin Garton[2]

[1]*Department of Environmental Science and Engineering, Kathmandu University, Dhulikhel, Nepal*

[2]*The Glacier Trust, Wiltshire, UK*

ABSTRACT

This research work records the baseline inventory of selected ions in running waters of Gokyo Valley, Nepal and tries to establish relationship of these ions' concentration with elevation gradient. Among these selected hydro-chemical parameters, the average concentration of Ca^{2+}, Mg^{2+}, SO_4^{2-} and total silica were determined to be 3.25 (±2.02), 0.64 (±0.47), 207.2 (±423.01) and 0.36 (±0.17) mg/L respectively. Concentration of these parameters was low compared to earlier studies except SO_4^{2-} that could be justified too since the sampling was performed only once in drier period with low atmospheric input of acid protons required for weathering process. Moreover, Pearson Product-moment Correlation Coefficients (r) for elevation gradient and concentration of Ca^{2+}, Mg^{2+}, SO_4^{2-} and total silica were determined to be +0.350, +0.145, +0.101 and −0.275 respectively. Such weak relationship between these parameters and elevation gradient suggests that the concentration of these ions is related to weathering phenomena, glacier and watershed characteristics and snow cover dynamics in Gokyo Valley.

Keywords: Gokyo Valley, Himalayan Streams, Hydrochemistry, Sagarmatha National Park

1. INTRODUCTION

Limnological research in the Himalayas has been carried out since the beginning of the 20th century by Sars, 1903 and Hutchinson, 1932 (As cited in [1]). Present decades have seen more detailed approach with greater focus on morphometric, physiochemical and biological features, primary production and trophic status and particular emphasis for fish production [1] indicating very low concentration of dissolved minerals at altitude above 4000 m a.s.l. Nevertheless, the studies in the higher Himalayas of Nepal are scanty irrespective to the studies done in high altitude water bodies in Kashmir and Sikkim in the northwest Himalayas [1] . The study carried out by [2] in 24 lakes at altitude between 4500 and 5600 m a.s.l. in the Mount Everest area provides the first data on morphometry, temperature, chemistry and biology. Concerning the overall lake studies performed in Nepal, more than 163 research studies have been reported including master's thesis from 1969 to 2010 [3] .

Lakes in Gokyo regions are studied systematically by assigning their names as Lake Cadstre Numbers (LCN) and the results indicate that these lakes are oligotrophic and alkaline and have low ionic content [1] . More recently, various limnological studies have been carried out in Gokyo region specially focusing on lakes [4] . The studies concerning physical-chemical including heavy metals, diatom and macroinvertebrate assemblages have also been carried out [4] [5] .

The impact on water chemistry due to anthropogenic pollution from the industrial and agricultural sources has concerned environmentalists and scientists for past decades [6] . However, due to their remoteness high altitude lakes are less often affected by local sources of pollution. But their catchments receive deposition of pollutants due to long distance atmospheric transport [7] . Moreover, the chemistry of running waters draining the high mountain areas of Asia is largely unknown nevertheless; some studies conclude that the degree of pollution of these rivers increases with the distance from the high mountain first order streams and that pollution is increasing through time [6] . Geologically different catchments differ in both the release rates of elements to streams and lakes, and the relative proportion that are released. Yet, for the major ions at least, the relative proportion does not greatly differ in well-watered portion of the temperate, arctic and sub-tropical zones. The most common ion sequences as equivalent weight according to [8] are as follows:

$$Ca^{2+} > Mg^{2+} > Na^{+} > K^{+}$$

$$HCO_3^{-} > SO_4^{2-} > Cl^{-}$$

In this study, two cations (Ca^{2+}, Mg^{2+}), sulphate and total silica are taken and their concentrations (mg/L) have been identified and plotted against elevation if there exists any relationship between these ions with elevation gradient.

2. MATERIALS AND METHODS

2.1. Study Area

Gokyo Valley is situated in Sagarmatha National Park, one of the World Heritage Sites and is a designated Ramsar Site. It is located within 27°52'N to 28°05'N and 86°40'E to 86°43'E. Total population of Khumjung VDC is 1912 with 913 males and 999 females, and has 551 households [9] . For the sampling of water for hydrochemistry, area from Phortse Thanga (3570 m asl) to Gokyo Lake Six (5125 m a.s.l) has been taken as shown in Figure 1. Several first order streams that are directly fed to Dudh Koshi River have been sampled. The sampling of water was done in the third week of May, 2013.

2.2. Sampling and Analysis

20 sampling stations were selected systematically and numbered from 1 to 20 as shown in Figure 2. Water samples for the analysis of calcium ion, magnesium ion, sulphate ion and total silica were analyzed in Aquatic Ecology Center, Kathmandu University. For the determination of calcium and magnesium ion concentration determination, Atomic Absorption Spectrophotometer (Thermo electron corporation; Model; M5MK2AA system), was used. Similarly for the determination of sulphate ion concentration and total silica concentration, Turbidimetric Spectrophotometric and Molybdosilicate Spectrophotometric (Model: Thermo Spectronic GRNESYS 10 UV) methods were used respectively. The guideline provided by APHA, AWWA and WEF (1998) [10] was strictly followed for the concentration determination of mentioned parameters. Microsoft Office Excel-2007 was used for the processing and analysis of data. This software was also used to construct Trend Graphs and to compute Pearson Product-moment Correlation Coefficient for selected parameters under study.

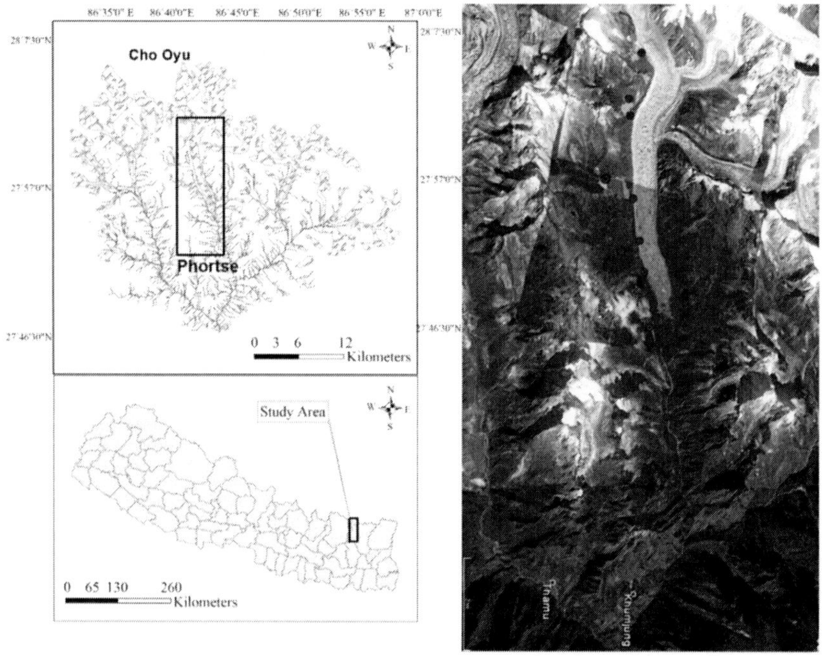

Figure 1. Study area.

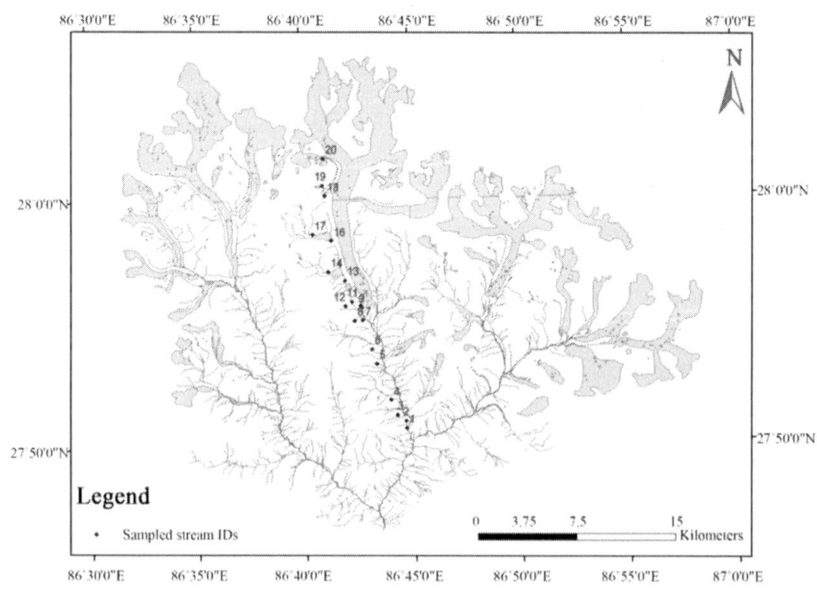

Figure 2. Sampling stations.

3. RESULTS AND DISCUSSION

One time sampling for two cations (Ca^{2+} and Mg^{2+}), total silica and sulphate was done in 3rd week of May, 2013 so as to examine whether there is any relationship of these parameters with elevation gradient. Sampling points 1 to 8 consist of first order streams which are devoid of lakes. Nevertheless, sampling points 9 to 15 are streams between Gokyo lake series. Sampling points 16, 17, 18, 19 and 20 are strictly located in lakes. Sampling points 16 and 17 are outlet and inlet of Lake 4 respectively. Similarly sampling points 18 and 19 are outlet and inlet of Lake 5 respectively. However, the sampling point 20 lies in Lake 6. The numbering of these streams is presented in Figure 2.

The average concentration of calcium ion in the sampling site is found to be 3.25 mg/L with standard deviation of 2.02, statistically as 3.25 (±2.02). Similarly, the average concentration of magnesium ion in mg/L with standard deviation is 0.64 (±0.47). The average concentration of total silica and sulphate in mg/L is 0.36 (±0.17) and 207.20 (±423.01) respectively. Owing to extremely high variability of result in the data sets of sulphate ion due to high concentration in four sites for which geometric mean has been calculated and is 15.16 which is more representative. The statistical results of four parameters that are analysed have been presented in Table 1.

Figures 3-6 demonstrate the concentration levels of analyzed four parameters of stream and lake waters in Gokyo region. The concentration of calcium (3.25 ± 2.02) is low compared to the former studies done in Everest region in 1995 and 1998 [6] [11] however, it is very low compared to the study done in 2010 [5] . In similar way, concentration of magnesium (0.64 ± 0.47) and total silica (0.36 ± 0.17) is also low compared to the results of study performed in 1995 and 1998 [6] [11] . However the total silica concentration is approximately similar to the study performed in 2010 [5] . Nevertheless, the sulphate concentration of this study is highly variable and very high compared to the previous studies [6] [11] .

Table 1. Mean concentration of selected parameters.

Parameters (mg/L)	Mean	Min	Max	Std. Dev	Geomean
Ca^{2+}	3.25	0.56	7.94	2.02	
Mg^{2+}	0.64	0.14	2.14	0.47	
Total silica	0.36	0.13	0.92	0.17	
SO_4^{2-}	207.2	1.46	1291	423.01	15.16

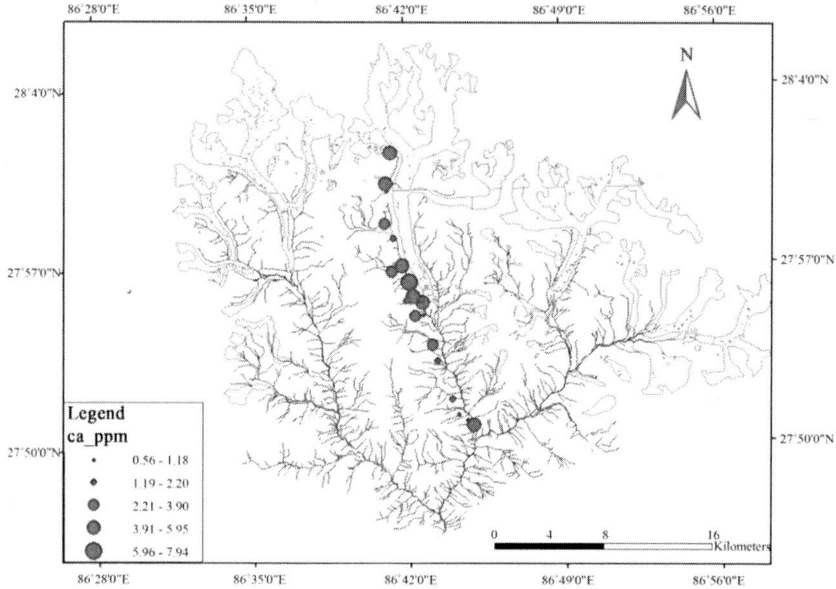

Figure 3. Calcium ion concentration.

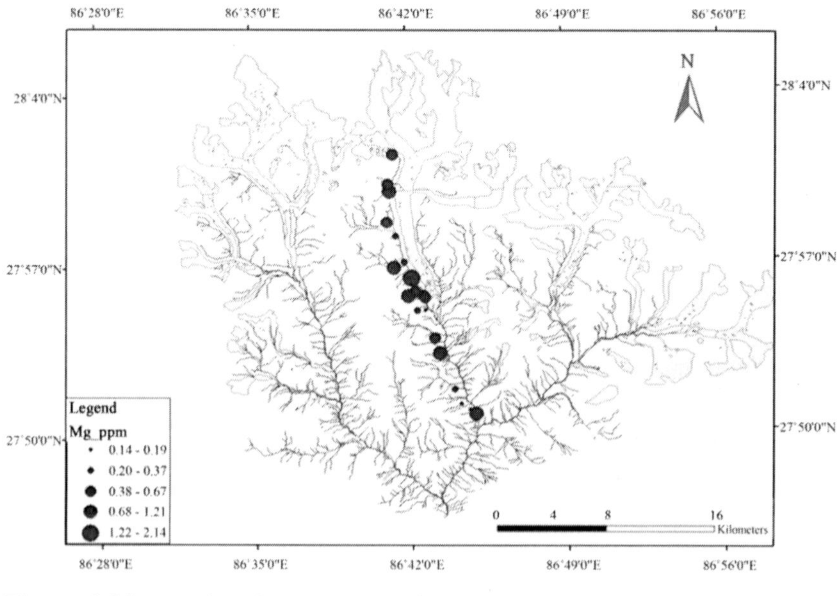

Figure 4. Magnessium ion concentration.

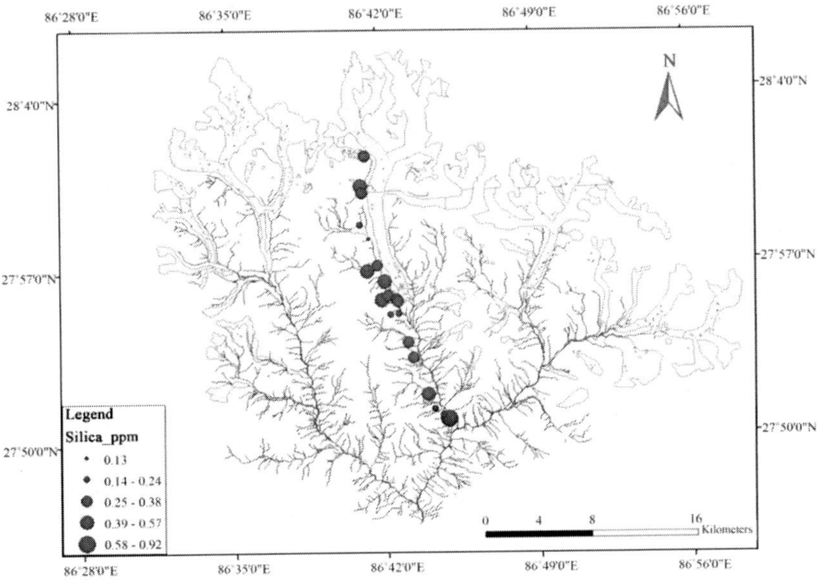

Figure 5. Silica ion concentration.

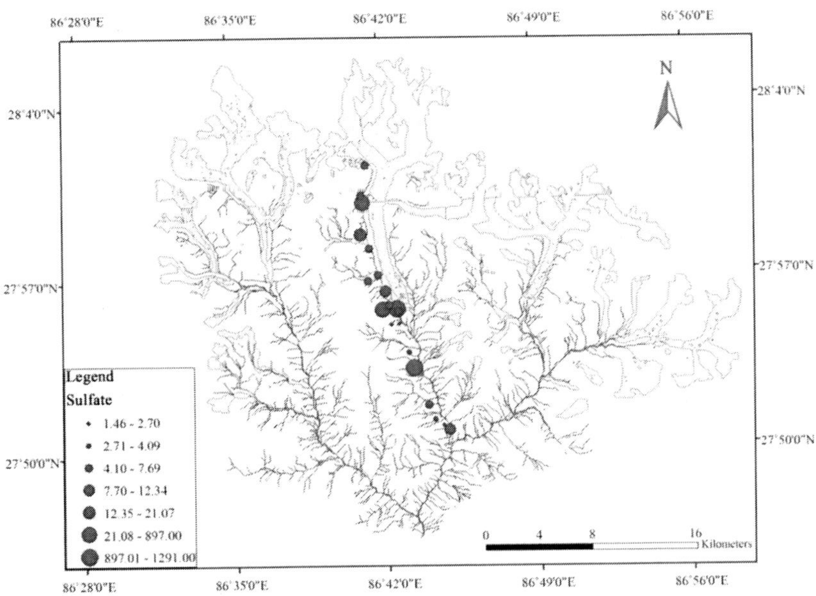

Figure 6. Sulphate ion concentration.

This study was done only in one season (3rd week, May 2013) during the dry period with low atmospheric inputs and low stream flows so that the major source of ions and silica to these streams and lakes is from chemical weathering. In the absence of significant atmospheric inputs of acid pollutants, the supply of acid protons required for weathering reactions in alpine glacial environments depends mainly on the dissolution of CO_2 and oxidation of pyrite, the processes which accounts for the dominance of HCO_3^- and SO_4^{2-} amongst anions solution [12] . And the readily weatherable source of Ca^{2+} and Mg^{2+} are alumino-silicates with carbonates that are easily identified in a number of geological formations in these regions [11] .

The final objective of this study was to determine whether the ions concentration is related to the elevation gradient or not. For this, scatter chart of elevation withthe concentration of calcium, magnesium and total silica were plotted as shown in Figures 7-9 respectively.

The Pearson product-moment correlation coefficient was calculated using Microsoft, Excel Workbook for four parameters as mentioned above. Owing to extremely high variability of sulphate concentration the chart is not constructed. The Pearson Product-moment Correlation Coefficients for elevation gradient and concentration of calcium, magnesium, total silica and sulphate are +0.350, +0.145, −0.275 and +0.101 respectively. All the values showed very weak relationship between elevation gradient and ions concentration for Gokyo region. Hence, this study derives the finding that the ions concentrations in Gokyo region are not strictly correlated to elevation gradient. Nevertheless, these are related tothe weathering phenomena, glacier and watershed characte- ristics and snow cover dynamics [1] .

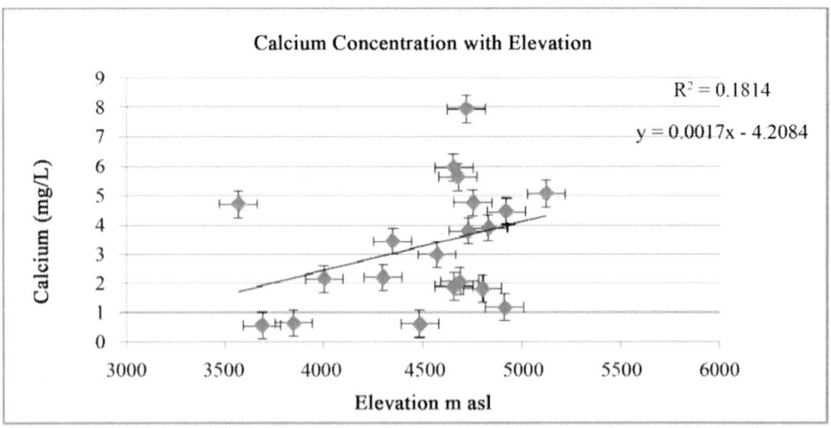

Figure 7. Scatter chart of calcium concentration.

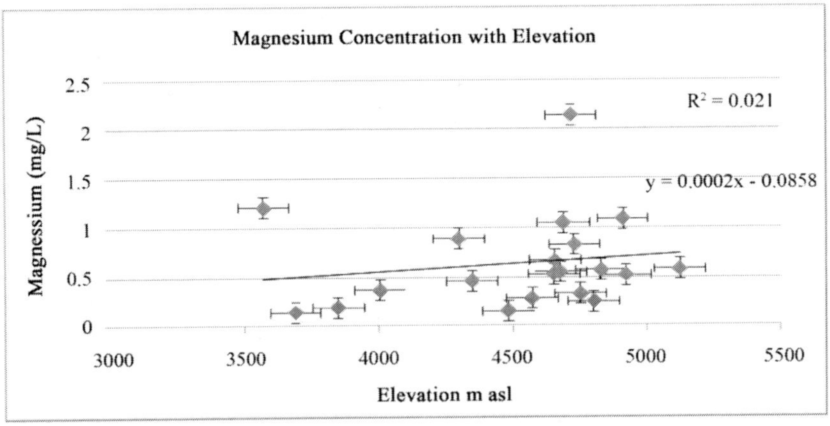

Figure 8. Scatter chart of magnessium concentration.

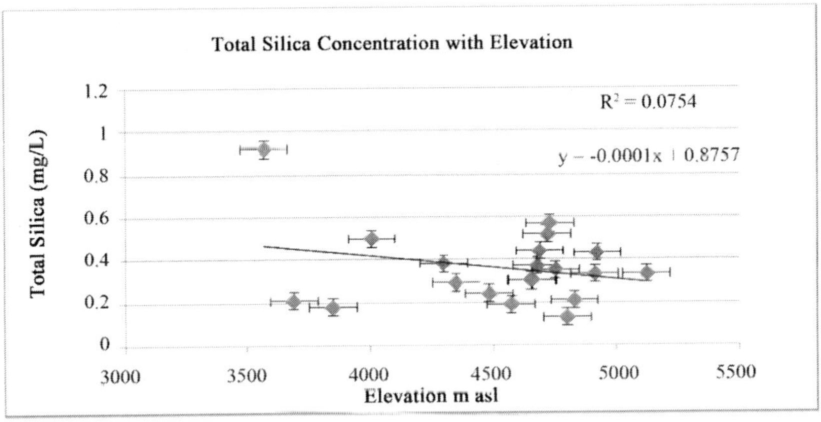

Figure 9. Scatter chart of total silica concentration.

4. CONCLUSION

The low concentration of calcium and magnesium ions in average compared to past studies was usual in the study site. However, the quantitative presence of total silica has been seen similar in temporal studies. Nevertheless, highly fluctuating concentration of sulphate ions could not be well explained and future studies are suggested. Pearson Product-moment Correlation Coefficient for elevation gradient and concentration of these four parameters showed very weak relationship which proves their disassociation.

ACKNOWLEDGEMENTS

We are very grateful to The Glacier Trust, UK for their financial assistance under Climate Change Education Project in Nepal supporting this research work. We are also thankful to the Department of National Park and Wildlife Conservation, Government of Nepal, and villagers in Gokyo Valley for their kind cooperation while conducting this research.

CITE THIS PAPER

Rajendra Bhandari, Subodh Sharma, Juerg Merz, Robin Garton (2015) Hydro-Chemistry of Gokyo Valley, Sagarmatha (Everest) National Park, Nepal. *Journal of Geoscience and Environment Protection*,**03**,74-81. doi: 10.4236/gep.2015.39007

REFERENCES

1. Tartari, G.A., Tartari, G. and Mosello, R. (1998) Water Chemistry of High Altitude Lakes in the Khumbu and Imja Kola Valleys (Nepalese Himalayas). Memorie dell'lstituto Italiano di Idrolnologia, 57, 51-76.
2. Loffler, H. (1969) High Altitude Lakes in Mt Everest Region. Verhandlungen Internationale Vereinigung Limnologie, 17, 373-385.
3. Bhuju, D.R., Sharma, S., Jha, P.K. and Gaire, N.P. (2012) Scientific Discourse of Lakes in Nepal. Nepal Journal of Science and Technology, 13, 147-158.
4. Sharma, C.M., Sharma, S., Bajracharya, R.M., Gurung, S., Juttner, I., Kang, S., Zhang, Q. and Li, Q. (2012) First Results on Bathymetry and Limnology of High-Altitude Lakes in the Gokyo Valley, Sagarmatha (Everest) National Park, Nepal. Limnology, 13, 181-192.
5. Gurung, S. (2012) High Altitude Aquatic Biodiversity of Gokyo Lake Series in Sagarmatha National Park, Nepal. A Dissertation, Submitted for the Partial Fulfillment of the Requirements for the Doctoral Degree (Ph.D.) in Environmental Science. Central Library, Kathmandu University, Lalitpur.
6. Reynolds, B., Jenkins, A., Chapman, P.J. and Wilkinson, J. (1998) Stream hydrochemistry of the Khumbu, Annapurna and Langtang Regions of Nepal. Ecovision World Monograph Series, 123-141.
7. Battarbee, R.W. (2005) Mountain Lakes, Pristine or Polluted? Limnetica, 24, 1-8.
8. Kallf, J. (2002) Limnology. Prentice Hall. Upper Saddle River.
9. National Population and Housing Census 2011. Volume 02, NPHC 2011, Government of Nepal National Planning Commission Secretariat Central Bureau of Statistics Kathmandu, Nepal.
10. APHA, AWWA and WEF (1998) Standard Methods for the Examination of Water and Wastewater. 20th Edition, American Public Health Association, Washington DC.
11. Reynolds, B., Chapman, B.J., French, M.C., Jenkins, A. and Wheater, H.S. (1995) Major, Minor and Trace Element Chemistry of Surface Waters in the Everest region of Nepal. Proceedings of a Boulder Symposium, IAHS, Publication No. 228.
12. Tratner, M., Brown, G., Raiswell, R., Sharp, M. and Gurnell, A. (1993) A Conceptual Model of Solute Acquisition by Alpine Glacial Meltwaters. Journal of Glaciology, 39, 573-581.

CHAPTER 11

Revisiting the Environmental Interactions of Pesticides

Phebe Asantewaa Owusu, Samuel Asumadu-Sarkodie*

Sustainable Environment and Energy Systems, Middle East Technical University—Northern Cyprus Campus, Kalkanli, Guzelyurt, Turkey

ABSTRACT

As population increases, the demand for safe and nutritious food, safe water and air quality increases. The emergence of pesticides has played as an essential role in preventing pests from destroying various valued properties in our homes, offices and farmlands. The usage of pesticides has a role to play in the achievement of a sustainable developed earth. The earth today needs to be sustainable for future generations, and hence the need to develop it sustainably. The sustainable development aspect consist of the social, economic and environment. The benefits of pesticide in these recent years seem to have been overlooked since the risks associated with its exposure to our health and environment is fatal. As a result, this study revisits the benefits (primary and secondary) of pesticides to man, its economy and environment. The negative effects of pesticide residue exposure in the environment: the atmosphere, soil and water, and how pesticide exposures affect the human system are revisited. Despite the evidence of the adverse effects of the use of pesticides, its usage has increased in the past decades. This paper ascertains the reasons behind the use of pesticides and propose some recommendations to reduce the effects of its exposure.

Keywords: Pesticides, Sustainable Development, Climate Change, Agriculture, Environment

1. INTRODUCTION

Global interest in mitigating climate change and its impact has increased within the past decades because of the long-term effect of the greenhouse gas emissions. Scientific contribution to climate change mitigation has increased in the field of energy, agriculture, environment and natural resources [1-12]. Our environment consists of living and non-living components, which interact with each other. Pests have their role they play in the environment, but the majority is destructive in nature. Pesticides are inorganic or organic toxic compounds that are formulated to prevent the destruction of the vegetation of plants (food produce) by pests (weeds, fungi, insects, etc.). Pesticides continue to play a critical role in food production and environmental pollution around the globe. According to the Sustainable Development goal (SDG) 12, target 12.4, it is essential to achieve an environmentally sound management of chemicals and all wastes throughout their lifecycle. In accordance with the agreed international frameworks, and significantly reduce their release into the air, water and soil in order to minimize their adverse effects on human health and the environment by the year 2020 [13-16].

The identification of pests as posing threats to the environment spans from 2500 BC. In attempts to find a solution to the menace, Sulfur was used to repel insects and prevent fungus on crops. In the 15th century, arsenic, mineral oil, mercury and lead were applied to food produce as insecticides. Food producers (Agriculturalists) began the extraction of chemicals from some plants that had developed resistance against some insects that were destroying crops in the 1700's. In the 1900's, Paul Mueller discovered DDT an inorganic chemical produced in the laboratory which proved to be effective in killing insects as an insecticide [17,18].

The use of pesticide around the globe continues to increase as shown in in Table 1. Table 1 shows the major grouping of pesticides [19]. Pesticide applications were found to have an effect on the non-targeted components (environment and humans) with only about 0.3% of pesticide administered getting to the target pest, which caused concerns to view the other 99.7% pesticide interactions with the non-target components [20].

The paper attempts to examine the various relationships and interactions between the environment at large and pesticides. As a contribution to literature, the study will increase the global debate on ensuring a sustainable consumption and production patterns in the chemical industry. A qualitative research methodology, which includes information, gathered from the actors in the field of the study, journals, articles, and other literature relevant to the study are analyzed and reviewed. As a contribution, the study increases the global debate on the role of pesticides in the environment.

Table 1. Major Pesticide Groups.

Type	Used Against
Insecticide	Insects and related animals
Herbicide	Weeds
Fungicide	Fungi (Fungal diseases, molds and mushrooms)
Bactericide	Bacteria
Rodenticide	Mice, rats and other rodents
Algaecide	Algae
Nematocide	Nematodes
Acaricide (miticide)	Mites

The remainder of the study include; Benefits of pesticides to the environment, the various aspects of the environment (atmosphere, soil and water) and pesticides, pesticides effect in the human body, continual use of pesticides despite negative effects, Integrated Pest management, ways for reducing pesticides negative effects and Conclusion.

2. PESTICIDES

Pesticides were formulated to aid in the prevention, destruction, deterring of pests in the environment, which were beginning to be a cause for concern. Later it was also realised that in its use trying to help improve the situation it also came with its own concerns of pollution, of which some are banned (Deldrin, DDT, etc.). Pesticides can be grouped in various ways; the following shows the classification of pesticides according to their chemical structure [21]:

- Organochlorine
- Organophosphate
- Carbanates
- Pyrethriods

- Others (Nitro compounds, bipyridyl compounds, sulfur containing compounds and many more)

The table (Table 2) below also shows the classification of pesticides in terms of the toxic level, expressed as its Lethal dose (LD_{50}) in mg/kg.

Table 2. Classification of Pesticides (Toxicity).

Divisions	Toxic level	Examples
Class IA	Extremely dangerous	Parathion, Dieldrin
Class IB	Highly dangerous	Eldrin, Dichlorvos
Class II	Moderately hazardous	DDT, Chlordane
Class III	Slightly hazardous	Malathion,

Pesticides have both primary and secondary benefits to our environment. The primary benefits are the most obvious ones and the main reasons why pesticides were manufactured in the first place, while the secondary benefits are indirect and are less noticed or observed since it happens over a long period.

2.1. Primary Benefits

Increase in crop yield: The use of pesticides primarily was to prevent pests and pest diseases from destroying the crop produce. This has helped a lot in reducing crop infestation and inversely increasing crop yield and putting an end to many hunger situations.

Disease control: The use of insecticides has aided in the prevention of many diseases such as Malaria, which is caused by the female anopheles mosquito, the black fly also causing onchocerciasis (river blindness) and others.

Improving food quality and availability: Food is essential for the survival of humankind, hence the need for adequate quality and consumption by man. The crops that are produced when treated with pesticides attain the maximum nutrient content improving food quality. The use of herbicides also prevents weeds from competing with crops for nutrients [13,22].

2.2. Secondary Benefits

Pesticides in the long term improve productivity by increasing revenue and profit of farmers. It also increases macro-economic benefits of societies, countries and the globe at large by improving export of produce, labor force and employment opportunities. Globally the following are some of the benefits; decline in civil unrest due to food shortage, reducing the pressure on uncropped land, reduction in the emission of greenhouse gases, reduction in the spread of diseases internationally and trees reducing global warming [22].

Human health and life expectancy are improved due to the increase in the nutritional content of food crops. Also, herbicides use to aid in reducing manual and mechanical labor, as well as reducing soil disturbance and moisture from digging.

In the end, pesticides also help in conserving biodiversity. They also help in sports by destroying pests on the pitches and in buildings by protecting building materials, especially wood from being destroyed by pest infestation [13,22]. Metabolites are produced after pesticides in the environment go through some chemical (photochemical) reactions, which are comparatively non-hazardous to the environment or man [23].

3. NEGATIVE EFFECTS OF PESTICIDES ON THE ENVIRONMENT AND MANKIND

3.1. Effects in the Soil

Most pesticides are target oriented but sometimes ends up destructing non-target areas like the soil. The soil supports many life forms. Pesticides in soils also do not remain in the soil, but moves through while some stick to the soil. Although the use of pesticides in some researches has shown to aid in biodiversity, others have also reduced biodiversity in the soil by killing soil invertebrates such as earthworms. These residues also disrupt nutrients in the soil by killing of earthworms and losing their activity benefits as well as soils losing their water retention ability [24,25].

3.2. Effects on the Atmosphere

The atmosphere is probably an important medium for long-distance dispersal of pesticides. Pesticides can be referred to as semi-volatile organic compounds, which can be transported owing to their aerial or particulate distribution into the atmosphere over wider distances.

Pesticide residue that moves into the atmosphere mixes with other chemicals and may return to the soil surface by rainfall. It gets into the atmosphere by spray application, volatilization from crops or soils, physiochemical properties of pesticides and environmental conditions (humidity, higher air temperature, wind speed and soil temperature) [24,25].

3.3. Effects on Water

Water supports many life forms like humans, animals, plants and aquatic life [15,26-28]. Water also serves for recreational purposes. Water pollution by pesticides is mostly by anthropogenic factors. Pesticide residue is a severe source of pollution to water, which affects both surface water and groundwater.

Pesticides gets into these water sources by surface runoff and leaching. The leaching of pesticides to reach groundwater sources depends on soil characteristics, properties of pesticide, depth of water table and drainage rate, while macropores in soil enhances leaching of pesticide residue. If these water sources are affected, it disrupts many life forms that depend on them [20,24].

Pesticides residue in water bodies such as rivers or lakes, cannot be really measured due to the flow and velocity of water in these bodies, hence they can be measured by assessing its persistence in fish in these sources [29].

3.4. Human System and Pesticides

Humanity benefits a lot from the use of pesticides by various means: the control of vector-borne disease causing organisms, both in man and its livestocks as well as plants that enable man to have adequate food for consumption. However, the negative effects of the use of these pesticides cannot be overlooked or ignored because to a greater extent, it causes a lot of harm [22].

Humans mostly get into contact with pesticides through non-targeted areas unless it is accidental or suicidal. It usually occurs by ingestion through exposure of foods that have pesticide residue, water contaminated with pesticide residue as well as breathing it in and getting onto the skin from atmospheric exposure to pesticides. Food intake is the major medium through which pesticide gets into the human system. The most immediate contact persons to pesticide exposure are farmers and their workers [20,30]. Aside these major group pesticides affect in one way or the other the wider population. Figure 1 depicts a brief summary of the population groups at risk [31].

The effects of pesticides affect various parts of the human system. A study by Bassil, Vakil [32] revealed a significant relationship between pesticides exposure and the development of some types of cancers namely; brain cancer, prostate cancer and kidney cancers. In addition, Non-Hodgkin Lymphoma (NHL) and Leukemia were prevalent in pesticide exposures. Aside causing cancer, there are other acute and chronic diseases related to a man's exposure depending on the dose of exposure and the duration, it also disrupts sex hormone actions in both men and women [33].

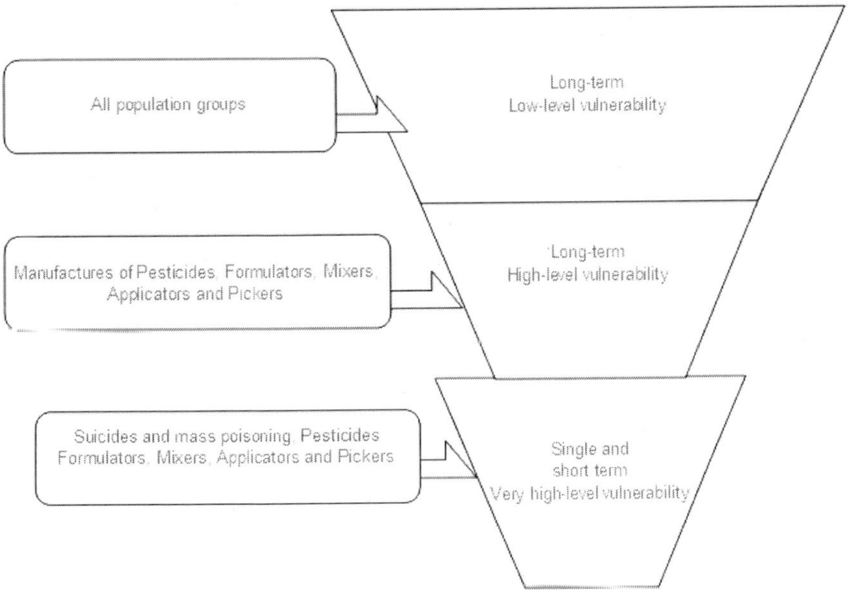

Figure 1. Population at risk of pesticide vulnerability.

3.5. The Continual Use of Pesticide by Farmers in Spite of Its Negative Effects

The evidence of the effects on our environment and health of pesticide in most cases is substantiated, but still its use is still ongoing. These are some reasons why farmers continue to use pesticides [34]:

- Cost involved in switching to a more sustainable approach
- Positive profits from pesticide use by increasing yield and productivity

- Market systems, in the notion of farmers that if I do not join in the use of pesticide, I stand to lose a lot. Hence, I should partake in sharing the proceeds.
- Ignorance of some uses
- Underestimation of the various negative effects of pesticide use
- No immediate solution to the prevention of the destruction of food produce by pests.

4. INTEGRATED PEST MANAGEMENT (IPM)

There are several definitions to the explanation of Integrated Pest Management (IPM), but Prokopy defined it as "a decision-based process involving the coordinated use of multiple tactics for optimizing the control of all classes of pests (insects, pathogens, weeds, vertebrates) in an ecologically and economically sound manner" [35]. IPM was a concept that its implementation or adoption will be able to offer users (farmers) with some degree of economic development and keep both our environment and human health safe.

Integrated Pest Management goals as initially intended, is not being implemented, but has been done by looking at its entity in various aspects. It is in the form of a policy framework being definite and is recognized globally for the management of pest. The fact that it cannot be implemented in its entity means that there are some limitations, which can be as follows [36]:

- Most agriculturalist finds the implementation of Integrated Pest Management as tiring or demanding and complex.
- There is also a conflict of interest on the part of pest control specialists since pesticide manufacturers mostly employ them and they do not have enough time for the monitoring of pest and their natural adversaries.
- Scientist in the pest field of various institutions have being interested in the various individual segments of IPM but not wholly.

5. REDUCING THE NEGATIVE EFFECTS OF PESTICIDE USE

The negative effects of pesticides are reduced through various approaches, innovations and by a collective action being employed. There are bans on some pesticide use such as DDT, and a host of others, which is still in our environment. To really deal with this situation there should be strict measures or policies and its enforcement put into action by opinion leaders and politicians (governments) to help reduce the effects of pesticide in our environment and on humanity.

There should be a broad risk communication and education programs in communities or societies, countries, especially in developing countries, to the dangers that pesticides exposure poses and to the extent of the damaged caused especially in human populations. Aside the education on risks, education on use and safe handling of pesticides needs to be given a lot of attention since most of the exposure are through these (unsafe handing and misuse).

Exposure is reduced by the discontinuation of calendar spraying to reduce the amounts of pesticide in our environment. In addition, the development of sustainable approaches to the use of pesticides, by substituting non-resistant seeds with pest-resistant ones, crop rotation in agricultural farmlands is essential in the control of pests, in order to avoid the use of pesticides.

Farmers or pesticide users have to read and follow label instructions adequately. Before the application of pesticides, protective clothes (gloves, long pants and respirators) have to be used. Leaking equipment should not be used as well as the right measuring equipment used and quantities not over-dosed. Pesticides should be kept safely and most importantly to be kept away from food and hard to reach areas in the home. Biological control of insects or predators, which are harmless to the crops, needs to be adopted. In case of rodent traps can be used instead of the use of pesticide [37-39].

6. CONCLUSION

Everybody (citizens, politicians, economists and environmentalists) in our world today wants a better future for the generations yet unborn, hence the need to protect our environment and health. Sustainable development has become a relevant issue to be discussed when it was realized by some scientists with evidence, that by the way human populations were increasing and the rate of degradation of our environment there needed to be a sustainable approach being employed.

The use of pesticides became necessary when pests were destroying food crops and some vector disease pests also causing some illness in man and animals that were in most cases detrimental. During these periods pesticides use was sure solution to the problems at hand. It had both primary and secondary benefits. The primary benefits that was immediately noticed as increasing yield, profits to farmers, curbing hunger situations around the globe and preventing some vector-borne diseases such as malaria. In the long run, its secondary benefits also were realized in various sectors like improving human health due to the nutritional quality of food, improving productivity and economic benefits of countries and improving the social well-being of people by killing pests in recreational areas such as pitches and also the killing and prevention of insects in buildings.

In the centuries that followed its negative impact became evident in the environment (soil, atmosphere, water and biota) and health. Pesticide residues in our environment cause decline in biodiversity and the killing of non-target species such as birds. In human health, its exposure is becoming detrimental causing cancers, diabetes, birth defects and a host of other diseases. Various attempts employed include the Integrated Pest Management approach, which was to improve the situation but had its own constraints on the implementation process, with efforts toward its absolute adoption.

In spite of its negative effects, pesticides are still being used extensively and to help improve the situation there are some measures that needs to be adopted to reduce the impact: education, measuring to reduce exposure and researches into substitutes that can help remedy the situation aside using pesticides. In doing this the world can be sustainably developed in the economic, social and environmental aspects. We (humans) can have positive attitudes (behavior change) towards the use of pesticides, which can go a long way to improve the situation.

REFERENCES

1. Asumadu-Sarkodie S, Owusu P. The Casual Nexus between Child Mortality Rate, Fertility Rate, GDP, Household Final Consumption Expenditure, and Food Production Index. Cogent Economics & Finance. 2016;4:1191985.
2. Asumadu-Sarkodie S, Owusu PA.Carbon dioxide emissions, GDP, energy use and population growth: a multivariate and causality analysis for Ghana, 1971-2013. Environmental Science and Pollution Research International. 2016;23:13508–20.

3. Owusu PA, Asumadu-Sarkodie S. Is there a causal effect between agricultural production and carbon dioxide emissions in Ghana? Environmental Engineering Research. 2016;0:0.
4. Asumadu-Sarkodie S, Owusu PA.The relationship between carbon dioxide and agriculture in Ghana: a comparison of VECM and ARDL model. Environmental Science and Pollution Research International. 2016;23:10968-82.
5. Asumadu-Sarkodie S, Owusu PA.Multivariate co-integration analysis of the Kaya factors in Ghana. Environmental Science and Pollution Research International. 2016;23:9934-43.
6. Asumadu-Sarkodie S, Owusu PA.Carbon Dioxide Emission, Electricity Consumption, Industrialization and Economic Growth Nexus: The Beninese Case. Energy Sources, Part B: Economics, Planning, and Policy. 2016.
7. Asumadu-Sarkodie S, Owusu PA.A Multivariate Analysis of Carbon Dioxide Emissions, Electricity Consumption, Economic Growth, Financial Development, Industrialization and Urbanization in Senegal. Energy Sources, Part B: Economics, Planning, and Policy. 2016.
8. Asumadu-Sarkodie S, Owusu PA.The Causal Nexus between Carbon Dioxide Emissions and Agricultural Ecosystem, an Econometric Approach. Environmental Science and Pollution Research International. 2016. http://dx.doi.org/10.1007/s11356-016-7908-2.
9. Asumadu-Sarkodie S, Owusu PA.Energy use, carbon dioxide emissions, GDP, industrialization, financial development, and population, a causal nexus in Sri Lanka: With a subsequent prediction of energy use using neural network. Energy Sources, Part B: Economics, Planning, and Policy. 2016;11:889-99.
10. Asumadu-Sarkodie S, Owusu PA.The Causal Nexus between Energy Use, Carbon Dioxide Emissions and Macroeconomic Variables in Ghana. Energy Sources, Part B: Economics, Planning, and Policy. 2016.
11. Asumadu-Sarkodie S, Owusu PA.The Causal Effect of Carbon Dioxide Emissions, Electricity Consumption, Economic Growth and Industrialization in Sierra Leone. Energy Sources, Part B: Economics, Planning, and Policy. 2016.
12. Asumadu-Sarkodie S, Owusu PA.The Relationship between Carbon Dioxide Emissions, Electricity Production and Consumption in Ghana. Energy Sources, Part B: Economics, Planning, and Policy. 2016.
13. Aktar W, Sengupta D, Chowdhury A. Impact of pesticides use in agriculture: their benefits and hazards. Interdisciplinary toxicology. 2009;2:1-12.
14. United Nations. Sustainable Development Goals. 2016.
15. Owusu PA, Asumadu-Sarkodie S, Ameyo P. A review of Ghana's water resource management and the future prospect. Cogent Engineering. 2016;3:1164275.
16. Owusu P, Asumadu-Sarkodie S. A Review of Renewable Energy Sources, Sustainability Issues and Climate Change Mitigation. Cogent Engineering. 2016;3:1167990.
17. Jain H. Green revolution: history, impact and future: THE GREEN REVOLUTION: HISTOR; 2010.
18. Costa LG. Toxicology of pesticides: A brief history. Toxicology of Pesticides: Springer; 1987. p. 1-10.
19. The University of Arizona. Pesticide Types and Formulations. Chapter 3: Arizona Agricultural Pesticide Applicator Training Manual, College of Agriculture and Life Sciences2000.
20. Van der Werf HM. Assessing the impact of pesticides on the environment. Agriculture, Ecosystems & Environment. 1996;60:81-96.
21. Prieto F, Cortés S, Gaytán J, Ceruelos A, Vázquez P. Pesticides: Classification, Uses and Toxicity. Measures of Exposure and Genotoxic Risks. Journal of Research in Environmental Science and Toxicology. 2012;1:3-23.
22. Cooper J, Dobson H. The benefits of pesticides to mankind and the environment. Crop Protection. 2007;26:1337-48.
23. Kole R, Banerjee H, Bhattacharyya A, Chowdhury A, AdityaChaudhury N. Phototransformation of some pesticides. ChemInform. 2000;31.
24. Edwards C. Environmental pollution by pesticides: Springer Science & Business Media; 2013.
25. Polyrakis IT. Environmental pollution from pesticides. Predictive Modeling and RiskAssessment: Springer; 2009. p. 201-24.

26. Asumadu-Sarkodie S, Owusu PA, Rufangura P. Impact analysis of flood in Accra, Ghana. Advances in Applied Science Research. 2015;6:53-78.
27. Asumadu-Sarkodie S, Rufangura P, Jayaweera HM, Owusu PA. Situational Analysis of Flood and Drought in Rwanda. International Journal of Scientific and Engineering Research. 2015;6:960-70. http://dx.doi.org/10.14299/ijser.2015.08.013
28. Asumadu-Sarkodie S, Owusu PA, Jayaweera HM. Flood risk management in Ghana: A case study in Accra. Advances in Applied Science Research. 2015;6:196-201. http://doi.org/10.13140/2.1.4945.7440
29. Khan M. Pesticides in aquatic environments: Springer Science & Business Media; 2013.
30. Margni M, Rossier D, Crettaz P, Jolliet O. Life cycle impact assessment of pesticides on human health and ecosystems. Agriculture, Ecosystems & Environment. 2002;93:379-92.
31. Jeyaratnam J. Acute pesticide poisoning: a major global health problem. World Health Stat Q. 1990;43:139-44.
32. Bassil K, Vakil C, Sanborn M, Cole D, Kaur J, Kerr K. Cancer health effects of pesticides Systematic review. Canadian Family Physician. 2007;53:1704-11.
33. Andersen HR, Vinggaard AM, Rasmussen TH, Gjermandsen IM, Bonefeld-Jørgensen EC. Effects of currently used pesticides in assays for estrogenicity, androgenicity, and aromatase activity in vitro. Toxicology and applied pharmacology. 2002;179:1-12.
34. Wilson C, Tisdell C. Why farmers continue to use pesticides despite environmental, health and sustainability costs. Ecological economics. 2001;39:449-62.
35. Prokopy RJ. Two decades of bottom-up, ecologically based pest management in a small commercial apple orchard in Massachusetts. Agriculture, Ecosystems & Environment. 2003;94:299-309.
36. Ehler LE. Integrated pest management (IPM): definition, historical development and implementation, and the other IPM. Pest management science. 2006;62:787-9.
37. Ibitayo O, Monosson E. Agricultural pesticide contamination.Encyclopedia of Earth(Washington, DC: Environmental Information Coalition, National Council for Science and the Environment). 2007.
38. Fait A, Iversen B, Tiramani M, Visentin S, Maroni M, He F. Preventing health risks from the use of pesticides in agriculture: WHO; 2001.
39. Subramanyam B, Hagstrum DW. Alternatives to pesticides in stored-product IPM: Springer Science & Business Media; 2012.

CHAPTER 12

Hyporheic Zone Hydrochemistry of the Mine-Polluted River

D. Ciszewski, P. Bijata

AGH-University of Sciences and Technology, Kraków, Poland

ABSTRACT

Intensity of stream waters mixing with groundwaters and lateral extent of these processes in the hyporheic zone were investigated in a near-bank sandbar and an adjacent floodplain through the comparison of groundwaters and stream water chemistry of the Biała Przemsza River in southern Poland. The stream waters were polluted by the discharge of mine waters from "Bolesław" lead and zinc mine. The investigated waters were several times more mineralized than the natural spring waters of the river valley. The concentration of: potassium, sodium, and the pH, as well as cadmium, lead, and zinc decreased in the hyporheic zone towards the stream bank, whereas conductance, calcium, magnesium, sulphates, as well as silica contents were the highest on the floodplain, diminishing towards the stream. The changes observed in the chemical composition of groundwaters were apparent in mixing stream waters below the depth of 2 m with shallow groundwaters draining the valley slope. Hyporheic mixing also takes place in the 10-meter-wide, marginal zone of the sandbar, whereas in the 5-meter-wide stream-side zone of the sandbar groundwaters represent weakly transformed stream water.

Keywords: Hyporheic Zone, Groundwaters, Pollution, Water Chemistry, Metal Mining

1. INTRODUCTION

The hyporheic zone is an area beneath and adjacent to the stream where stream waters mix with groundwaters. This zone contains from several per cent up to 100% of river waters which penetrate alluvia over distances from centimetres to kilometres from the channel and several meters below the channel bottom (Woessner, 2000) [1]. Intensive investigations of the hyporheic zone during two most recent decades revealed their important ecological role. It was found that a channel bottom constitutes a shelter for hyporheic organisms which are unique for this specific ecotone. This ecotone can act as a physical, chemical and biological filter capable of transforming nutrients and pollutants (Bourg & Bertin, 1993) [2]. Microbial biofilms, in particular, mediate the transformation of water chemistry and affect the nitrogen cycling within the river system. Typically, along the hyporheic flow path well-oxygenated stream waters infiltrating through the bottom lose oxygen during nitrification, followed by denitrification in anoxic conditions, before unaerated waters reenter the stream (Hancock, Boulton & Humphreys, 2005) [3]. High oxygen, temperature and flow velocity gradients in the hyporheic zone, together with the microbial activity, favour the transformation of contaminants transported by the streams affected by mining. Studies of gravel alluvia also recognized the precipitation of manganese and iron oxides on coarse grains as an important mechanism controlling the removal of heavy metals from river waters (Fuller & Bargar, 2014) [4].

Whereas the hyporheic zone is extensively studied on gravel-bed rivers and gravel bars (considered as part of this zone), the processes of contaminant transformation are much less recognized in the sandy alluvia. Sandy river beds are more mobile and this mobility was found to affect the storage and transformation of solutes and fine particles (Harvey et al., 2012) [5]. The aim of the study was to recognize the relative importance of the hyporheic flow in the sand bar of the Biała Przemsza River channel, cut in the fine grained alluvia, which are typical for the Pleistocene sediments in Central Europe. The river channel is polluted by the discharge of mine waters from "Bolesław" lead and zinc mine in southern Poland. The chemistry of the river water is markedly altered due to the discharged waters, thus their comparison to groundwater chemistry was used to reveal the lateral extent of the hyporheic zone.

2. METHOD

The Biała Przemsza River is the left-bank tributary of the Vistula River, which is the longest and largest river of Poland. The Biała Przemsza, with the length of over 60 kilometres, drains the upland area of the most industrialized region of Poland called Upper Silesia. The river valley is several hundred meters wide when it crosses cuesta sections, built of carboniferous rocks, and is several kilometres wide in inter-cuesta basins filled with fluvioglacial sandy deposits, with a thickness of as much as 25 m. In the studied middle river reach the valley gradient decreases and the channel reaches the width of up to 50 m because the sandy sediments accumulate, forming mid- and side channel bars (Figure 1). The tendency for the accretion of bed material has been observed here since the middle of the 19th century when weirs protecting the railway bridge and later the water supply plant had been constructed about 2 km downstream. These constructions, together with the valley narrowing by the high Pleistocene terrace in this river section, induced the lateral channel stabilization over the course of the 20th century. This stabilization favoured the development of the turf grass cover over most of the sand bars in this river reach as their surface became closer to the average level of the water table. The average discharge of the Biala Przemsza River in this reach equalled 4 m^3/s with a half of it originating from one of the largest lead and zinc mines in Europe—the "Bolesław" Mine and Metallurgical Plant (ZGH). The plant discharges mine water with higher-than-natural mineralization, the content of zinc, lead, cadmium, and turbidity, which has decreased since the beginning of the mine operation (Ciszewski, 1998) [6].

Seven sediment cores were collected from the 20-metre-long transect across the side channel bar and the floodplain of the Biała Przemsza River at Maczki (Figure 1). Four sediment cores were drilled to the depth of 4 meters and three profiles—up to 1 meter deep. In the core sites, 4- and 1-meter long piezometers were installed. Sediment cores were split into sections about 30 cm long and transported to a laboratory for analyses. Samples of groundwaters and stream waters were collected 10 times over the period of one year. The values of the pH and conductance were established using a Thermo Scientific device. The samples were filtered before the analysis through 0.45 μm filters and concentrations of Cl^-, NO_3^-, SO_4^{2-}, PO_4^{3-}, CO_3^{2-} and macroions (Ca, Mg, Na, K) were determined within 48 h using Ion Chromatography (DIONEX 1000). Standard reference materials (Canadian waters Hamilton-20) were employed to determine the accuracy of anion analyses. For analyses of Zn, Cd, Pb, Fe, Ba and Si, the waters were filtered on site through a syringe filter 0.45 μm, acidified to a pH below 2 with ultrapure HNO_3 and measured with an inductively coupled plasma-mass spectrometer (Perkin Elmer ELAN 6100) in the certified

Hydrogeochemical Laboratory according to the standard certified analytical quality control procedure (PN-EN ISO 17294-1:2007). In the collected sediments, losses on ignition at 550°C and content of 0.2 mm fraction were analyzed after drying at 105°C.

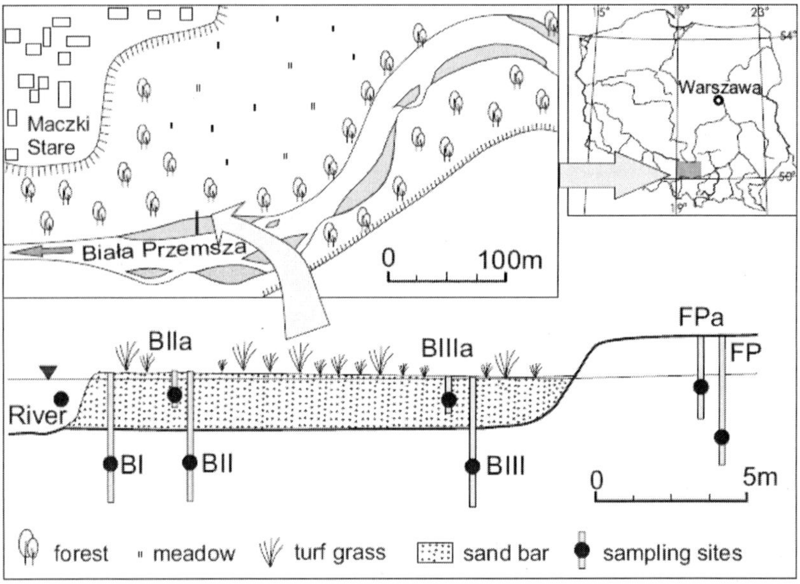

Figure 1. Location of sampling sites of river water and groundwater in the channel bar (B) and in the floodplain (FP) (black dots).

3. RESULTS

Groundwater levels in the sandbar varied within less than 10 cm over the study period and were never recorded below 20 cm of the bar surface. The groundwater depths on the floodplain varied between 68 and 78 cm within the year. No head gradient was observed within the channel bar. Generally, the average conductance of stream waters was high and equalled 885 µS (Table 1). Very similar values were in groundwaters within 5 meters from the bar edge and they increased markedly at the river bank. These values were higher at the surface than at greater depths and on the floodplain.

The opposite distribution pattern was observed for the pH. Whereas stream waters were alkaline, with the pH of 8, groundwaters became more acidified further away from the stream, reaching 6.5 on the floodplain

(Figure 2). In the same sampling site (II, III or FP) more acidified waters occurred closer to the surface than at the depth of 3 m. Very similar distribution pattern was observed for potassium as the maximum concentrations of this element occurred in the stream waters and there were markedly lower values in the floodplain; the decrease of this element content from the stream was progressive (Figure 2). Higher concentrations in stream waters than in floodplain groundwaters were observed also for sodium and its changes were closely associated with those of potassium (Table 1). Particularly low sodium values were recorded in shallow floodplain waters, whereas at the depth of 3 m the content of both elements was much higher or even similar to that in the river.

Table 1. River and ground waters characteristics from the sampling sites in the channel bar (B) and in the floodplain (FP) of the Biała Przemsza river at Maczki (Median values) [mg/dm³]; n.d.: not detected.

Water chemistry characteristics	River	B I depth 3 m	B II depth 3 m	B IIA depth 0.6 m	B III depth 3 m	B IIIA depth 0.6 m	FP I depth 3 m	FP IA depth 1 m	Natural spring
Conductivity (μS)	885	880	875	904	831	979	961	915	228
pH	8.1	7.5	7.5	6.8	7.3	6.9	7.0	6.5	6.5
Ca	118	120	115	129	118	141	135	134	30
Mg	33	32	32	30	33	35	37	36	7
Na	22	26.3	23	21	17	11	21	9	3
K	5.6	6.8	5.1	4.0	4.8	4.5	3.0	0.8	2
NH_4	0.72	0.02	0.47	0.16	0.72	0.51	0.19	0.44	0.01
HCO_3	278	268	273	341	273	390	311	297	76
SO_4	198	199	210	181	179	191	248	211	35
Cl	35	31	33	28	28	29	33	31	8
NO_3	8.4	1.53	nd.	nd.	nd.	nd.	nd.	5.2	2.4
NO_2	0.22	0.21	nd.	nd.	nd.	nd.	nd.	0.67	nd.
PO_4	0.39	0.01	nd.	nd.	nd.	nd.	nd.	nd.	nd.
F	0.74	0.269	0.37	0.35	0.46	0.270	0.16	0.308	0.1
Ba (μg/dm³)	91	111	211	129	290	127	259	148	80
Cd (μg/dm³)	3.5	<0.1	<0.1	<0.1	<0.1	<0.1	<0.1	3.3	nd.
Pb (μg/dm³)	36	5.4	2.1	5.0	2.8	6.2	11.5	32	3.1
Zn (μg/dm³)	878	284	14	681	11	21	11	950	30
Fe (μg/dm³)	148	59	798	6795	3042	12,006	4804	940	300
Si (μg/dm³)	8792	7982	7890	9470	8437	10,471	8357	16,231	12,650

In contrast with sodium and potassium, magnesium exhibited the distribution pattern similar to that of the conductance. Differences between waters of the floodplain and those of the channel bar were not very high, but statistically significant. Intermediate values characterized shallow waters from the marginal zone of the channel but they varied markedly over the investigated period (BIIIA, Figure 2). Variability of calcium was also similar to that of magnesium; calcium content in the surface waters was generally lower than in the sandbar close to the stream (BI, BII) and increased at the channel margin and on the floodplain.

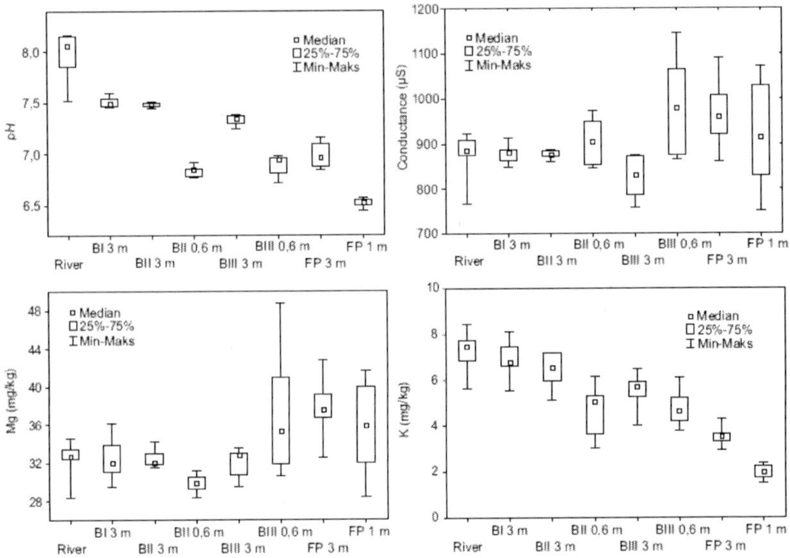

Figure 2. Variability of the pH, conductance and Mg and K concentrations in waters sampled from the river and from different depths of the channel bar (B) and of the floodplain (FP).

River waters were markedly polluted with sulphates to the level of about 200 mg/l, but their content did not differ significantly from the values for the channel bar. Only values at the deep horizon of the floodplain were markedly higher than the rest of the investigated points. Hydrocarbonates with the content of about 280 - 300 mg/l were more uniformly distributed in the investigated transect with significantly (statistically) higher values only in shallow waters of the sandbar. Their concentration in floodplain waters was only slightly higher than that in the river. Concentration of chlorides was also weakly differentiated with the highest values, on average, in the stream. Detectable values of nitrates, nitrites and phosphates occurred solely in the stream, in close proximity to the stream (BI) and at the floodplain surface (FPI). Concentrations of all these compounds were at least several times higher than in the natural spring water draining the Pleistocene terrace about 3 km upstream the sampling site (Table 1).

The sediments of the sandbar were composed almost exclusively of uniform, medium grained sands. The content of the fine fraction of sand and silt did not exceed several per cent in almost all strata of sediment cores. The highest content of the fine fraction was found at the surface of the sandbar. The surface 0.3 m strata contained also up to 10% of organic matter. However, some lenses of more organic and finer grained deposits were also observed at lower depths of the profiles.

4. DISCUSSION

There are several methods of investigation of groundwater and surface waters interactions in the hyporheic zone. In addition to the direct flow measurements, pumping tests, temperature measurements or solute tracing, contaminants may be used as markers for stream water or groundwater which mix beneath the stream (Kalbus, Reinstorf & Schirmer, 2006) [7]. Hydrochemical methods are particularly effective where chemical properties of stream water and groundwater differ markedly (Woessner, 2000; Hoehn & Scholtis, 2011) [1] [8]. In the presented studies this difference was used in order to identify the relative intensity of mixing of the two types of water. Two different distribution patterns of chemical compounds were identified. In one of them the concentrations decreased from the highest in the stream towards the lowest in the floodplain. It is represented by potassium, sodium and the pH, as well as by cadmium, lead, and zinc. The opposite distribution, with overall higher values in the floodplain and lower in the stream and in the sandbar, is represented by conductance, calcium, magnesium, and sulphates, as well as silica. These differences are less visible for chlorides, hydrocarbonates, barium, and iron and seem to represent an intermediate distribution pattern. The observed distribution patterns indicate that shallow groundwaters in the floodplain differ markedly from the rest of the waters because they drain a large area of the valley slope and have a prolonged contact with the subsurface. This is particularly visible in the highest content of silica. The groundwaters at lower levels were for some compounds more closely related to groundwaters which drain the floodplain and some concentrations were more similar to that in the sandbar at the channel margin under the influence of water from the stream. High variability also characterizes concentrations of many compounds in the subsurface waters at the channel margin (BIIIA). The influence of water draining the floodplain is higher during rainfall or snow-melting periods and only the channel margins are the zone of most intensive mixing of ground and stream waters. The proportion of stream waters at the depth of 3 m on the floodplain seems to be quite high, considering the chlorides content is as high as in the stream. Nevertheless, it is clearly visible that the proportion of stream waters is the highest closer to the stream. In particular, the concentrations of most of the compounds in BI and BII sites i.e. in the marginal zone of the sandbar about 5 metres wide are almost the same as in the stream. It seems that in this zone 100% of stream water flow is parallel or the intensive exchange with the stream and lateral groundwater inflow from the floodplain occurs at the channel margin, where stream waters mix with groundwaters in variable proportions. This zone is characterized by the highest average conductance. The data enabled the identification of the hyporheic zone extending over the channel bar and outside the river channel below the depth of 2 m.

Investigations of hyporheic exchange processes in partially submerged bed structures use observations of stream stage, head data in piezometers, artificial tracer additions or variations in natural electrical conductivity and were attempted mainly in gravel bed rivers (Trauth et al., 2015) [9]. The electrical conductivity and the other characteristics of water in the sandbar were used to show high complexity of the morphological factors controlling the exchange in the sandy alluvia. These investigations indicated that sand bars, particularly at channel margins, not actively reworked in the channel, may not have a uniform structure. The sediments accreted during numerous flood episodes contain some stratification which modify the flow velocity in the hyporheic zone. The bars formed at the channel margin shift out of the bank both the flow and the zone of the flow parallel to the channel, sometimes called the parafluvial zone (Boulton et al. 1998) [10]. As a result, hyporheic mixing processes occur as well within the channel in a zone proportional to the width of the channel bar and their grain size variability.

The mixing processes seem to be connected to the conservative behaviour of chemical compounds. However, some substances are biologically transformed in the hyporheic zone. Nitrates and nitrites, abundant in the stream water, had to have already disappeared in the BI point, localized a few decimetres from the bar edge (Table 1). The same substances are also present only in the surface waters of the floodplain. This suggests that microbial reduction of nitrogen load is intensive and takes place within the decimetres of the flow in the hyporheic zone. The widespread absence of oxygen in the sandbar or its very low content (<0.5 mg/l) only in the BI site at the bar edge, demonstrates its rapid drop in the fully oxygenated (10 mg/l) river waters infiltrating the sand bar. Some of the other chemical compounds can also be transformed within the channel bar. The loss of phosphates from the stream waters is suggested by its rapid decrease already at the bar margin. Moreover, processes of dissolution of iron hydroxides are responsible for very high iron content in the subsurface waters on the sandbar. These processes are characteristic of the reducible dissolution of iron compounds in the presence of organic matter and also occur in the swampy areas of the Biała Przemsza River valley.

5. CONCLUSION

Stream waters polluted by the discharge from "Bolesław" lead and zinc mine are more abundant in most of the chemical compounds investigated than in the natural spring waters of the Biała Przemsza River valley. The concentrations of potassium, sodium, and the pH, as well as cadmium, lead, and zinc decreased in the hyporheic zone towards the stream banks, whereas conductance, calcium, magnesium, and sulphates, as well as silica contents were higher on the floodplain and decreased towards the stream. The observed changes in the chemical composition of the groundwater exhibit the inflow of shallow groundwater from the catchment mixing with stream water below the depth of 2 m. Hyporheic mixing extends along the bank and within the approximately 10-metre-wide marginal zone of the sandbar, whereas in the 5-metre-wide marginal zone of the sandbar, groundwaters represent weakly transformed stream waters flowing along the stream.

ACKNOWLEDGEMENTS

The authors gratefully acknowledge the National Science Center of Poland for funding the project No. 2012/05/B/ST10/00235.

CITE THIS PAPER

D. Ciszewski, P. Bijata (2015) Hyporheic Zone Hydrochemistry of the Mine-Polluted River. *Journal of Geoscience and Environment Protection*,**03**,47-52.

REFERENCES

1. Woessner, W.W. (2000) Streams and Fluvial Plain Ground Water Interactions: Rescaling Hydrogeologic Thought. Ground Water, 38, 423-429.
2. Bourg, A.C.M. and Bertine, C. (1993) Biogeochemical Processes during the Infiltration of River Water into an Alluvial Aquifer. Environmental Science and Technology, 27, 661-666.

3. Hancock, P.J., Boulton, A.J. and Humphreys, W.F. (2005) Aquifers and Hyporheic Zones: Towards an Ecological Understanding of Groundwater. Hydrogeological Journal, 13, 98-111.
4. Fuller, Ch.C. and Bargar, J.C. (2014) Processes of Zinc Attenuation by Biogenic Manganese Oxides Forming in Hyporheic Zone of Pinal Creek, Arizona. Environmental Sciences and Technology, 48, 2165-2172.
5. Harvey, J.W., Drummond, J.D., Martin, R.L., McPhillips, L.E., Packman, A.I., Jerolmack, D.J., Stonedahl, S.H., Aubenau, A.F., Sawyer, A.H., Larsen, L.G. and Tobias, C.R. (2012) Hydrogeomorphology of the Hyporheic Zone: Stream Solute and Fine Particle Interactions with Dynamic Streambed. Journal of Geophysical Research, 117, 1-20.
6. Ciszewski, D. (1998) Channel Processes as a Factor Controlling Accumulation of Heavy Metals in River Bottom Sediments: Consequences for Pollution Monitoring (Upper Silesia, Poland). Environmental Geology, 36, 45-54.
7. Kalbus, E., Reinstorf, F. and Schirmer, M. (2006) Measuring Meth-ods for Groundwater-Surface Water Interactions: A Review. Hydrology and Earth System Science, 10, 873-887. http://dx.doi.org/10.5194/hess-10-873-2006
8. Hoehn, E. and Scholtis, A. (2011) Exchange between a River and Groundwater, Assessed with Hydrochemical Data. Hydrology and Earth System Science, 7, 9023-9042.
9. Trauth, N., Schmidt, Ch., Vieweg, M., Oswald, S.E. and Flecken-stein, J.H. (2015) Hydraulic Controls of In-Stream Gravel Bar Hyporheic Exchange and Reactions. Water Resources Research, 51, 2243-2263.
10. Boulton, A.J., Findlay, S., Marmonier, P., Stanley, E.H. and Valet, H.M. (1998) The Functional Significance of the Hyporheic Zone in Streams and Rivers. Annual Review of Ecological Systems, 29, 59-81.

CHAPTER 13

Soil C and N models that integrate microbial diversity

Benjamin P. Louis1,2,3, Pierre-Alain Maron4, Vale´rie Viaud2, Philippe Leterme1,2,3, Safya Menasseri-Aubry1,2,3

1 UMR 1069 SAS, Agrocampus-Ouest, 65, Rue de Saint Brieuc, 35042 Rennes Cedex, France

2 UMR 1069 SAS, INRA, 65, Rue de Saint Brieuc, 35042 Rennes Cedex, France

3 Universite´ Europe´enne de Bretagne, Rennes, France

4 INRA, AgroSup Dijon, UMR 1347 Agroecology, Universite´Bourgogne Franche Comte´, UMR Agroe´cologie 17, Rue

ABSTRACT

Industrial agriculture is yearly responsible for the loss of 55–100 Pg of historical soil carbon and 9.9 Tg of reactive nitrogen worldwide. Therefore, management practices should be adapted to preserve ecological processes and reduce inputs and environmental impacts. In particular, the management of soil organic matter (SOM) is a key factor influencing C and N cycles. Soil microorganisms play a central role in SOM dynamics. For instance, microbial diversity may explain up to 77 % of carbon mineralisation activities. However, soil microbial diversity is actually rarely taken into account in models of C and N dynamics. Here, we review the influence of microbial diversity on C and N dynamics, and the integration of microbial diversity in soil C and N models. We found that a gain of microbial richness and evenness enhances soil C and N dynamics on the average, though the improvement of C and N dynamics depends on the composition of microbial community. We reviewed 50 models integrating soil microbial diversity. More than 90 % of models integrate microbial diversity with discrete compartments representing conceptual functional groups (64 %) or identified taxonomic groups interacting in a food web (28 %). Half of the models have not been tested against an empirical dataset while the other half mainly consider fixed parameters. This is due to the difficulty to link taxonomic and functional diversity.

1. INTRODUCTION

Massive use of mineral fertilisers for agricultural production has altered C, N and P cycles at the global scale. Historic global C loss due to agricultural activities is estimated at 55–100 Pg from soil C pool (Lal 1999). In Europe, the annual reactive nitrogen surplus of 9.9 Tg contributes to substantial N leaching to water bodies, volatilisation as NH_3 and emission of N_2O and NO (Sutton et al. 2011). Since the twentieth century, agriculture intensification has also been a major factor of land use change (Reidsma et al. 2006), and its impact is particularly noticeable for soil biodiversity (Creamer et al. 2016). A current major challenge for agriculture therefore is to adapt management practices to preserve ecological processes and reduce inputs and environmental impacts on soil, water and atmosphere.

Management of soil organic matter (SOM) is a key element of agroecological transitions because it affects both agricultural production and soil environmental functions, and conversely agricultural management affects SOM (McLauchlan 2006): SOM is highly involved in soil chemical quality by supplying nutrients for plant production, in soil physical quality by supporting structural stability and in soil biological quality by supporting biodiversity reserves (Reeves 1997). In addition to having a direct impact on the soil itself, SOM has implications in pollutant transfers to air and water. It has a strong impact on the local and global C cycles and the interest in the potential of agricultural soils to sequester C is increasing (Chambers et al. 2016). SOM dynamics and stocks partly control the emission and discharge of dissolved nutrients, such as nitrates and dissolved organic C.

Dynamics of soil C and N are greatly driven by soil microorganisms (Paul and Clark 1989; Sylvia et al. 2005). New developments in molecular biology methods provided new evidence that not only microbial abundance but also microbial diversity affects C and N cycling in the soil (e.g. Baumann et al. 2013; Juarez et al. 2013) in different experimental and agricultural contexts. Agricultural management practices interact with pedoclimatic factors and can change soil physical properties via soil tillage, soil chemical properties via organic and inorganic inputs and soil biological properties via the type of plant cultivated for instance. All three influence microbial community diversity and differences are often observed in soil microbial abundance and composition between agricultural land and less-disturbed areas such as natural areas, forest or grassland (Rampelotto et al. 2013). Other studies emphasise differences in microbial community diversity for contrasting agricultural practices: soil tillage and/or fertilisation and/or weed management (Doran et al. 1998; Feng et al. 2003; Zhong et al. 2010; Berthrong et al. 2013; Lazcano et al. 2013; Sul et al. 2013) in interaction with pedoclimatic factors (Kaisermann et al. 2013; Lienhard et al. 2013). It has been shown that microbial diversity could explain up to 77 % of carbon dynamic activities (Tardy et al. 2015). Thus, predictive tools of C and N dynamics which integrate microbial diversity should make it possible to fully evaluate the impacts of change in agricultural practises induced by agroecological transition.

Modelling of carbon (C) and nitrogen (N) dynamics is an essential tool to predict SOM dynamics in response to pedoclimatic and agricultural systems (Manzoni and Porporato 2009) to adapt relevant management practises. Today C and N dynamics models are predominantly dynamic deterministic compartment models (Manzoni and Porporato 2009), in which SOM transformation is assumed to follow first-order kinetics, and where SOM is conceptually divided into several homogeneous pools such as pool of organic matter, microbial biomass or crop residue (Halfon 1979, cited in Manzoni and Porporato 2009), each with its unique decomposition rate, and more or less resistant to decomposition. Many examples of models that integrate microbial biomass are found in the literature, which confirms the interest of including these microorganisms in soil C and N dynamics models (Schimel 2001; Neill and Gignoux 2006). However, questions about incorporating information about microbial diversity continue to enliven the debate (e.g. Nannipieri et al. 2003; McGuire and Treseder 2010; Nielsen et al. 2011; Treseder et al. 2012; Graham et al. 2014). The diversity of soil microorganisms, i.e. number of different species and relative contribution of each species to the community composition, is extensive, with a high level of functional redundancy in C and N transformations (Nannipieri et al. 2003); consequently, it is legitimate to question whether integrating diversity parameters into models could better predict soil C and N dynamics. It has long been assumed that soil microbial diversity has no functional significance because of the high functional redundancy of soil microorganisms (Nannipieri et al. 2003). This assumption was partly justified by the lack of a demonstrated diversity–function relationship, which was mainly attributed to technical limitations, which often restricted analysis of the diversity to indigenous soil microbial communities. This makes difficult to explicitly integrate the microbial community diversity in soil C and N models (Maron et al. 2011). Progress in analytical methods over the two last decades, such as in molecular biology, has generated new methods and has provided the necessary tools to reconsider this assumption (Maron et al. 2007).

In this context, the objective of this paper was to identify the various ways to further develop predictive models of C and N dynamics integrating soil microbial diversity. Focusing on the agricultural management context, we first summarise existing knowledge about the role of microbial diversity in C and N dynamics. We then analyse advantages and limits of how microorganisms have been incorporated into current models. Finally, we discuss methodological approaches towards a better integration of microbial diversity in soil C and N dynamics models.

2. ROLE OF MICROBIAL DIVERSITY IN SOIL CARBON AND NITROGEN DYNAMICS

Diversity is a generic term which can be described in different ways. In this review, we refer as *taxonomic indexes*, the diversity as described by synthetic indexes taking into account a number of entities such as genotypes or species, and their relative distribution in a community: richness, evenness, Shannon index etc. To refer to the identity of the present entities as well as their abundances in the community, we used the common term *taxonomic composition*. Eventually, diversity describing functional traits of the community is referred as *functional diversity*. Considering these different ways of describing microbial community diversity, different approaches have been developed to demonstrate its role in soil carbon and nitrogen dynamics.

2.1 Influence of reduced soil microbial diversity

Several experimental studies have enabled to assess the impact of soil microbial diversity on C and N dynamics by manipulating diversity and mainly by artificially reducing soil microbial diversity (Nielsen et al. 2011). These studies created gradients of microbial diversity in the soil while keeping a constant biomass, and measured one or several soil functions for each level of diversity. Three main methods exist, each with advantages and disadvantages (Table 1). The "construction" method is performed by creating different microbial consortia (i.e. number of species and/or taxonomic groups) by combining microbial species after isolating them in culture media (Deacon 1985). This helps understand the link between taxonomic and functional diversity by testing hypotheses of functioning in the presence or absence of specifically chosen microorganisms, such as species that use specific substrates, to study facilitation mechanisms among microorganisms (Wohl et al. 2004). The number of taxa is generally small (<100) compared to the diversity of indigenous communities, and only culturable microorganisms can be isolated, which represent less than 10 % of all existing microorganisms (Maron et al. 2011). These taxa often have particular traits such as a fast growth rates, and studies using this method mainly focus on bacterial or fungal taxa, suppressing possible interactions between them (Romaní et al. 2006). The other two methods, called "destruction" and "erosion" (Table 1), create different levels of microbial diversity by (1) eliminating different quantities of soil microorganisms through different durations of fumigation with chloroform (Jenkinson and Powlson 1976) or by (2) inoculating sterile soil with different dilutions of a solution containing microorganisms in suspension (Griffiths et al. 2000). Both approaches provide more realistic diversity levels regarding the number of taxa and coexistence of bacteria and fungi compared to the construction method. However, they also have major disadvantages or "hidden effects" (Huston 1997) represented by a possible increase in soil nutrients from dead cells in the destruction method, and the non-random selection of taxa due to the preferential elimination of rare ones in the erosion method. Reduction of diversity is assessed by measurement of taxonomic indexes, but the effect on taxonomic and functional composition is not controlled.

Most construction method results find a positive relationship between microbial diversity and soil processes involved in soil C and N cycles when there are fewer than 10 species (Deacon 1985; Robinson et al. 1993; Wohl et al. 2004), but they also reveal functional redundancy (Setälä and McLean 2004; Bell et al. 2005; Tiunov and Scheu 2005) in microbial communities when there are more than 10 species (except for Tiunov and Scheu 2005). The unrealistically low level of diversity obtained from the construction method can result in low niche complementarity (Loreau 2001). It is likely that certain non-cultivable microorganisms are able to perform functions that the cultivable ones cannot (Hättenschwiler et al. 2011). Genetic diversity increases functional diversity and helps microorganisms exploit a vast range of organic C compounds (Burns et al. 2013). This unattained resource partitioning could lead to an apparent redundancy in studies using complex substrates (SOM or litter) as a C resource (Setälä and McLean 2004; Bell et al. 2005), meaning that mineralisation stops because certain compounds cannot decompose or interactions (competition and facilitation) between microorganisms (e.g. bacteria vs. fungi) are not possible.

Table 1. Experimental methods used to create different microbial diversity levels

Method	References	Principles	Advantage	Disadvantage
Construction	Deacon (1985), Robinson et al. (1993), Setälä and McLean (2004), Wohl et al. (2004), Bell et al. (2005) and Tiunov and Scheu (2005)	Combinations of different microorganisms previously isolated in culture media	Taxa selection	Culture-dependency unrealistic diversity levels
Destruction	Degens (1998), Griffiths et al. (2000), and Griffiths et al. (2004)	Different duration of fumigation events	Realistic diversity levels	Hidden effect (Huston 1997): increase in soil nutrients
Erosion	Griffiths et al. (2001), Griffiths et al. (2004), Wertz et al. (2006, 2007); Baumann et al. (2013), Juarez et al. (2013) and Philippot et al. (2013)	Successive dilutions of a soil suspension	Realistic diversity levels	Hidden effect (Huston, 1997): non-random taxa selection = elimination of rare taxa

A hierarchy of functioning is often reported between different levels of diversity in more species-rich communities: the higher the level of diversity (i.e. higher richness or Shannon index), the higher the process, even if differences are not always statistically significant. We focused on nine studies based on destruction and erosion methods. Four studies (Degens 1998; Griffiths et al. 2001; Wertz et al. 2006, 2007) found no statistically significant differences in the measured processes between diversity levels and concluded that no relationship exists between microbial diversity and soil processes. The results of three studies (Griffiths et al. 2000; Baumann et al. 2013; Juarez et al. 2013) indicate a relationship between microbial diversity and soil processes such as C decomposition and denitrification, which are highly redundant functions in the microbial world (Griffiths et al. 2000), and also more specialised processes such as nitrification or methane oxidation (Griffiths et al. 2000). This suggests that decreased diversity initially might not affect soil processes; however, beyond a given threshold, diversity could become a limiting factor affecting soil processes. Another hypothesis to explain the complex relationship between diversity and soil processes is that community composition could be a determinant parameter rather than richness per se (Cox et al. 2001; Griffiths et al. 2004; Philippot et al. 2013). In this case, both diversity and community composition could have an impact on soil processes, but the specific effect of each taxa might depend on environmental conditions and biotic interactions between microorganisms (Nielsen et al. 2011) causing an increase (facilitation) or a decrease (inhibition) in functioning. Under this hypothesis, the effect of change in diversity on C and N dynamics may be more variable.

2.2 Relationship between microbial community composition and soil processes

Recent advances in molecular biology make it possible to track and analyse specific functional genes, which provides the opportunity to compare the presence, abundance and diversity of genes in different communities (Colloff et al. 2008). This method is only possible for a few processes, most of them involved in the N cycle, such as denitrification or nitrification, because of the well-known genetic coding of the enzymes responsible for transforming N (Maron et al. 2007). For other processes, such as C decomposition, that involve highly complex and mostly unknown sets of functional genes, an approach that focuses on functional genes is highly problematic making very difficult so far to describe functional diversity. An alternative approach is to study the link between taxonomic and functional composition by determining whether taxonomic groups are related to the processes. Relationships between bacterial phyla and functional traits suggest that the phylum level has ecological coherence (Philippot et al. 2010) meaning that members of one phylum share similar life strategy or traits that make them different from members of another phylum. Certain bacterial phyla that only grow and reproduce in nutrient-rich environments (Langer et al. 2004) are classified as "copiotrophs" (Fierer et al. 2007). They also are identified as "r-strategists" or opportunists because they grow rapidly after the addition of easily available C sources in the environment and are highly competitive for this kind of substrate. Other phyla, grown in nutrient-poor environments (Langer et al. 2004), are classified as "oligotrophs" (Fierer et al. 2007). They are mainly "K-strategists" because their metabolism is slower and synthesises more complex enzymes, making possible the decomposition of recalcitrant compounds. These functional traits of phyla have been observed in different studies, such as those studying decomposition of plant residues (Pascault et al. 2010) or comparing it to that of SOM (Pascault et al. 2013) or in a N fertilisation gradient (Fierer et al. 2012). Even when these classifications are identified, the phylum remains a high taxonomic rank, meaning that members of one phylum can exhibit different physiological traits. Overall, the literature shows the importance of community composition in soil processes. However, knowledge remains limited, and further study is required to understand the influence of microbial composition on C and N dynamics and maybe identify key taxa as indicators of microbial community functional traits (Nannipieri et al. 2003).

The overall view of the cited references bring evidence that soil microbial diversity does not have a neutral effect on the soil processes involved in C and N dynamics. But we are still lacking a full understanding of the underlying changes in soil community functioning with shifts in soil diversity (Nielsen et al. 2011; Graaff et al. 2015). It seems that a change in taxonomic diversity could lead to a change in the functional traits of the whole community and thus to a change in soil functioning.

To summarise, we propose a conceptual model (Figure 1) that assumes an increase in soil functioning with increased diversity. At low levels of diversity, the redundant relationship found in construction method studies could be an artefact of the method. However, low levels of diversity have been shown to develop under real conditions when a major perturbation occurs, such as fire or severe pollution, and are often associated with a reduction in soil C and N transformations (e.g. Griffiths et al. 2000, 2001, 2004). It is likely that at low microbial diversity soil processes reach a minimum, which lies a certain distance from a potential maximum (Figure 1, section A), but grows rapidly with an increase in diversity (Figure 1, section B). Conversely, at high levels of diversity, we suggest that a threshold exists at which processes are not affected by an increase in diversity (Figure 1, section C) because functional redundancy is reached or because diversity is no longer the determinant factor. This relationship is similar to the well-known mass-ratio hypotheses in plant community diversity (Grime 1998). On average (Figure 1, solid line), an augmentation of dominant species (i.e. augmentation of diversity) controls the ecosystem functioning. However, the rate at which this plateau is reached (Figure 1, section B) is controlled by the functional traits hypothesis (Díaz et al. 2007) which states that ecosystem functioning depends on functional diversity determined by the composition of the community. Besides, this relationship depends on the redundancy in studied processes or whether another parameter becomes more determinant. For instance, the relationship between diversity and processes seems to be expressed more significantly when substrate availability is higher (Philippot et al. 2013; Tardy et al. 2014). This suggests that the general relationship could be "situation-dependent" (e.g. substrate availability-dependent). More knowledge is needed to quantify this relationship and its potential variability in different situations.

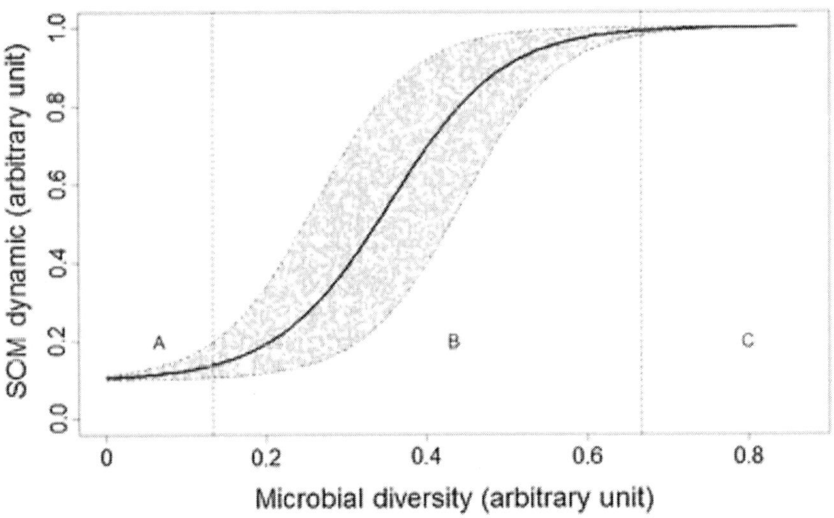

Figure 1. Conceptual diagram of the relationship between microbial diversity and soil organic matter (SOM) dynamics. *Solid line* general relationship; *dashed lines* area of variability in the relationship; *grey points* variability of pathways in the relationship; *A* area in which the relationship is minimum; *B* area of potential highest variability; *C* area with no relationship

3. SOIL MICROBIAL DIVERSITY IN CURRENT SOIL CARBON AND NITROGEN DYNAMICS MODELS

Models are characterised by state variables corresponding to C or N compartments and kinetics parameters. According to Manzoni and Porporato (2009) reviewing approximately 250 models of C and N dynamics developed in the last 80 years, thirty per cent of the models have no microbial compartment. Thirty per cent represent microbial biomass as an additional organic matter compartment. Among these, few models represent explicitly microbial biomass (linearly or nonlinearly) in first-order kinetics of C and N dynamics (Wutzler and Reichstein 2008). Forty per cent of the models use at least two state variables to represent microorganisms, suggesting a more explicit representation of their diversity. Although models might represent the observed reality, the initialisation of such models remains critical as each soil condition requires an optimisation analysis to initialise the OM compartments and their decomposition rates. This section focuses on the latter model group plus five other models developed since 2009.

3.1 Representation of microbial diversity in existing C and N dynamics models

We distinguished two model categories (Table 2 and Figure 2): models representing the functional diversity of soil microorganisms (FD), and soil food web models representing taxonomic diversity (TD).

References in bold were published after Manzoni and Porporato (2009). The class are FD: representation of functional diversity and TD: Soil Food Web models representing taxonomic diversity. Number of pools is equal to the infinite (Inf) for the models presenting a continuous of microbial diversity

The FD models are designed to explicitly model the diversity of substrate decomposition kinetics, by representing different affinities of microbial functional pools for substrates. Defining multiple functional pools enables multiple kinetic components in the model, such as decomposition, growth and decay rates, depending on the organic matter/microbial pool combination considered (e.g. Hansen et al. 1991; Garnier et al. 2001; Moorhead and Sinsabaugh 2006; Miki et al. 2010; Perveen et al. 2014). Functional pools are represented as discrete compartments in 89 % of the FD models. In the guild-based model developed by Moorhead and Sinsabaugh (2006), the "opportunist pool" represents microorganisms that compete highly for easily degradable material and have fast growth and decay rates when the resource is missing. The "decomposer pool" represents microorganisms that degrade cellulosic and lignocellulosic compounds. The "miner pool" represents microorganisms that degrade humified organic matter. In a more recent model, Perveen et al. (2014) distinguished two pools of microorganisms to better simulate the priming effect. The "SOM builders" pool is composed of microorganisms that degrade only fresh organic matter. The "SOM decomposers" pool represents microorganisms that degrade fresh organic matter and SOM. More detailed models with a larger number of microbial pools have been developed to represent complex interactions between functional pools or interactions between soil microorganisms and plant diversity (Loreau 2001). Grant (2001) distinguished nine microbial pools based on nine different metabolisms differing in the source of energy and nutrients mobilised; this representation enables different structural and kinetic components in substrate-microbe complexes.

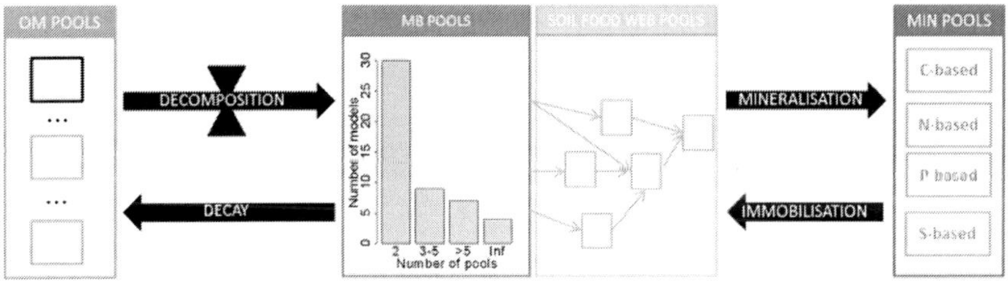

Figure 2. Diagram of microbial diversity in current carbon and nitrogen dynamics models. *OM* organic matter, *MB* microbial biomass, *MIN* mineral compounds. *Black* pool always encountered in models; *grey* pool specific to certain models

TD models (28 % of those listed) distinguish taxonomic groups of microorganisms and focus on interactions along the food web. They are used to analyse contributions of these groups, specifically on N mineralisation (Hunt et al. 1987; Ruiter et al. 1993; Zheng et al. 1999) or C and N dynamics (Hunt et al. 1983; Kuijper et al. 2005). The microbial groups differ in their nutrient ratios, feeding preferences, life spans, assimilation efficiencies, production-to-assimilation ratios and decomposability.

Table 2. Classification of models representing microbial communities according to the class of microbial community (MC) representation and the number of pools associated with the class (from Manzoni and Porporato, 2009)

Model	Reference	MC	Pools
PWNEE	Patten (1972)	TD	6
PHOENIX	McGill et al. (1981)	FD	2
NCSOIL	Molina et al. (1983)	FD	2
–	Hunt et al. (1983)	TD	4
NCSOIL	Hadas et al. (1987)	FD	2
–	Hunt et al. (1987)	TD	11
–	Leffelaar (1986)	FD	2
–	Leffelaar and Wessel (1988)	FD	2
–	Robinson et al. (1989)	TD	2
DAISY	Hansen et al. (1991)	FD	2
GEM	Hunt et al. (1991)	TD	5
DNDC	Li et al. (1992a, b)	FD	4
–	Griffiths and Robinson (1992)	TD	2
–	Ruiter et al. (1993)	TD	9
Ecosys	Grant et al. (1993)	FD	4
–	Kersebaum and Richter (1994)	FD	2
Q-model	Bosatta and Agren (1994)	FD	Inf
Q-model	Bosatta and Agren (1995)	FD	Inf
Q-model	Bosatta and Agren (1996)	FD	Inf
–	Zheng et al. (1997)	TD	2
NCSOIL	Hadas et al. (1998)	FD	2
–	Zheng et al. (1999)	TD	2
–	Henriksen and Breland (1999)	FD	2
DNDC	Li et al. (2000) and Stange et al. (2000)	FD	4
SOILN-NO	Korsaeth et al. (2001)	FD	2
CANTIS	Garnier et al. (2001)	FD	2
–	Loreau (2001)	FD	m
Ecosys	Grant (2001)	FD	9
–	Kravchenko et al. (2004)	FD	2
–	Moore et al. (2004)	TD	2
–	Foereid and Yearsley (2004)	TD	2
–	Long and Or (2005)	FD	2
CN-SIM	Petersen et al. (2005a, b)	FD	2
INDISIM-S	Ginovart et al. (2005)	FD	2
–	Kuijper et al. (2005)	TD	5
–	Moore et al. (2005)	TD	10
EnzModel	Allison (2005)	FD	2
–	Fontaine and Barot (2005)	FD	2
–	Raynaud et al. (2006)	FD	2
BACWAVE-WEB	Zelenev et al. (2006)	TD	5
GDM	Moorhead and Sinsabaugh (2006)	FD	3
–	Roy et al. (2008)	FD	2
TOUGHREACT-N	Maggi et al. (2008)	FD	4
NICA	Ingwersen et al. (2008)	FD	2
CEM	d'Annunzio et al. (2008)	FD	Inf

3.2 Advantages and limitations of existing modelling approaches

The models representing functional microbial diversity are mainly theoretical and based on conceptual microbial pools characterised by distinct functional traits.

They represent selected groups of microorganisms with distinct functional traits, mainly to test hypotheses on the processes involved in C and N dynamics. The corresponding pools might not be explicitly identified (e.g. Hansen et al. 1991) or distinguished according to known functional traits, as in the distinction between autochthonous microorganisms mainly found in environments rich in recalcitrant material and zymogenous microorganisms found in environments rich in easily decomposable material (Kersebaum and Richter 1994; Hadas et al. 1998; Garnier et al. 2001). Few models identify microbial functional pools in explicit taxonomic microbial groups. In these models, bacteria and fungi are often separate pools because their physiological differences induce contrasting C and N stoichiometries, and their relative abundance influences C and N dynamics and net primary production at the ecosystem scale (Waring et al. 2013). All food web models have explicit microbial pools because they study nutrient and energy transfer between taxonomic groups in an ecosystem. Two groups of microorganisms are found in soil food web models: decomposers, composed of bacteria and fungi, and predators of decomposers, mainly composed of protozoa and nematodes.

Almost half of the models that consider microbial diversity (FD and TD) have not been assessed with empirical datasets, i.e. data from an experiment or from field observations, at least in the reviewed articles. These heuristic models are essentially used to test hypotheses (Figure 3, arrows 4a and 5) about the coexistence of species in heterogeneous soil (Long and Or 2005) or effects on C and N dynamics of species and enzyme diversities (Allison 2012), rhizodeposition and microbial interactions (Robinson et al. 1989; Griffiths and Robinson 1992; Foereid and Yearsley 2004), global change (Kuijper et al. 2005) or anthropogenic and natural disturbances (Moore et al. 2005). Some were only analysed at equilibrium to assess the model's capacity to simulate processes such as the priming effect (Perveen et al. 2014), plant-soil feedback in plant communities (Miki et al. 2010), plant persistence and C storage (Fontaine and Barot 2005), or the effect of microbial diversity on ecosystems (Loreau 2001). The main reason not to evaluate such models by comparing predicted and observed data is that they are in essence theoretical; therefore, data are not available to calibrate the parameters (e.g. Moorhead and Sinsabaugh 2006). At most, the realism of model behaviour is assessed by comparing model outputs to general patterns from the literature or to those observed in real situations. These models are appropriate tools to assess ways to translate a hypotheses or a process into mathematical language, but are less suitable for prediction in the current state of knowledge (Figure 3, arrow 4b). The main exceptions are models focusing on N dynamic to whom genes implicated in the different processes are well known and where microbial biomass can be subdivided in measurable functional groups involved in these processes such as nitrifiers and denitrifiers (Leffelaar 1986; Leffelaar and Wessel 1988; Li et al. 1992a, b, 2000; Ginovart et al. 2005; Raynaud et al. 2006; Maggi et al. 2008).

Other models were evaluated by comparing their predictions to experimental data (Figure 3, arrow 3). However, a limit to these models is that initial values and parameters related to microbial communities are often assumed fixed or calibrated values. For instance, the initial biomass of each model compartment is often non-measurable. When compartments represent known taxonomic groups (e.g. bacteria vs. fungi; most compartments in TD), the initial biomass of each group is measurable, but when compartments represent conceptual functional groups, theoretical values are assumed. In the first case, compartments represent a high level of taxonomic classification, and we question whether this subdivision is always sufficient. This representation is often used at an ecosystem scale, but at smaller scales it may be relevant for better explaining bacterial diversity. In the second case, initial values are most often defined from the literature (Leffelaar and Wessel 1988; Li et al. 1992b; Henriksen and Breland 1999; Korsaeth et al. 2001; Foereid and Yearsley 2004; Ginovart et al. 2005), but they can also be fitted from a control incubation (Hadas et al. 1998) or estimated during model calibration (Maggi et al. 2008). The main issue of using assumed values compared to real measures is that making correct assumptions is not guaranteed, and the values may not be accurate. Yet, this information can be of great value. For instance, sensitivity analysis showed that the fraction of the late-stage decomposer pool in initial microbial biomass at time $t = 0$ in the model of Ingwersen et al. (2008) has the highest sensitivity among all fitted parameters. However, authors rarely discussed this aspect, and the way initial values are chosen is not always documented.

Efforts have been made to more explicitly represent microbial communities in C and N dynamics models, but the overview of the existing models shows that they multiply theoretical or conceptual functional compartments of microorganisms. This approach comes with difficulties in getting data to evaluate the models, creating a lack of predictive models.

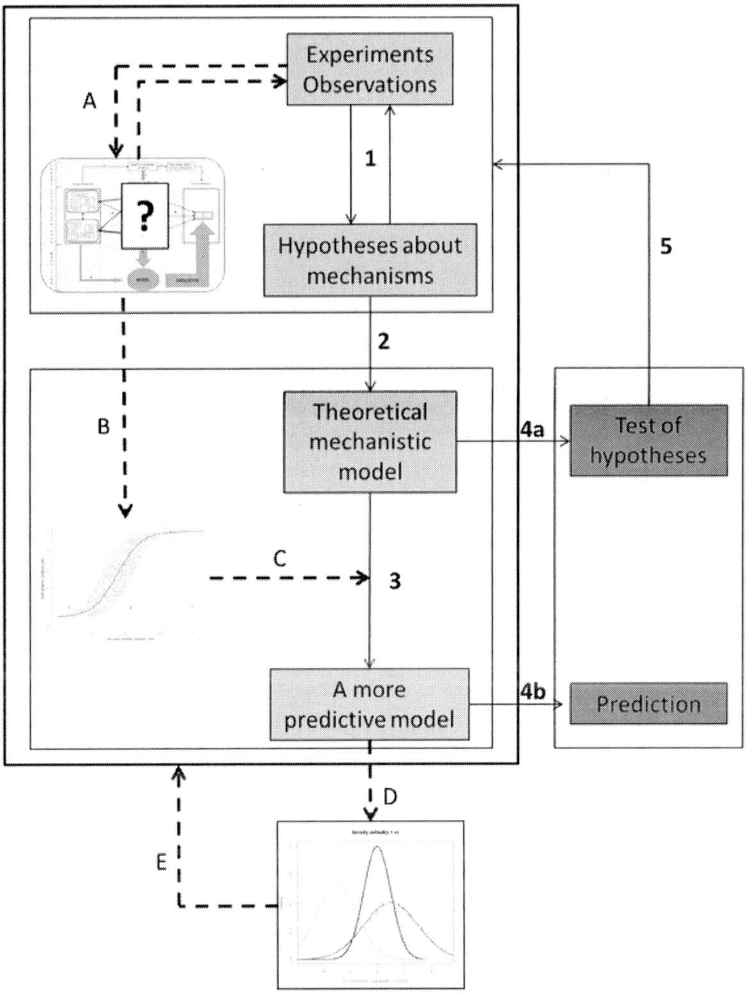

Figure 3. Diagram of model design (*solid arrows with numbers*) and stages for integrating microbial community descriptors (*dashed arrows with letters*). Stages correspond to (*1*) interactions between analysis of experimental/observed data that enable making hypotheses and hypotheses that influence future experiments, (*2*) translation of hypotheses into mathematical language, (*3*) model calibration and validation, (*4a*) simulations for testing hypotheses enabling (*5*) experiment/hypothesis interactions or (*4b*) for predictions, (*A*) search for best microbial community descriptors, (*B*) statistical learning, (*C*) coupling mechanistic modelling with statistical modelling, (*D*) sensitivity and uncertainty analyses, which help (*E*) in all model design

4. WHAT WAY TOWARDS PREDICTIVE MODELLING APPROACHES INCLUDING SOIL MICROBIAL DIVERSITY?

4.1 Getting additional datasets to inform and improve explicit models

As mentioned above, a lot of data are available on C and N dynamics and microbial diversity in a lot of context (different soil types, studied functions, used substrates etc.). Metaanalyses could enable to summarise these results and find general relationships between microbial diversity and C and N dynamics. However, considering the differences in the protocols measuring soil microbial diversity, deriving generic quantitative relationships between soil microbial diversity and C and N dynamics suitable to calibrate models remain challenging (Graaff et al. 2015). Consequently, additional datasets are still necessary (1) to further deepen our knowledge of soil microbial diversity impact on C and N dynamics, (2) to improve and evaluate the current models by testing with numerous quantitative diversity—C,N dynamics relationship, (3) and eventually provide more predictive modelling approaches.

To develop models with a relatively wide validity domain, these datasets must include data on soil microbial communities covering a large range of diversity and presenting enough intermediate levels to capture the likely nonlinearity of the relationship (Figure 1). The data must be representative of a range of pedoclimatic conditions and agricultural systems. As reviewed in "Role of microbial diversity in soil carbon and nitrogen dynamics" section, the aim of experimental studies creating a gradient of diversity is to detect potential difference of functioning when the diversity decreases. Consequently, they work with few levels of diversity (around 5) and seldom if ever build a relationship between diversity descriptors and soil functions. We suggest that the number of levels should be doubled (around 10) and range between a high rate of decrease and a non-decrease in the diversity. In addition, in these incubation experiments determination of microbial diversity should not be restricted to a unique initial point in time (as often done), but should be assessed parallelly to functions measurements since both are known to be strongly dynamic variables.

As microbial diversity is highly impacted by soil physical and chemical properties, climate and land use (Maron et al. 2011), the natural gradients of diversity found in soils could be used to study the relationship between diversity and C and N dynamics (Tardy et al. 2015). The major limit is the unavoidable presence of confusion between factors influencing C and N dynamics. Being capable of determine if different dynamics is a direct consequence of the gradient or an indirect consequence of the reason of the gradient is not that obvious. For instance, Rousk et al. (2010) found a strong link between a pH gradient and microbial diversity. In this case, it would be difficult to separate the biological effect (microbial diversity) from the chemical effect (pH) on C and N dynamics. However, using natural gradient, with uncontrolled factors, remains complementary to created gradient where others factors can be controlled.

In most of the available datasets based on incubation under controlled conditions to link soil microbial diversity to C and N dynamics, only the total mineralisation and decomposition fluxes (e.g. total CO_2 emissions) are measured. We argue that the use of isotopic labelling technics and the implementation of experiments, in which soil samples will be incubated with or without added labelled plant residues, are of high interest to better understand the processes driven by microbial diversity and quantify the relationship between microbial diversity and each C and N flux (decomposition, basal respiration, priming effect). These technics combined with microbial diversity characterisation are still used in ongoing research projects (ANR DIMIMOS ANR-08-STRA-06, and ANR MOSAIC ANR-12 AGRO-0005) and should provide key datasets to further develop and parameterise mechanistic models.

4.2 A complementary modelling approach based on taxonomic indexes

Although studies about the relationship between microbial diversity and SOM dynamics compute synthetic taxonomic diversity indexes (e.g. richness, Shannon index), none of the reviewed models use those available indexes. This can be explained by the lack of understanding of the relationship between taxonomic diversity and functional traits of a community (Heijden and Wagg 2013). Given the potential and availability of taxonomic indexes, additional effort is needed to better understand this relationship and assess the relevance of using them in predictive models.

As shown above, considering microbial diversity in C and N dynamics models significantly increases their complexity and consequently almost restricts their applicability to theoretical issues. While adding compartments helps represent functional diversity, it is limited by the exponential growth in the number of parameters introduced and the quantity of data needed to calibrate them. The resulting models with many compartments would be conceptually interesting but not highly applicable because it would be challenging to calibrate them. For prediction purposes, alternative modelling approaches based on empirical functions could be tested to account for impacts of microbial diversity in C and N dynamics models, even non-explicitly.

One possible alternative approach would consider soil microbial diversity descriptors in a function that modifies parameters in the first-order kinetics equation describing SOM decomposition/mineralisation: for example, simple representation of decomposition of a substrate by microbial biomass (Figure 4). Microbial biomass would be represented by a state variable and its activity by two parameters: decomposition rate (k) and efficiency yield (Y) (Figure 4).

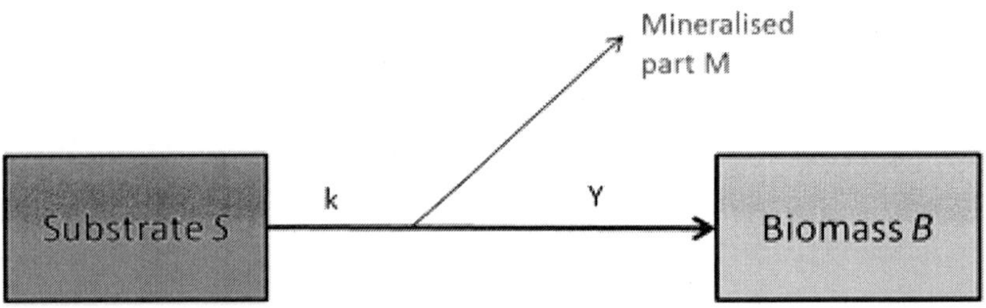

Figure 4. Diagram of a simple model of decomposition of a substrate

The decomposition rate and efficiency yield can be corrected by functions f and g, respectively, depending on groups θ and θ' of key taxonomic indexes, respectively:

$$k \xrightarrow{f(\theta)} k_{corr}$$
$$Y \xrightarrow{g(\theta')} Y_{corr}$$

This approach is similar to the traditional method of including the effect of pedoclimatic factors such as temperature, soil moisture or soil texture. Parameters are a function of pedoclimatic factors mainly in a multiplicative form (e.g. Hansen et al. 1991; Garnier et al. 2001). This illustrates why soil microbial communities, represented by the groups of taxonomic indexes, should be quantitatively linked to SOM dynamics, represented by decomposition rate and efficiency yield (Figure 3, arrow B).

Since many methods measure descriptors of microbial communities and knowledge about how the latter might influence these parameters is lacking, such empirical functions are difficult to define. The need exists for better integration of empirical knowledge through statistical learning in models (Figure 4, arrow C). In particular, statistical tools have three advantages:

1) Many exploratory tools exist (e.g. clustering, tree-based method, factorial analyses) to help identify descriptors of microbial communities that control processes of soil C and N dynamics and that are faster than mechanistic modelling and hypothesis testing.

2) Statistical modelling can be useful to link model parameters such as the decomposition rate to microbial community descriptors. Considering current limited knowledge, tools with little a priori (e.g. generalised additive models) are useful to understand complex systems in ecology and help in exploring underlying mechanisms (Guisan et al. 2002).

3) Statistical modelling comes with estimated uncertainties, which could be incorporated into traditional models by developing stochastic models, which have demonstrated their relevance for ecological modelling (Clark 2007; Katul et al. 2007).

4.3 Uncertainties and sensitivity: assessing for progressing

Assessing the accuracy of predictions should include comparing model predictions to independent observed data. The relevance of methods used to represent microbial communities more explicitly should be assessed with sensitivity and uncertainty analysis tools (Figure 3, arrow D), which are not used sufficiently in models of C and N dynamics. They have demonstrated their ability to assess the behaviour of complex dynamic system models, such as those simulating ground temperature (Gubler et al. 2013) and soil moisture (Wolf et al. 2008) or in some C and N dynamics models (Hunt et al. 1987; Raynaud et al. 2006; Roy et al. 2008). Sensitivity analysis can improve models and adapt microbial descriptors to include different scales of study (from micro- to global scales), mechanisms or systems (Figure 3, arrow E). These tools could be a way to improve the generality of a model (sensu Levins 1966) by adapting or modifying it according to the situation rather than developing a model applicable to all situations. By combining these tools, experiments would help develop models, and sensitivity analysis would help design experiments.

5. CONCLUSION

From our review, the increasing use of advanced tools from microbiology and molecular biology appears to provide evidence of the influence of microbial diversity in SOM dynamics. This diversity is driven by the combination of edaphic factors and agricultural management practices, suggesting that better understanding of its effects on SOM dynamics and its integration into models could increase the accuracy of the latter. Our review shows that a body of modelling studies already confirms that considering microbial diversity in C and N dynamics models should improve the ability to accurately predict SOM dynamics. Microbial diversity is mainly incorporated by adding new biomass compartments to represent groups of microorganisms. These studies allowed the development of mechanistic models that explicitly represent soil processes. They are needed to test new scientific hypothesis and help understand underlying mechanisms and the impact of the environment and human activity. With certain exceptions, the mechanistic approach seems to be the only one used to model soil C and N dynamics in recent decades. However, they are essentially theoretical and are not operational for prospective simulation. Combining the mechanistic approach with statistical tools could facilitate incorporation of explicit taxonomic diversity indexes and improve models that simulate real situations, which should follow a realistic quantitative approach to provide predictions conditioned by real data.

ACKNOWLEDGMENTS

Financial support for this research was provided by the Chair for "Ecologically Intensive Agriculture" and the Brittany Regional Council.

REFERENCES

1. Allison SD (2005) Cheaters, diffusion and nutrients constrain decomposition by microbial enzymes in spatially structured environments. Ecol Lett 8:626–635.
2. Allison SD (2012) A trait-based approach for modelling microbial litter decomposition. Ecol Lett 15:1058–1070.
3. Baumann K, Dignac M-F, Rumpel C et al (2013) Soil microbial diversity affects soil organic matter decomposition in a silty grassland soil. Biogeochemistry 114:201–212
4. Bell T, Newman JA, Silverman BW et al (2005) The contribution of species richness and composition to bacterial services. Nature 436:1157–1160.
5. Berthrong ST, Buckley DH, Drinkwater LE (2013) Agricultural management and labile carbon additions affect soil microbial community structure and interact with carbon and nitrogen cycling. Microb Ecol 66:158–170
6. Bosatta E, Agren GI (1994) Theoretical analysis of microbial biomass dynamics in soils. Soil Biol Biochem 26:143–148.
7. Bosatta E, Agren GI (1995) The power and reactive continuum models as particular cases of the q-theory of organic matter dynamics. Geochim Cosmochim Acta 59:3833–3835.
8. Bosatta E, Agren GI (1996) Theoretical analyses of carbon and nutrient dynamics in soil profiles. Soil Biol Biochem 28:1523–1531.

9. Burns RG, DeForest JL, Marxsen J et al (2013) Soil enzymes in a changing environment: current knowledge and future directions. Soil Biol Biochem 58:216–234.
10. Chambers A, Lal R, Paustian K (2016) Soil carbon sequestration potential of US croplands and grasslands: implementing the 4 per thousand initiative. J Soil Water Conserv 71:68A–74A
11. Clark JS (2007) Models for ecological data: an introduction. Princeton University Press Princeton, New Jersey
12. Colloff MJ, Wakelin SA, Gomez D, Rogers SL (2008) Detection of nitrogen cycle genes in soils for measuring the effects of changes in land use and management. Soil Biol Biochem 40:1637–1645. doi:10.1016/j.soilbio.2008.01.019CrossRefGoogle Scholar
13. Cox P, Wilkinson SP, Anderson JM (2001) Effects of fungal inocula on the decomposition of lignin and structural polysaccharides in Pinus sylvestris litter. Biol Fertil Soils 33:246–251CrossRefGoogle Scholar
14. Creamer R, Hannula S, Van Leeuwen J et al (2016) Ecological network analysis reveals the inter-connection between soil biodiversity and ecosystem function as affected by land use across Europe. Appl Soil Ecol 97:112–124
15. d'Annunzio R, Zeller B, Nicolas M et al (2008) Decomposition of European beech (*Fagus sylvatica*) litter: combining quality theory and 15 N labelling experiments. Soil Biol Biochem 40:322–333
16. de Graaff M-A, Adkins J, Kardol P, Throop HL (2015) A meta-analysis of soil biodiversity impacts on the carbon cycle. SOIL 1:257–271.
17. Deacon JW (1985) Decomposition of filter paper cellulose by thermophilic fungi acting singly, in combination, and in sequence. Trans Br Mycol Soc 85:663–669
18. Degens BP (1998) Decreases in microbial functional diversity do not result in corresponding changes in decomposition under different moisture conditions. Soil Biol Biochem 30:1989–2000
19. Díaz S, Lavorel S, de Bello F et al (2007) Incorporating plant functional diversity effects in ecosystem service assessments. Proc Natl Acad Sci 104:20684–20689.
20. Doran JW, Elliott ET, Paustian K (1998) Soil microbial activity, nitrogen cycling, and long-term changes in organic carbon pools as related to fallow tillage management. Soil Tillage Res 49:3–18
21. Feng Y, Motta AC, Reeves DW et al (2003) Soil microbial communities under conventional-till and no-till continuous cotton systems. Soil Biol Biochem 35:1693–1703
22. Fierer N, Bradford MA, Jackson RB (2007) Toward an ecological classification of soil bacteria. Ecology 88:1354–1364
23. Fierer N, Lauber CL, Ramirez KS et al (2012) Comparative metagenomic, phylogenetic and physiological analyses of soil microbial communities across nitrogen gradients. ISME J 6:1007–101
24. Foereid B, Yearsley JM (2004) Modelling the impact of microbial grazers on soluble rhizodeposit turnover. Plant Soil 267:329–342
25. Fontaine S, Barot S (2005) Size and functional diversity of microbe populations control plant persistence and long-term soil carbon accumulation. Ecol Lett 8:1075–1087
26. Garnier P, Néel C, Mary B, Lafolie F (2001) Evaluation of a nitrogen transport and transformation model in a bare soil. Eur J Soil Sci 52:253–268.
27. Ginovart M, López D, Gras A (2005) Individual-based modelling of microbial activity to study mineralization of C and N and nitrification process in soil. Nonlinear Anal Real world Appl 6:773–795
28. Graham EB, Wieder WR, Leff JW et al (2014) Do we need to understand microbial communities to predict ecosystem function? A comparison of statistical models of nitrogen cycling processes. Soil Biol Biochem 68:279–282
29. Grant RF (2001) A review of the Canadian ecosystem model ecosys. In: Shaffer M (ed) Modeling carbon and nitrogen dynamics for soil management, CRC Press, Boca Raton, pp. 173–264
30. Grant R, Juma N, McGill W (1993) Simulation of carbon and nitrogen transformations in soil: mineralization. Soil Biol Biochem 25:1317–1329

31. Griffiths B, Robinson D (1992) Root-induced nitrogen mineralisation: a nitrogen balance model. Plant Soil 139:253–263
32. Griffiths B, Ritz K, Bardgett RD et al (2000) Ecosystem response of pasture soil communities to fumigation-induced microbial diversity reductions: an examination of the biodiversity–ecosystem function relationship. Oikos 90:279–294CrossRefGoogle Scholar
33. Griffiths BS, Ritz K, Wheatley R et al (2001) An examinataion of the biodiversity-ecosystem function relationship in arable soil microbial communaities. Soil Biol Biochem 33:1713–1722
34. Griffiths BS, Kuan HL, Ritz K et al (2004) The Relationship between Microbial Community Structure and Functional Stability, Tested Experimentally in an Upland Pasture Soil. Microb Ecol 47:104–113
35. Grime JP (1998) Benefits of plant diversity to ecosystems: immediate, filter and founder effects. J Ecol 86:902–910.
36. Gubler S, Endrizzi S, Gruber S, Purves R (2013) Sensitivities and uncertainties of modeled ground temperatures in mountain environments. Geosci Model Dev 6:1319–1336
37. Guisan A, Edwards TC, Hastie T (2002) Generalized linear and generalized additive models in studies of species distributions: setting the scene. Ecol Model 157:89–100
38. Hadas A, Molina J, Feigenbaum S, Clapp C (1987) Simulation of nitrogen-15 immobilization by the model NCSOIL. Soil Sci Soc Am J 51:102–106
39. Hadas A, Parkin T, Stahl P (1998) Reduced CO_2 release from decomposing wheat straw under N-limiting conditions: simulation of carbon turnover. Eur J Soil Sci 49:487–494
40. Halfon E (1979) Preview: theory in ecosystem analysis. Theor Syst Ecol Adv Case Stud 4:1
41. Hansen S, Jensen H, Nielsen N, Svendsen H (1991) Simulation of nitrogen dynamics and biomass production in winter wheat using the Danish simulation model DAISY. Fertil Res 27:245–259
42. Hättenschwiler S, Fromin N, Barantal S (2011) Functional diversity of terrestrial microbial decomposers and their substrates. CR Biol 334:393–402
43. Henriksen T, Breland T (1999) Nitrogen availability effects on carbon mineralization, fungal and bacterial growth, and enzyme activities during decomposition of wheat straw in soil. Soil Biol Biochem 31:1121–1134.
44. Hunt H, Stewart J, Cole C (1983) A conceptual model for interactions among carbon, nitrogen, sulphur, and phosphorus in grasslands. The Major Biogeochemical Cycles and Their Interactions Wiley, New York, pp 303–325
45. Hunt H, Coleman D, Ingham E et al (1987) The detrital food web in a shortgrass prairie. Biol Fertil Soils 3:57–6
46. Hunt H, Trlica M, Redente E et al (1991) Simulation model for the effects of climate change on temperate grassland ecosystems. Ecol Model 53:205–246
47. Huston MA (1997) Hidden tretments in ecological expermiments: re-evaluating the ecosystem function of biodiversity. Oecologia 110:449–460
48. Ingwersen J, Poll C, Streck T, Kandeler E (2008) Micro-scale modelling of carbon turnover driven by microbial succession at a biogeochemical interface. Soil Biol Biochem 40:864–878
49. Jenkinson D, Powlson DS (1976) The effects of biocidal treatments on metabolism in soil-V: a method for measuring soil biomass. Soil Biol Biochem 8:209–213
50. Juarez S, Nunan N, Duday A-C et al (2013) Soil carbon mineralisation responses to alterations of microbial diversity and soil structure. Biol Fertil Soils 49:939–948
51. Kaisermann A, Roguet A, Nunan N et al (2013) Agricultural management affects the response of soil bacterial community structure and respiration to water-stress. Soil Biol Biochem 66:69–77.
52. Katul G, Porporato A, Oren R (2007) Stochastic dynamics of plant-water interactions. Ann Rev Ecol Evol Syst 38:767–791
53. Kersebaum K, Richter O (1994) A model approach to simulate C and N transformations through microbial biomass. Eur J Agron 3:355–360
54. Korsaeth A, Molstad L, Bakken L (2001) Modelling the competition for nitrogen between plants and microflora as a function of soil heterogeneity. Soil Biol Biochem 33:215–226

55. Kravchenko L, Shvytov I et al (2004) Mathematical simulation of the dynamics of interacting populations of rhizosphere microorganisms. Microbiology 73:189–195
56. Kuijper LD, Berg MP, Morriën E et al (2005) Global change effects on a mechanistic decomposer food web model. Glob Change Biol 11:249–265
57. Lal R (1999) Soil management and restoration for C sequestration to mitigate the accelerated greenhouse effect. Prog Environ Sci 1:307–326
58. Langer U, Böhme L, Böhme F (2004) Classification of soil microorganisms based on growth properties: a critical view of some commonly used terms. J Plant Nutr Soil Sci 167:267–269
59. Lazcano C, Gómez-Brandón M, Revilla P, Domíguez J (2013) Short-term effects of organic and inorganic fertilizers on soil microbial community structure and function. Biol Fertil Soils 49:723–733
60. Leffelaar P (1986) Dynamics of partial anaerobiosis, denitrification, and water in a soil aggregate: experimental. Soil Sci 142:352–366
61. Leffelaar P, Wessel W (1988) Denitrification in a homogeneous, closed system: experiment and simulation. Soil Sci 146:335–349
62. Levins R (1966) The strategy of model building in population biology. Am Sci 54:421–431
63. Li C, Frolking S, Frolking TA (1992a) A model of nitrous oxide evolution from soil driven by rainfall events: 1. Model structure and sensitivity. J Geophy Res Atmos (1984–2012) 97:9759–9776
64. Li C, Frolking S, Frolking TA (1992b) A model of nitrous oxide evolution from soil driven by rainfall events 2. Model applications. J Geophys Res 97:9777–9783
65. Li C, Aber J, Stange F et al (2000) A process-oriented model of N2O and NO emissions from forest soils: 1. Model development. J Geophy Res Atmos (1984–2012) 105:4369–4384
66. Lienhard P, Tivet F, Chabanne A et al (2013) No-till and cover crops shift soil microbial abundance and diversity in Laos tropical grasslands. Agron Sust Dev 33:375–384
67. Long T, Or D (2005) Aquatic habitats and diffusion constraints affecting microbial coexistence in unsaturated porous media. Water Resour Res 41:W08408.
68. Loreau M (2001) Microbial diversity, producer–decomposer interactions and ecosystem processes: a theoretical model. Proc R Soc Lond B Biol Sci 268:303–309
69. Maggi F, Gu C, Riley WJ, Hornberger GM, Venterea RT, Xu T, Spycher N, Steefel C, Miller NL, Oldenburg CM (2008) A mechanistic treatment of the dominant soil nitrogen cycling processes: model development, testing, and application. J Geophys Res 113:G02016.
70. Manzoni S, Porporato A (2009) Soil carbon and nitrogen mineralization: theory and models across scales. Soil Biol Biochem 41:1355–1379.
71. Maron P-A, Ranjard L, Mougel C, Lemanceau P (2007) Metaproteomics: a new approach for studying functional microbial ecology. Microb Ecol 53:486–493
72. Maron P-A, Mougel C, Ranjard L (2011) Soil microbial diversity: methodological strategy, spatial overview and functional interest. C R Biol 334:403–411.
73. McGill WB, Hunt HW, Woodmansee RG, Reuss JO (1981) Phoenix, a model of the dynamics of carbon and nitrogen in grassland soils. In: Clark FE, Rosswall T (eds) Terrestrial nitrogen cycles, processes, ecosystem and management impacts, vol 33, pp. 49–116
74. McGuire KL, Treseder K (2010) Microbial communities and their relevance for ecosystem models: decomposition as a case study. Soil Biol Biochem 42:529–535
75. McLauchlan K (2006) The nature and longevity of agricultural impacts on soil carbon and nutrients: a review. Ecosystems 9:1364–1382
76. Miki T, Ushio M, Fukui S, Kondoh M (2010) Functional diversity of microbial decomposers facilitates plant coexistence in a plant–microbe–soil feedback model. Proc Natl Acad Sci 107:14251–14256
77. Molina J, Clapp C, Shaffer M et al (1983) NCSOIL, a model of nitrogen and carbon transformations in soil: description, calibration, and behavior. Soil Sci Soc Am J 47:85–91
78. Moore JC, Berlow EL, Coleman DC et al (2004) Detritus, trophic dynamics and biodiversity. Ecol Lett 7:584–600

79. Moore JC, McCann K, de Ruiter PC (2005) Modeling trophic pathways, nutrient cycling, and dynamic stability in soils. Pedobiologia 49:499–510
80. Moorhead DL, Sinsabaugh RL (2006) A theorical model of litter decay and microbial interacions. Ecol Monogr 76:151–174
81. Nannipieri P, Ascher J, Ceccherini M et al (2003) Microbial diversity and soil functions. Eur J Soil Sci 54:655–670
82. Neill C, Gignoux J (2006) Soil organic matter decomposition driven by microbial growth: a simple model for a complex network of interactions. Soil Biol Biochem 38:803–811
83. Nielsen U, Ayres E, Wall D, Bardgett R (2011) Soil biodiversity and carbon cycling: a review and synthesis of studies examining diversity–function relationships. Eur J Soil Sci 62:105–116
84. Pascault N, Cécillon L, Mathieu O et al (2010) In situ dynamics of microbial communities during decomposition of wheat, rape, and alfalfa residues. Microb Ecol 60:816–828
85. Pascault N, Ranjard L, Kaisermann A et al (2013) Stimulation of different functional groups of bacteria by various plant residues as a driver of soil priming effect. Ecosystems 16:810–822
86. Patten BC (1972) A simulation of the shortgrass prairie ecosystem. Simulation 19:177–186
87. Paul EA, Clark FE (1989) Soil Microbiology and Biochemistry. Academic Press Limited, Cambridg
88. Perveen N, Barot S, Alvarez G, Klumpp K et al (2014) Priming effect and microbial diversity in ecosystem functoining and response to global change: a modeling approach using the SYMPHONY model. Glob Change Biol 20:1174–119
89. Petersen BM, Berntsen J, Hansen S, Jensen LS (2005a) CN-SIM: a model for the turnover of soil organic matter. I. Long-term carbon and radiocarbon development. Soil Biol Biochem 37:359–374
90. Petersen BM, Jensen LS, Hansen S et al (2005b) CN-SIM: a model for the turnover of soil organic matter. II. Short-term carbon and nitrogen development. Soil Biol Biochem 37:375–393
91. Philippot L, Andersson SGE, Battin TJ et al (2010) The ecological coherence of high bacterial taxonomic ranks. Nat Rev Microbiol 8:523–529
92. Philippot L, Spor A, Hénault C et al (2013) Loss in microbial diversity affects nitrogen cycling in soil. ISME J 7:1609–1619
93. Rampelotto PH, de Siquiera FA, Barboza ADM, Roesch LFW (2013) Changes in Diversity, Abundance, and Structure of Soil Bacterial Communities in Brazilian Savanna Under Different Land Use Systems. Microb Ecol 66:593–607
94. Raynaud X, Lata J-C, Leadley PW (2006) Soil microbial loop and nutrient uptake by plants: a test using a coupled C:N model of plant–microbial interactions. Plant Soil 287:95–116
95. Reeves D (1997) The role of soil organic matter in maintaining soil quality in continuous cropping systems. Soil Tillage Res 43:131–167
96. Reidsma P, Tekelenburg T, Van den Berg M, Alkemade R (2006) Impacts of land-use change on biodiversity: an assessment of agricultural biodiversity in the European Union. Agric Ecosyst Environ 114:86–102
97. Robinson D, Griffiths B, Ritz K, Wheatley R (1989) Root-induced nitrogen mineralisation: a theoretical analysis. Plant Soil 117:185–193
98. Robinson CH, Dighton J, Frankland JC, Coward PA (1993) Nutrient and carbon dioxide release by interacting species of straw-decomposing fungi. Plant Soil 151:139–142
99. Romaní AM, Fischer H, Mille-Lindblom C, Tranvik LJ (2006) Interactions of bacteria and fungi on decomposing litter: differential extracellular enzyme activities. Ecology 87:2559–2569
100. Rousk J, Bååth E, Brookes PC et al (2010) Soil bacterial and fungal communities across a pH gradient in an arable soil. ISME J 4:1340–1351CrossRef
101. Roy M, Mandal S, Ray S (2008) Detrital ontogenic model including decomposer diversity. Ecol Model 215:200–206

102. Ruiter PCD, Moore JC, Zwart KB et al (1993) Simulation of Nitrogen Mineralization in the Below-Ground Food Webs of Two Winter Wheat Fields. J Appl Ecol 30:95–106
103. Schimel J (2001) 1.13—Biogeochemical models: implicit versus explicit microbiology. In: Schulze E-D, Heimann M, Harrison S et al (eds) Global biogeochemical cycles in the climate system. Academic Press, San Diego, pp 177–183
104. Setälä H, McLean MA (2004) Decomposition rate of organic substrates in relation to the species diversity of soil saprophytic fungi. Oecologia 139:98–107
105. Stange F, Butterbach-Bahl K, Papen H et al (2000) A process-oriented model of N2O and NO emissions from forest soils: 2. Sensitivity analysis and validation. J Geophy Res Atmos (1984–2012) 105:4385–4398
106. Sul WJ, Asuming-Brempong S, Wang Q et al (2013) Tropical agricultural land management influences on soil microbial communities through its effect on soil organic carbon. Soil Biol Biochem 65:33–38.
107. Sutton MA, Howard CM, Erisman JW et al (2011) The European nitrogen assessment: sources, effects and policy perspectives. Cambridge University Press, Cambridge
108. Sylvia DM, Fuhrmann JJ, Hertel PG, Zuberer DA (2005) Principles and applications of soil microbiology, 2nd edn. Pearson, London
109. Tardy V, Mathieu O, Lévêque J et al (2014) Stability of soil microbial structure and activity depends on microbial diversity. Environ Microbiol Rep 6:173–183CrossRef
110. Tardy V, Spor A, Mathieu O et al (2015) Shifts in microbial diversity through land use intensity as drivers of carbon mineralization in soil. Soil Biol Biochem 90:204–213
111. Tiunov AV, Scheu S (2005) Facilitative interactions rather than resource partitioning drive diversity-functioning relationships in laboratory fungal communities. Ecol Lett 8:618–625
112. Treseder KK, Balser TC, Bradford MA et al (2012) Integrating microbial ecology into ecosystem models: challenges and priorities. Biogeochemistry 109:7–18
113. van der Heijden MG, Wagg C (2013) Soil microbial diversity and agro-ecosystem functioning. Plant Soil 363:1–5
114. Waring BG, Averill C, Hawkes CV (2013) Differences in fungal and bacterial physiology alter soil carbon and nitrogen cycling: insights from meta-analysis and theoretical models. Ecol Lett 16:887–894
115. Wertz S, Degrange V, Prosser JI et al (2006) Maintenance of soil functioning following erosion of microbial diversity. Environ Microbiol 8:2162–2169
116. Wertz S, Degrange V, Prosser JI et al (2007) Decline of soil microbial diversity does not influence the resistance and resilience of key soil microbial functional groups following a model disturbance. Environ Microbiol 9:2211–2219
117. Wieder W, Grandy A, Kallenbach C, Bonan G (2014) Integrating microbial physiology and physio-chemical principles in soils with the MIcrobial-MIneral Carbon Stabilization (MIMICS) model. Biogeosciences 11:3899–3917
118. Wohl DL, Arora S, Gladstone JR (2004) Functional redundancy supports biodiversity and ecosystem function in a closed and constant environment. Ecology 85:1534–1540
119. Wolf A, Blyth E, Harding R et al (2008) Sensitivity of an ecosystem model to hydrology and temperature. Clim Change 87:75–89
120. Wutzler T, Reichstein M (2008) Colimitation of decomposition by substrate and decomposers? a comparison of model formulations. Biogeosciences 5:749–759
121. Zelenev V, Van Bruggen A, Leffelaar P et al (2006) Oscillating dynamics of bacterial populations and their predators in response to fresh organic matter added to soil: the simulation model BACWAVE-WEB. Soil Biol Biochem 38:1690–1711
122. Zheng DW, Bengtsson J, Agren GI (1997) Soil food webs and ecosystem processes: decomposition in donor-control and Lotka-Volterra systems. Am Nat 149:125–148
123. Zheng DW, Ågren GI, Bengtsson J (1999) How do soil organisms affect total organic nitrogen storage and substrate nitrogen to carbon ratio in soils? A theoretical analysis. Oikos 86:430–442
124. Zhong W, Gu T, Zhang Wang B W et al (2010) The effects of mineral fertilizer and organic manure on soil microbial community and diversity. Plant Soil 326:511–522

CHAPTER 14

Pb concentrations and isotopic compositions in the soil and sediments around the abandoned mine in southwest of Korea

Jong-Woo Choi1, Eun-Jin Yoo1, Jee-young Kim2, Jong-Yeon Hwang1, Khanghyun Lee1*, Won-Seok Lee1, Jin-Seok Han1, Kyu-Seung Lee3

[1]*Environmental Measurement & Analysis Center, Department of Environmental Infrastructure Research, National Institute of Environmental Research, Incheon, South Korea*

[2]*Department of Earth System Sciences, Yonsei University, Seoul, South Korea*

[3]*Department of Bio-Environmental Chemistry, Chungnam National University, Daejeon, South Korea*

ABSTRACT

This study was undertaken to investigate the influence of Pb contaminant dispersal from an abandoned mine. For the study, 23 samples of sediment and soil were collected around the abandoned mine located in southwest of Korea. Pb concentrations and isotope ratios of sediment samples were analyzed through ICP-OES (Optima 5300 DV, Perkin Elmer) and MC-ICP-MS (Plasma II, Nu), respectively. The sediment samples collected in the main stream from the mine entrance showed high Pb concentrations of 139 -

1079 mg/kg, while Pb concentrations of the other sites were 19 - 29 mg/kg. Pb isotope ratios of the main stream sediments were 2.4854 - 2.4883 for $^{208}Pb/^{207}Pb$ and 1.1693 - 1.1740 for $^{206}Pb/^{207}Pb$. On the other hand, the Pb isotopic compositions of sediments collected from the other sites were clearly different from those of ore deposits. The Pb isotope ratios versus inverse concentrations plot showed high linear correlation between the main stream, the tributary stream and one of the downstream sites representing the two end member mixing system between these sites. According to the binary mixing equation, the relative contribution of Pb pollutants from the abandoned mine to the downstream area were approximately 33% - 36%.

Keywords: Pb Isotope Ratios; MC-ICP-MS; Abandoned Mine; Stream Sediment

1. INTRODUCTION

Pb contamination in mine areas followed by ore digging and smelting has been reported through various environmental samples such as mine waste water, soils and plants [1-6]. Heavy metal pollutants released from mines are known to keep affecting the neighboring areas not only during the mine operation but even after the mines became disused. For example, the Pb concentrations of surface soils near the inactive mine in southwest of Spain were found 100 times as high as those of natural background [4]. Surface sediments of streams near the abandoned mine in Scotland were also reported to have a Pb concentration level similar to that of mine wastes [1]. All these indicate that the release of Pb pollutants from mines continues to have impact on the environment even after finishing the mine operation.

Over the past few decades, Pb pollution in mine areas has been evaluated based on concentrations and crustal enrichment factors. However, the total concentrations are not sufficient for a precise evaluation of contamination sources especially in the case of multiple sources for contaminants. Pb isotope ratios have thus been introduced as effective tracer for environmental pollution. Each source of Pb has its own isotopic ratio range. The isotopic composition of Pb in environmental samples reflects the mixing of these sources, and source apportionment can be quantified in cases where all potential sources of Pb are characterized [7]. For example, the surface soils in the Swiss National Park are influenced from natural Pb, Pb ores in mines and air pollutants, and that the relative contributions of these three sources are 43%, 21% and 36%, respectively [8].

In Korea, about 900 abandoned mines are known to be distributed around the nation. Since 1992, the Korean government has investigated environmental impact in the abandoned mine areas. The investigation showed that more than 90% of these areas suffered from serious soil contamination by heavy metal pollutants [9]. According to [10], the soils around Imcheon mine, which has been disused since the 1970s, have an average Pb concentration of 410 mg/kg, which reaches 40% of the maximum allow levels of heavy metals in factory and industrial areas provided by the Soil Conservation Act of Korea. Also, the Pb concentrations of soils near Duckum mine, which has been stopped its operation since 1994, are also found up to 200 times as high as those of natural soil [11]. However, these studies mostly focus on monitoring pollutants, and thus they do not effectively present contributions of pollutants necessary for compensation or reparation when environmental disputes arise. In this study, we investigated Pb isotopic compositions as well as Pb concentrations of soils and sediments around an abandoned mine. This provided the first application of Pb isotope ratios in the Korean abandoned mine area to identify the source for Pb pollutant and assess the relative contribution from that source.

2. MATERIAL AND METHODS

2.1. Sample Collection and Acid Digestion

For the research, surface sediment and soil samples were collected around an abandoned mine located in the southwest province of Korea. The alluvial system of this area consists of many tributary streams from surrounding mountains with one stream from the entrance of the abandoned mine (**Figure 1**). For the comparison of Pb isotope ratios between Pb contaminant from the abandoned mine and natural Pb, we collected twelve sediment samples at the main stream from the mine entrance (M1 ~ M12), two samples at the tributary stream coming from other direction (T1, T2) and two samples at the downstream area which goes down past the confluence (D1, D2). In addition, seven sediment and soil samples were collected from mountainside upper than the mine entrance (U1, U2), creeks flowed from a farm house (F) and artificial water reservoir (W1, W2) and agricultural fields nearby the main stream (A1, A2) to investigate the influences from potential contaminant sources.

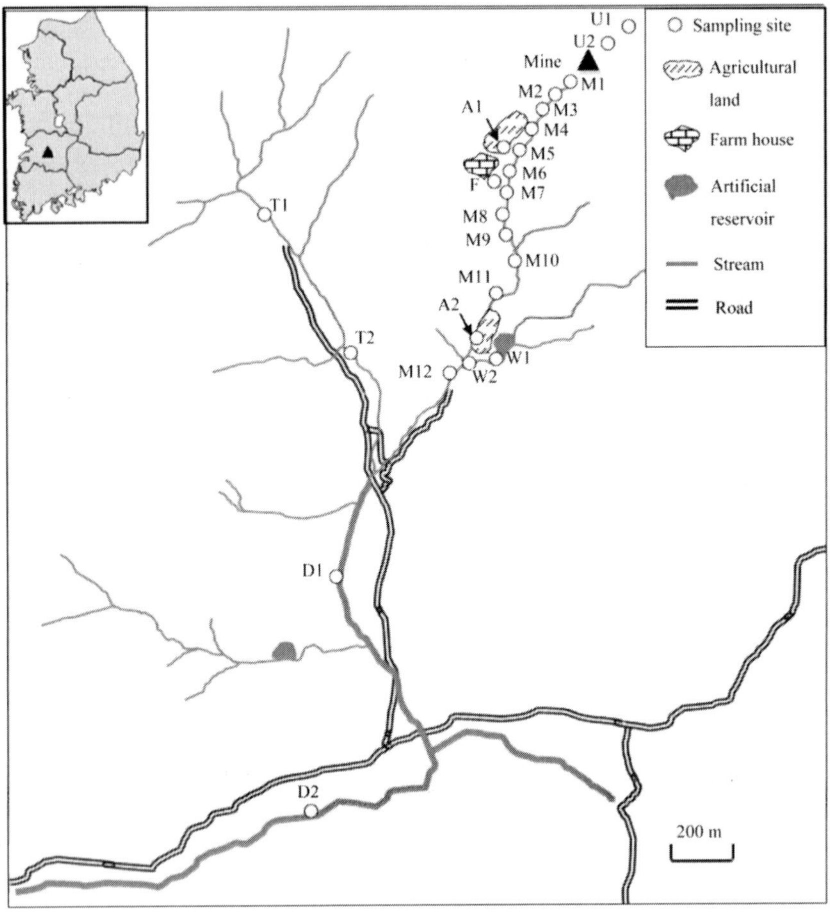

Figure 1. Location of sampling sites near the abandoned mine in southwest of Korea. M1 - M12 are the main stream sediments; U1 and U2 are the mountain soils upper than the mine entrance; A1 and A2 are the agricultural soils; F is the sediments of a creek from the farm house; W1 and W2 are the sediments of the creek from the artificial reservoir; T1 and T2 are the tributary sediments; and D1 and D2 are the sediments of the downstream area.

Each sediment sample was dried at room temperature in the class 1000 clean booth for one week. The dried samples were sieved with non-metallic sieves and acid-digested. The acid digestion carried out in this study followed the procedure modified from the EPA method 3050, 3051 and 3052. Approximately 200 mg of each sample powder was loaded into an acid-cleaned Teflon vessel with 5.0 mL of the mixture of HNO_3 and $HClO_4$

(4:1 v/v), and then dissolved at 175°C for 12 hours. The digested samples were dried down on a graphite heating block at 175°C, reconstituted in 5.0 mL of the mixture of HF and $HClO_4$ (3:1 v/v), and re-dissolved at 175°C for 12 hours. The samples were then dried again and sequentially re-dissolved by 1.0 mL of HCl and 5.0 mL of the mixture of $HClO_4$ and H_3BO_3 (1:3 v/v). Finally, the samples were diluted with 10 mL of 1% HNO_3 and then centrifuged at 3000 rpm for 10 minutes. About 8 mL of supernatant was collected from each sample for the measurement of Pb concentrations and isotope ratios.

2.2. Chemical Separation

Since MC-ICP-MS performance can be affected by various spectral interferences, a targeted element should be separated from the sample matrix [12]. The Pb in each sample was separated through anion exchange column separation. The column purification was done as below: 2 mL of Biorad® AG-MP1 resin was packed in Savillex Teflon micro column and cleaned using HNO_3, HCl and HF [13]. Before column separation, the resin was conditioned with 10 mL of 7 N HCl and 0.001% H_2O_2 solution mixture; and 1 mL aliquot of each digested sample was dried and redissolved in 1 mL of that solution. Then, it was loaded into a column. The elution was done in the following sequence: 10 mL of 7 N HCl + 0.001 % H_2O_2 (for Pb and the matrix), 24 mL of 7 N HCl + 0.001 % H_2O_2 (for Cu), 12 mL of 1 M HCl (for Fe), 8 mL of 0.01 M HCl (for Zn), 8 mL of HCl (for the matrix) and 10 mL of 0.001 M HCl (for Cd). Because the modified procedure aimed to perform multi-separation of Pb, Cu, Fe, Zn and Cd through one stage, Pb and the matrix were separated together from other elements such as Cu, Fe, Zn and Cd at the first elution. In order to separate Pb from the sample matrix, second elution was done using 4 mL of 1 N HCl (for the matrix) and 8 mL of 0.1 N HCl (for Pb).

2.3. Analysis of Pb Concentrations and Isotope Ratios

Pb concentrations and isotope ratios were analyzed with the ICP-OES (Optima 5300 DV, Perkin Elmer) and the MC-ICP-MS (Nu Plasma II, Nu), respectively, at the National Institute of Environmental Research (NIER). The DSN-100 desolvating system including a micromist nebulizer was used for sample introduction of the MCICP-MS. Typical instrumental parameters of the MCICP-MS are presented in **Table 1**. The mass bias during the analysis of Pb isotope ratios was corrected using NIST SRM 997 Tl spiked in samples. In addition, the Pb isotope ratios of NIST SRM 981 Pb spiked with NIST SRM 997 Tl were measured after every five samples to check the instrumental drift. The $^{206}Pb/^{207}Pb$ and $^{208}Pb/^{207}Pb$ ratios of NIST SRM 981 determined in this research were 1.0932 ± 0.0001 and 2.3688 ± 0.0003, respectively. These values well agreed with the previous research [14]. The variations in Pb isotope ratios of NIST SRM 981 were less than 130 ppm, and thus the influence from the instrumental drift was not corrected.

Table 1. Instrument operating conditions of the MC-ICP-MS.

Instrument settings	
RF power	1300 W
Coolant gas flow	13.0 L/min
Auxiliary gas flow	0.8 L/min
Cone (Sampler + Skimmer)	Ni
Analyzer pressure	6.0×10^{-9} mbar
Signal intensity	32 ~ 354 V/ppm
Sample introduction	
Desolvating system	DSN-100
Nebulizer type	GE micromist
Nebulizer pressure	30.5 psi
Sample uptake rate	100 μL/min
Spray chamber temperature	110°C
Membrane temperature	110°C
Hot gas flow	0.3 L/min
Membrane gas flow	2.30 - 2.43 L/min
Sample analysis	
Scans	25 scans per block
Block	2 blocks
Cup configuration	H2 (^{208}Pb), H1 (^{207}Pb), Ax (^{206}Pb), L1 (^{205}Tl), L2 (^{204}Pb), L3 (^{203}Tl), L4 (^{202}Hg)
Mass bias correction	Tl internal standard normalization

3. RESULTS AND DISCUSSION

3.1. Pb Concentrations of Stream Sediments

The Pb concentrations and isotopic compositions of the sediment and soil samples are summarized in **Table 2**. The Pb concentration data clearly showed the enrichment of Pb in the main stream sediments compared with the other sites (**Figure 2**). The Pb concentrations determined in the main stream sediments ranged from 139.0 - 1079.43 mg·kg^{-1}. These values were similar to the Pb concentrations in the soils and sediments at various mine impacted areas in Korea [10,11,15]. On the other hand, the soil and sediment samples collected at the other sites showed relatively low Pb concentrations (18.56 - 29.52 mg·kg^{-1}) similar to the mean concentration of Pb in the upper continental crust (17 mg·kg^{-1}) [16]. Since the agricultural soil near to the mine entrance (A1) showed low Pb concentrations, it can be inferred that the Pb pollutant from the mine was transported only through the main stream, and thus atmospheric pollutants deposition and/ or dispersal of soluble Pb was negligible in this area. Among the other sites, D1 displayed slightly increased concentration, and this might result from confluence of the main stream.

3.2. Spatial Distribution of Pb Isotope Ratios

The most striking features in Pb isotopic compositions were high $^{208}Pb/^{207}Pb$ ratios and low $^{206}Pb/^{207}Pb$ ratios of the main stream sediments compared with the other sites (**Table 2**, **Figure 2**). The $^{208}Pb/^{207}Pb$ ratios of the main stream sediments ranged from 2.4854 to 2.4883, and only three sites (D2, W1 and W2) among the other sites showed higher $^{208}Pb/^{207}Pb$ ratios than those of the main stream (**Table 2**, **Figure 2**). According to [17], the higher abundance in ^{208}Pb was the characteristic feature of Korean Pb ores. Thus high $^{208}Pb/^{207}Pb$ ratios observed in the main stream sediments suggest the influence of Pb from the closed mine over these sites. The $^{206}Pb/^{207}Pb$ ratios of the main stream sediment ranged from 1.1693 to 1.1740 were the lowest values determined in this research. Unfortunately, Korean Pb ore are known to have a large variation in $^{206}Pb/^{207}Pb$ ratios [17-19]. Therefore it is difficult to characterize $^{206}Pb/^{207}Pb$ ratios of Pb ore. However, the $^{206}Pb/^{207}Pb$ ratios of the main stream sediment samples quite well met those of Pb ores in several mines located in the southwest province of Korea (Figure 3). In addition, the $^{206}Pb/^{207}Pb$ ratios of the marine sediment in the Southern East Sea decreased during 1930~1990 indicating low $^{206}Pb/^{207}Pb$ ratios of anthropogenic Pb in Korea [19]. Therefore it can be inferred that low $^{206}Pb/^{207}Pb$ ratios of main stream sediments represent the contributions of Pb from the closed mine to these sites.

Table 2. The concentrations and isotopic compositions determined in soil and sediment samples. The uncertainties of data (95% significance level) are 13.5% for Pb concentrations, 0.1% for $^{208}Pb/^{207}Pb$ and 0.2% for $^{206}Pb/^{207}Pb$ based on duplicate measurements of M9 sample.

Site	Pb concentration ($mg \cdot kg^{-1}$)	$^{208}Pb/^{207}Pb$	$^{206}Pb/^{207}Pb$
The main stream			
M1	189.88	2.4867	1.1710
M2	1079.43	2.4881	1.1697
M3	185.23	2.4877	1.1716
M4	809.78	2.4883	1.1693
M5	253.80	2.4880	1.1720
M6	465.32	2.4867	1.1700
M7	261.44	2.4875	1.1713
M8	150.51	2.4873	1.1729
M9	169.61	2.4869	1.1731
M10	243.43	2.4874	1.1726
M11	225.53	2.4873	1.1740
M12	139.00	2.4854	1.1737
The tributary stream			
T1	20.65	2.4798	1.1838
T2	21.52	2.4803	1.1865
The downstream area			
D1	29.52	2.4825	1.1804
D2	23.93	2.4951	1.1947

Upper mountain			
U1	26.73	2.4763	1.1835
U2	26.41	2.4749	1.1856
Agricultural soil			
A1	18.66	2.4684	1.2299
A2	22.78	2.4808	1.2040
Sewage from farmhouse			
F	18.55	2.4654	1.2173
Artificial pool			
W1	24.96	2.4950	1.1864
W2	23.38	2.5033	1.1843

The Pb isotope ratios of sediments collected from the other sites such as tributary stream, the downstream area and the creeks showed large variations depending on sites (**Figure 2**). Since no significant pollutions were found in corresponding sites, the differences between the sites might result from speciation of natural Pb after deposition. In this research, since all samples were total digested using HF, it was impossible to investigate the speciation of Pb in these sites.

3.3. Source Identification and Binary Mixing System

Figure 3 shows a three-isotope plot of $^{206}Pb/^{207}Pb$ versus $^{208}Pb/^{207}Pb$ ratios for all soil and sediment samples. The Pb isotopic compositions of the main stream sediments (M1 - M12) were close to the isotopic signatures of the Pb ores in the mines located in the southwest of Korea [18], while the Pb isotope ratios of the mountain soils upper than the mine entrance (U1 and U2) were clearly different from those of the main stream (**Figure 3**). This indicates that the Pb in the main stream sediments mainly originated from the closed mine.

In **Figure 3**, the Pb isotope ratios of the main stream sediments (M1 - M12), the tributary stream sediments (T1 and T2) and the downstream sediment sample (D1) display good linear correlation ($R^2 = 0.9185$, $p < 0.001$). In addition, the isotopic composition of D1 is plotted between those of the main stream and the tributary stream sediments. This suggests that the isotopic composition of the Pb in the downstream area resulted from mixing of the main stream and tributary stream. As mentioned above, the Pb in the main stream sediments dominantly came from the closed mine.

On the other hand, the Pb in the tributary stream sediments were probably natural Pb, because there were no pollution sources around the tributary stream. The low concentrations of Pb at T1 and T2 similar to the mean concentration of global continental crust also support little contribution of anthropogenic Pb in the tributary stream. Therefore, it can be inferred that the Pb in the downstream area were the mixture of anthropogenic Pb from the closed mine and natural Pb from the tributary stream.

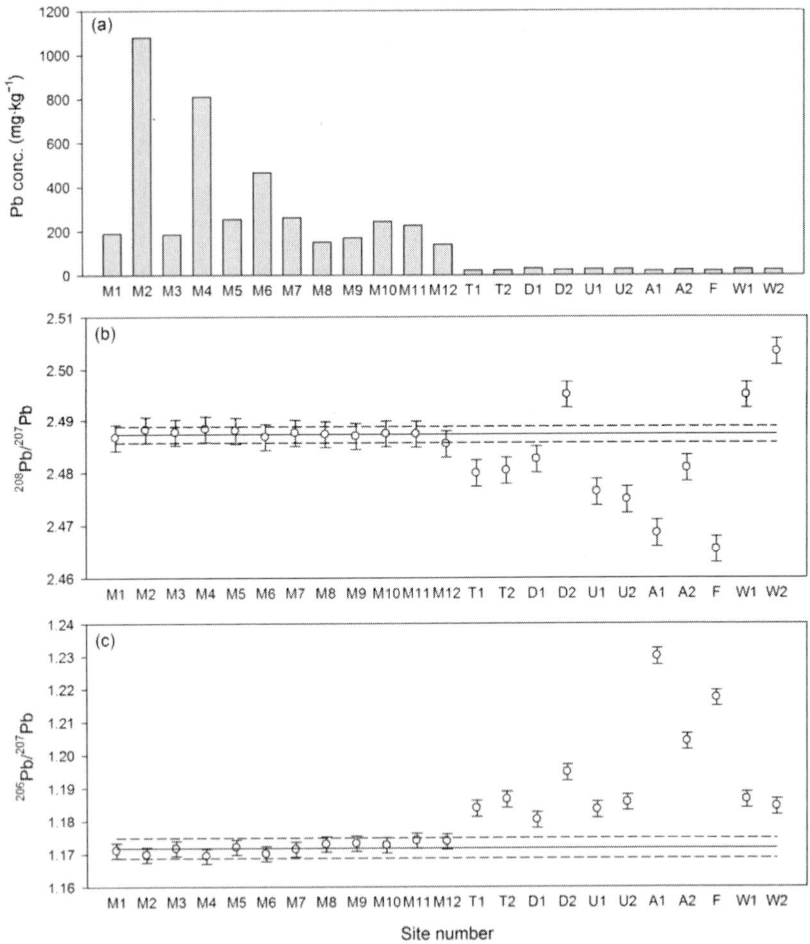

Figure 2. Pb concentrations and isotope ratios determined in sediment and soil samples. Each error bar represents uncertainty based on the duplicate measurements of M9 sample.. The solid lines and dashed lines represent the mean values and 2 σ values of whole Pb isotope ratios of the main stream sediments, respectively.

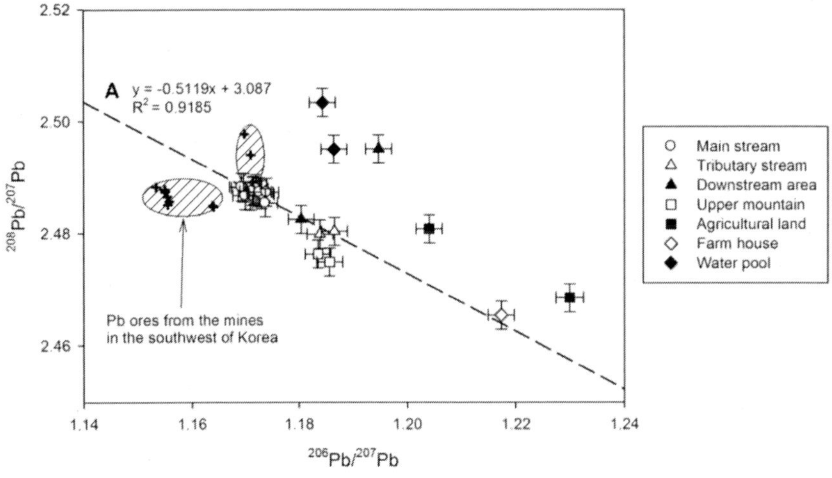

Figure 3. The three-isotope plot for $^{206}Pb/^{207}Pb$ vs. $^{208}Pb/^{207}Pb$ in soil and sediment samples. The dashed line (A) represents the linear regression of Pb isotopic composition of the main stream sediments (M1-M12), the tributary stream sediments (T1-T2) and one of the downstream area (D1), while the dash-dotted line (B) represents that of artificial water pool, (W1, W2), one of the agricultural soil (A2) and sewage from the farmhouse (F). The crosses in the shaded area were the Pb isotope ratios of Pb ores in several mines located in southwest of Korea [18].

According to [19], if samples were mixtures of two component end members such as natural and anthropogenic Pb, linear trend can be obtained from the isotope ratio and the inverse of concentration. In this research, high linear correlation ($R^2 = 0.9492$, $p < 0.001$) between M1 ~ M12, T1 ~ T2 and D1 was also observed in the isotope ratio versus inverse of concentration plot (**Figure 4**). In two end member mixing system, the relative contribution from each end member can be calculated by a simple binary mixing equation as below [20]:

$$F_M \times \left({}^{208}Pb/{}^{207}Pb \right)_M + F_T \times \left({}^{208}Pb/{}^{207}Pb \right)_T = \left({}^{208}Pb/{}^{207}Pb \right)_D \quad (1)$$

where F_M and F_T represent the relative contributions of anthropogenic Pb from the closed mine and natural Pb from the tributary stream, respectively; $(^{208}Pb/^{207}Pb)_{M,\ T,\ D}$ are the isotope ratio of each end member and mixture sample (D1). Since there were no additional sources, the boundary condition, $F_M + F_T = 1$, can be applied, and the equation (1) can be solved as below:

$$F_M = \frac{\left({}^{208}Pb/{}^{207}Pb\right)_D - \left({}^{208}Pb/{}^{207}Pb\right)_T}{\left({}^{208}Pb/{}^{207}Pb\right)_M - \left({}^{208}Pb/{}^{207}Pb\right)_T} \quad (2)$$

According to the above calculation with the mean $^{206}Pb/^{207}Pb$ and $^{208}Pb/^{207}Pb$ ratios, the relative contributions of the abandoned mine to the downstream area (D1) are about 36 and 33%, respectively. In our knowledge, this is the first assessment of a relative contributions from a closed mine in Korea. Thus, no data are available to compare with. In addition, the influence from a closed mine can be changed according to weather conditions and/or geochemical characteristics of a region [21]. However, the similar results calculated with different isotope ratios may represent authenticity of our assessment.

4. CONCLUSION

The high concentrations of Pb up to 1080 mg·kg^{-1} in the main stream sediments represent the contamination by the Pb pollutants from the closed mine. It is also supported by the Pb isotope ratios of the main stream sediments which are similar to Pb ores of mines in southwest of Korea. On the other hand, the influence of the closed mine is not appeared in the tributary stream and other creeks around the main stream. The Pb isotopic compositions of the stream sediment samples also suggest a twosource mixing system between the main stream, the tributary stream and the downstream area. The relative contribution of Pb from the mine to the downstream area reaches up to 36%. However, it can be changed according to the weather condition. Conclusively, this study reveals that the Pb isotope ratios are useful tracer not only for identification of pollutant source but also for assessment of its relative contribution.

ACKNOWLEDGEMENTS

The results presented in this paper were obtained in the framework of the project "Improvement in Precision and Accuracy of Isotope Jong-Woo Choi, Eun-Jin Yoo, Jee-young Kim, Jong-Yeon Hwang, Khanghyun Lee, Won-Seok Lee, Jin-Seok Han, Kyu-Seung Lee Analysis of Heavy Metals by MC-ICP-MS" which is financially supported by the National Institute of Environmental Research.

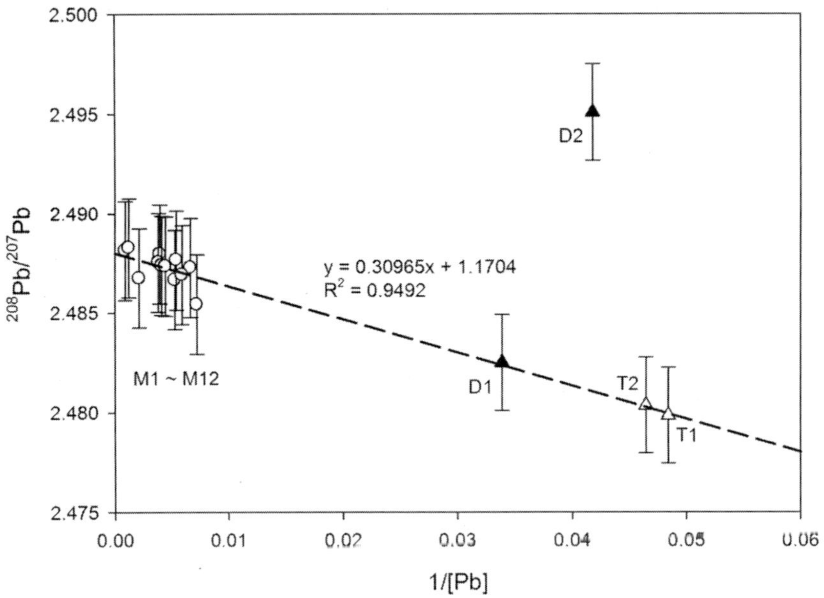

Figure 4. Inverse concentration plot versus $^{208}Pb/^{207}Pb$ ratios of the main stream, the tributary stream and downstream area sediment samples. The dashed line represents the linear regression between these sites except D2.

REFERENCES

1. MacKenzie, A.B. and Pulford, I.D. (2002) Investigation of contaminant metal dispersal from a disused mine site at Tyndrum, Scotland, using concentration gradients and stable Pb isotope ratios. Applied Geochemistry, 17, 1093- 1103.
2. Ettler, V., Mihaljevič, M., Šebek, O., Molek, M., Grygar, T. and Zeman, J. (2006) Geochemical and Pb isotopic evidence for sources and dispersal of metal contamination in stream sediments from the mining and smelting district of Příbram, Czech Republic. Environmental Pollution, 142, 409-417.
3. Arcega-Cabrera, F., Armienta, M.A., Daesslé, L.W., Castillo-Blum, S., Talavera O. and Dótor, A. (2009) Variations of Pb in a mine-impacted tropical river, Taxco, Mexico: Use of geochemical, isotopic and statistical tools. Applied Geochemistry, 24, 162-171.
4. Fernández-Caliani, J.C., Barba-Brioso, C., González, I. and Galán, E. (2009) Heavy metal pollution in soils around the abandoned mine sites of the Iberian pyrite belt (Southwest Spain). Water Air and Soil Pollution, 200, 211-226.

5. Rodríguez, L., Ruiz, E., Aloso-Azcárate, J. and Rincón, J. (2009) Heavy metal distribution and chemical speciation in tailing and soils around a Pb-Zn mine in Spain. Journal of Environmental Management, 90, 1106-1116.
6. Chrastný, V., Vaněk, A., Teper, L., Cabala, J., Procházka, J., Pechar, L., Drahota, P., Penížek, V., Komárek, M. and Novák, M. (2012) Geochemical position of Pb, Zn and Cd in soils near the Olkusz mine/smelter, South Poland: Effects of land use, type of contamination and distance from pollution source. Environmental Monitoring Assessment, 184, 2517-2536.
7. Komárek, M., Ettler, V., Chrastný, V. and Mihaljevic, M. (2008) Lead isotopes in environmental sciences: A review. Environment International, 34, 562-577.
8. Hansmann, W. and Köppel, V. (2000) Lead-isotopes as tracers of pollutants in soils. Chemical Geology, 171, 123- 144.
9. Korea Environment Corporation (2007) Investigation of soil pollution in closed mines. http://attfile.konetic.or.kr/konetic/xml/MARKET/51A4A0850074.pdf
10. Jung, M.C. (2001) Heavy metal contamination of soils and waters in and around the Imcheon Au-Ag mine, Korea. Applied Geochemistry, 16, 1369-1375.
11. Kim, J.-Y., Kim, K.-W., Lee, J.-U., Lee, J.-S. and Cook, J. (2001) Assessment of As and heavy metal contamination in the vicinity of Duckum Au-Ag mine, Korea. Environmental Geochemistry and Health, 24, 215-227.
12. Yang, L. (2009) Accurate and precise determination of isotopic ratios by MC-ICP-MS: A review. Mass Spectrometry Reviews, 28, 990-1011.
13. Lacan, F., Francois, R., Ji, Y. and Sherrell, R.M. (2006) Cadmium isotopic composition in the ocean. Geochimica et Cosmochimica Acta, 70, 5104-5118.
14. Cocherie, A. and Robert, M. (2007) Direct measurement of lead isotope ratios in low concentration environmental samples by MC-ICP-MS and multi-ion counting. Chemical Geology, 243, 90-104.
15. Jung, M.C. (2008) Heavy metal concentrations in soils and factors affecting metal uptake by plants in the vicinity of a Korean Cu-W mine. Sensors, 8, 2413-2423.
16. Wedepohl, K.H. (1995) The composition of the continental crust. Geochimica et Cosmochica Acta, 59, 1217- 1232.
17. Mukai, H., Furuta, N., Fujii, T., Ambe, Y., Sakamoto, K. and Hashimoto, Y. (1993) Characterization of sources of lead in the urban air of Asia using ratios of stable lead isotopes. Environmental Science and Technology, 27, 1347-1356.
18. Chung, J.-I. and Park, K.-H. (2006) Pb isotopic composition of the ore deposits distributed in Jeonbuk Province. Journal of Petrology Society of Korea, 15, 81-89.
19. Choi, M.-S., Yi, H.-I., Yang, S.Y., Lee, C.-B. and Cha, H.-J. (2007) Identification of Pb sources in Yellow sea sediments using stable Pb isotope ratios. Marine Chemistry, 107, 255-274.
20. Rosman, K.J.R. (2001) Natural isotopic variations in lead in polar snow and ice as indicators of source regions. In: Caroli, S., Cescon, P. and Walton, D.W.H., Eds., Environmental Contamination in Antarctica: A Challenge to Analytical Chemistry, Elsevier, Oxford, 87-106.
21. Lee, H., Kabir, M.I., Kwon, P.S., Kim, J.M., Kim, J.G., Hyun, S.H., Rim, Y.T., Bae, M.S., Ryu, E.R. and Jung, M.S. (2009) Contamination assessment of abandoned mines by integrated pollution index in the Han River watershed. The Open Environmental Pollution & Toxicology Journal, 1, 27-33.

CHAPTER 15

Improved soil carbonate determination by FT-IR and X-ray analysis

Viktor J. Bruckman, Karin Wriessnig

1 Austrian Academy of Sciences (O¨AW), Commission for Interdisciplinary Ecological Studies (KIOES), Dr. Ignaz Seipel-Platz 2, 1010 Vienna, Austria

2 Institute of Applied Geology, University of Natural Resources and Life Sciences, Peter Jordan-Strasse 70, 1190 Vienna, Austria

ABSTRACT

In forest soils on calcareous parent material, carbonate is a key component that influences both chemical and physical soil properties and thus fertility and productivity. At low organic carbon contents, it is difficult to distinguish between organic and inorganic carbon, e.g. carbonates, in soils. The common gas-volumetric method to determine carbonate has a number of disadvantages. We hypothesize that a combination of two spectroscopic methods, which account for different forms of carbonate, can be used to model soil carbonate in our region. Fourier transform mid-infrared spectroscopy was combined with X-ray diffraction to develop a model based on partial least squares regression. Results of the gas-volumetric Scheibler method were corrected for the calcite/dolomite ratio. The best model performance was achieved when we combined the two analytical methods using four principal components. The root mean squared error of prediction decreased from 13.07 to 11.57, while full cross-validation explained 94.5 % of the variance of the carbonate content. This is the first time that a combination of the proposed methods has been used to predict carbonate in forest soils, offering a simple method to precisely estimate soil carbonate contents while increasing accuracy in comparison with spectroscopic approaches proposed earlier. This approach has the potential to complement or substitute gas-volumetric methods, specifically in study areas with low soil heterogeneity and similar parent material or in long-term monitoring by consecutive sampling.

Keywords: Forest soil, Carbonate, Fourier, transform, mid-infrared spectroscopy, X-ray diffraction, Partial least squares regression Scheibler method

1. INTRODUCTION

Carbonate is a key component of soils on calcareous parent material, influencing both chemical and physical soil properties and hence fertility and productivity. It directly affects the soil pH and buffer capacity and is in strong reciprocation with biochemical cycles. Carbonate promotes the formation of stable soil aggregates and influences physical soil properties such as hydraulic characteristics and erodibility. It typically occurs as calcite ($CaCO_3$) and/or dolomite ($CaMg[CO_3]_2$), depending on the parent material and soil formation processes. A number of different methods of carbonate determination in soils were proposed over the last decades, principally based on approaches in which released CO_2 gas is directly measured either volumetrically or gravimetrically as a consequence of mass loss after treatment with acids. Other approaches are spectroscopic by means of Fourier transform infrared (FT-IR) or Raman spectroscopy or X-ray diffractometry (XRD) (Gunasekaran et al. 2006). A good overview of common methods is presented in Kamogawa et al. (2001). The gas-volumetric Scheibler method (ON L 1084-99, 1999) is still widely used, despite its weaknesses, for example, analytical precision, time consumption, and varying reaction times based on the present type of carbonates. Low solubility of dolomite in 10 % HCl leads to longer reaction times as compared to calcite (Schlichting et al. 1995), and consequently, total carbonate might be underestimated or organic carbon might be overestimated (after subtraction of inorganic carbon from total carbon). FT-IR spectroscopy, on the other hand, has proven to have the potential for accurate quantitative estimations (Janik and Skjemstad 1995; Tatzber et al. 2007, 2011; Ji et al. 2009). The current study aims at developing a model for carbonate quantification using a combination of FT-IR spectroscopy and XRD. Reference material is forest soil from a series of recently established long-term observation plots for studying woody biomass production. A similar approach was recently used in aquatic ecosystems for the identification of Heinrich events in marine deposits (Ji et al. 2009). We hypothesize that the determination of the form of carbonate actually occurring on our plots (calcite, dolomite) would increase the accuracy of predicting carbonate contents in our model.

2. EXPERIMENTAL

2.1 Soil sampling and preparation

The soil samples were collected in Eastern Austria, in a *Quercus* dominated high forest management system. The parent material consists of gravel, sand and silt built up during the Pannonium as a result of the early development of the Danube River. Parts of the deposited material originate from the northern limestone Alps, therefore containing considerable amounts of carbonate. Younger Aeolian deposits of loess (Pleistocene) from the same source region led to periglacial formation of Chernozems. A detailed site description and soil characterization as well as belowground organic carbon dynamics can be found in Bruckman et al. (2011). Only soils that obviously contained carbonates (determined by a pH [$CaCl_2$] value > 6 (Cools and de Vos 2010)), which is the case for approximately 10 % of all samples, were selected for this study. As a consequence of the geological background and diverse parent material, both calcite and dolomite are present in the soil matrix. Three soil reference groups were identified according to WRB (IUSS Working Group WRB 2006) across our study region, where five plots are classified as Cambisols, three as Luvisols and one as Chernozem. Mineral soil cores were divided into five geometric horizons, 0–5, 5–10, 10–20, 20–40 and 40–50 cm soil depth, and were analysed separately. Soil samples were sieved to 2 mm, and a subsample was ground to powder for Fourier transform mid-infrared (FT-MIR) and XRD analysis.

2.2 Gas-volumetric Scheibler method

The gas-volumetric determination of carbonate (Scheibler method) was conducted following ÖN L 1084-99 (1999), also described in Tatzber et al. (2007). Carbonate contents calculated from the Scheibler method were corrected for the dolomite content of the respective samples using a stoichiometric coefficient based on the calcite/dolomite XRD peak area ratios. Ambient air pressure and temperature were recorded for each sample in order to calculate the pT coefficient, which corrects the volume of CO_2 for current temperature and air pressure during the reaction. The reaction time was extended in order to allow complete dissolution of dolomite in the samples.

2.3 Fourier transform mid-infrared spectroscopy

KBr pellets were produced to measure samples in transmission mode in the mid-infrared area, that is, 400–4,000 cm^{-1}, using a Bruker Optics Tensor 27 spectrometer. Approximately 1 mg bulk soil sample, which was dried at 70 °C for at least 12 h prior to measurement, was mixed with 200 mg FT-IR grade KBr (both weighed exactly) and ground in an agate grinding mill. The mixed bulk soil and KBr powder was immediately pressed to obtain a transparent pellet and measured. According to our observation, it is crucial to store KBr and all manipulation tools such as press chambers and cylinders at 70 °C as this promotes sintering during pressing and results in completely transparent pellets of high quality. Sixteen scans were performed for each sample and averaged. The spectra were corrected against a pure KBr pellet and ambient air. A background spectrum was measured at least every 30 min. All spectra were baseline corrected before conducting further analysis. According to Beer's law, the magnitude of a signal is proportional to the concentration of the respective component causing IR absorption at its specific wavelength, allowing quantitative conclusions. For this reason, absorption spectra were used for analysis, rather than transmission spectra that are shown in Figure 1. Six indicative peak regions were selected for integration with base points at 2,686 and 2,460 cm^{-1} (Peak I), 1,850 and 1,784 cm^{-1} (Peak II), 1,567 and 1,295 cm^{-1} (Peak III), 889 and 867 cm^{-1} (Peak IV), 734 and 719 cm^{-1} (Peak V), and 719–708 cm^{-1} (Peak VI). Location of peaks was derived from Ji et al. (2009) and Tatzber et al. (2010). While peaks I–IV represent a combination of calcite and dolomite, which is not clearly separable, V is indicative of dolomite and VI of calcite. The exact location of the peaks and base points was set according to our own results; hence, they are not necessarily convergent to the cited papers.

2.4 Powder X-ray diffraction

Powder XRD was performed with ground bulk soil samples using a PANalytical X'pert Pro Diffractometer with CuKα radiation at a tension of 45 kV, and a current of 40 mA. Scanning was performed from 5° to 70°2θ at an interval of 0.017°, with a measuring time of 25 s per step. X-Ray diffraction allows a distinct differentiation between calcite and dolomite and calculation of respective ratios, which could be used to raise accuracy of carbonate determination. Approximately two grams of ground sample material is needed, which can be recycled and used for further analysis since the measurement is non-destructive. Sample preparation is simple and measurement is relatively fast if newer instrumentation is used (~15 min per sample) and no additional consumables are needed. However, the main limitations are its detection limit which is around 1 % by weight and relatively high investment costs for the analytical equipment (Ji et al. 2009). Two indicative peaks were selected for integration, representing calcite at 29.5°2θ (Peak VII) and dolomite at 31°2θ (Ji et al. 2009). Base points were set individually for each sample, representing local minima before and after the peak location.

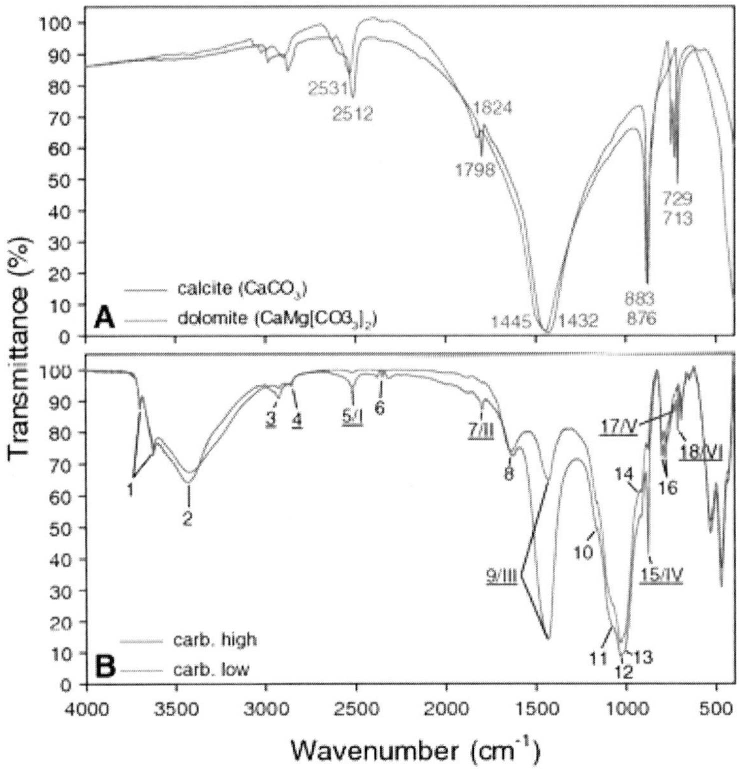

Figure 1. Pure carbonate and bulk soil infrared spectra. **A** Mid-infrared (MIR) range spectrum of pure calcite ($CaCO_3$) and dolomite ($CaMg[CO_3]_2$). Calcite and dolomite peaks used in this study are indicated by their respective wavenumber. **B** MIR spectra of bulk soil samples with the lowest (carb. low) and the highest (carb. high) carbonate concentrations of all samples. Carb. high: 19.6 % carbonate by weight, 40–50 cm soil depth, calcite/dolomite ratio = 1.52, Chernozem. Carb. low: 0.5 % carbonate by weight, 0–5 cm soil depth, calcite/dolomite ratio = 0.92, Chernozem. Peaks are indicated for most important infrared-active functional groups occurring in mineral soils (Tatzber et al. 2010).

Indicative peaks for carbonate are *underlined,* and *Roman numbers* represent peaks used in our model. 1 = Si–O–H vibrations of layer silicates; 2 = O–H (H bonded), N–H; 3 = asymmetrical CH_2 stretching, carbonate; 4 = symmetrical CH_2 stretching, carbonate; 5 = carbonate; 6 = ambient CO_2; 7 = carbonate (calcite dominated); 8 = C–O stretch (carboxylates, aromatic vibrations), O–H bending of water; 9 = carbonate, carboxylates; 10 = C–O stretch of OH deformation of COOH; 11 = Si–O–Si, sulphate; 12 = C–O stretch, carbohydrates; 13 = SiO_3^{2-}, CH vibrations; 14 = C–H deformation of aromatics, cyclopentane; 15 = carbonate; 16 = quartz; 17 = dolomite; 18 = calcite

2.5 Partial least squares regression

Partial least squares regression (PLSR) was used to develop a model for carbonate estimation at our study site. PLSR is among the most commonly used procedures for spectral chemometrics, which is also confirmed in the comprehensive review by Rossel et al. (2006). The first step is a reduction in dimensions. Principal components are computed from the predictor values, which are used to predict scores of the dependent values' components. Subsequently, the scores are used to compute the actual dependent values. Advantages of this method are its relative robustness with respect to missing data and noise and its ability to handle multicollinearity among the independent input values. The statistical software R (R Development Core Team 2011) in combination with the PLS package (Mevik and Wehrens 2007) was used to develop our model.

3. RESULTS AND DISCUSSION

3.1 Results obtained by the Scheibler method

Using the Scheibler method, we found a mean carbonate content of 5.4 % by weight, ranging from 0.5 to 19.6 %, which represents contents of 4.6–196.4 mg g^{-1}. The mean calcite/dolomite ratio, based on the integrated XRD bands VII and VIII, was 1.18, when excluding four samples with low dolomite contents which present an elevated mean ratio of 41.67. The four samples were from the plot where the soil was identified as Cambisol. In situ observations during soil sampling showed clear signs of secondary carbonates, that is, carbonates that are formed by biological processes at these rather dry sites and consist of calcite. Earthworm-secreted carbonate could be a significant component of biogeochemical C and Ca cycling (Lambkin et al. 2011). Based on this observation, grinding the soil samples for analysis is strongly suggested after sieving (2 mm).

3.2 Combination of Fourier transform mid-infrared and X-ray diffraction results

Pure calcite and dolomite spectra are shown in Figure 1A. Bands appeared at 2,984, 2,873, 2,512, 1,798, 1,432 cm^{-1} (ν_3), 876 cm^{-1} (ν_2), and 713 cm^{-1} (ν_4). Tatzber et al. (2007) produced very similar results on pure calcite, using a different instrument (Perkin Elmer Paragon 500 Spectrometer), and a detailed description of the underlying functional groups of the respective bands can be found in their study. It seems that dolomite peaks appear at shorter wavelengths, with a mean difference of 21.4 ± 11.1 cm^{-1} in comparison with calcite. Respective dolomite bands appeared at 3,024, 2,902, 2,531, 1,824, and 1,445 cm^{-1} (ν_3), 883 cm^{-1} (ν_2), and 729 cm^{-1}(ν_4). This comparison reveals potential difficulties in setting appropriate base points when integrating bands of bulk soil samples since carbonate typically occurs in different relative amounts of calcite and dolomite (sometimes also other carbonates, e.g. siderite) in soils. As a consequence, bands of bulk soil samples (Figure 1B) are less distinctive and usually broader as compared to pure carbonate spectra and in some cases overlapped by other non-calcareous constituents. This, for instance, in the 3,024–2,873 cm^{-1} band area, which is overlapped by aliphatic groups, causes CH_2 stretching signals. Hence, this area was excluded for carbonate determination based on FT-MIR spectroscopy. We selected an integration mode where the integrated peak area (corrected for the amount of sample) is delimited by the peak itself within the base points and a straight line between them. Integration up to the *x*-axis showed weaker model performances, which was most likely caused by slightly different amounts of sample in each pellet.

While most bands appear in the same magnitude at a given amount of sample, some bands tend to show weaker signals in case of dolomite, which is especially true for the 1,824 cm^{-1}/II band. If only this band is considered for predictive calculations and treated as a common carbonate band, one might underestimate carbonate if the soil contains considerable amounts of dolomite.

X-ray diffractograms (Figure 2) reveal a distinct pattern of mineral soil samples, with clear indicators of considerable amounts of clay minerals (e.g. chlorites) in general and to some extent, more specific details of other types of layer silicates. Being one of the fractions most resistant against chemical weathering in soils, quartz represents the most dominant peaks at 20.9°2θ and 26.6°2θ. Calcite (at 29.5°2θ/VII) and dolomite (at 31°2θ/VIII) are well separated and clearly no overlap occurs, making it comparatively simple to set base points for peak area integration and to determine the calcite/dolomite ratio to improve the Scheibler results. Using this approach, our results were corrected by up to 3.8 %, while the mean correction factor was 0.1 %, depending on the calcite/dolomite ratio and the pT factor during Scheibler measurement.

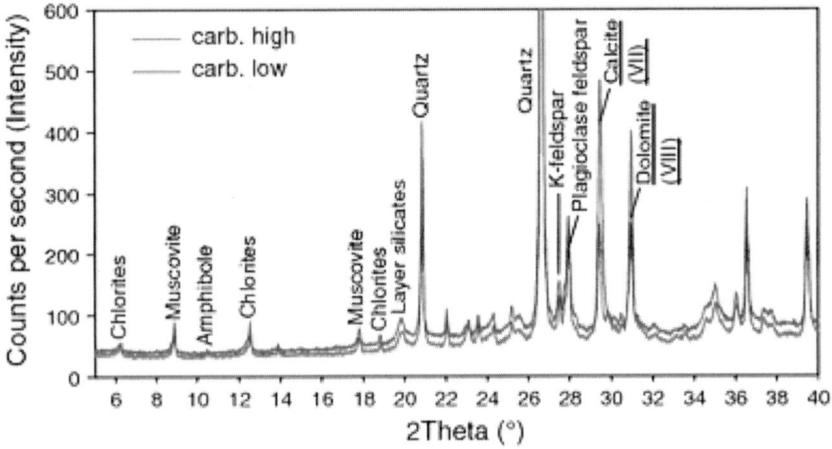

Figure 2. X-ray diffractogram of the same carb. high and carb. low bulk soil samples as shown in Figure 1B. The most important peaks to characterize the soil are indicated. Note the distinct separation of calcite and dolomite

Figure 3 shows the results of the carbonate prediction model including mid-infrared bands and the calcite and dolomite peaks from XRD as predictors and corrected Scheibler carbonate concentrations as the dependent variable; 94.65 % variance of the predictors was explained, employing a model that embraces four principal components. Including more components leads to a higher root mean squared error of prediction and hence a poorer model performance as a consequence of increased model complexity. Our model represents the data set satisfactory, and both inclusions of bands VII and VIII from XRD and the improved Scheibler estimates (stoichiometric coefficient) enhanced overall performance. As a result, the root mean squared error of prediction decreased from 13.07 to 11.57. The improvement was pronounced in samples with higher carbonate contents >100 mg g^{-1}. However, we had to exclude one sample with the highest carbonate content (196.4 mg g^{-1}) from our model calculation as it turned out to be an outlier.

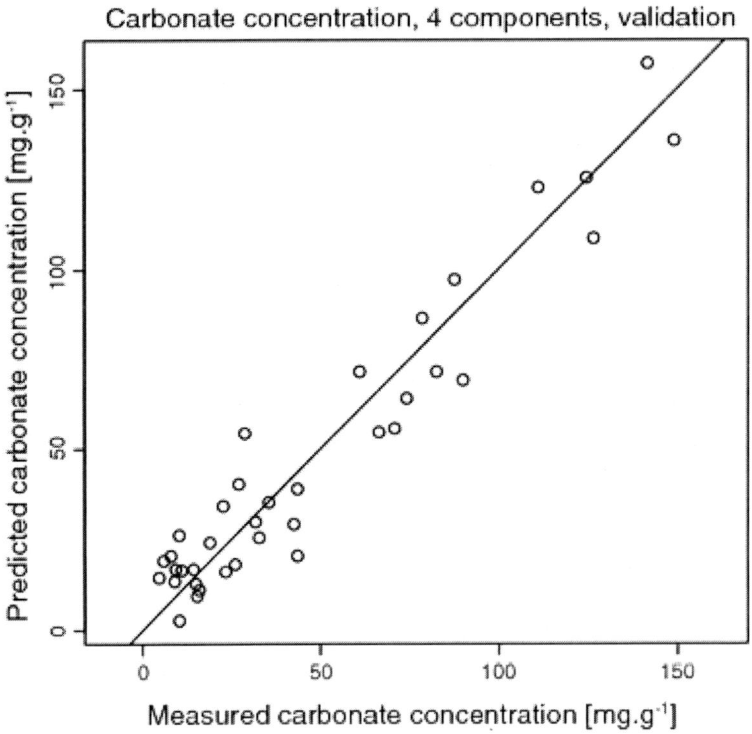

Figure 3. Cross-validated predictions for carbonate contents derived by Scheibler method, corrected for calcite/dolomite ratios. The lowest root mean squared error of prediction (RMSEP (r_{CV}^2) = 11.57) (adj. CV) was derived with a model including four principal components and XRD calcite and dolomite peaks in combination with improved Scheibler carbonate estimates. Cross-validation training explained 100 % of the variance of predictors and 94.5 % of the variance of the dependent (carbonate concentration)

3.3 Interpretation of main model components

Calcite and dolomite represent the first two main components, and a loading plot is shown in Figure 4. While bands I–IV, VI, and VII are indicative of calcite or calcite and dolomite and load predominantly on Component 1, bands V and VIII represent dolomite and predominantly load on Component 2. Depicted positions of soil horizons with their respective pH value in aqueous solution clearly show the trend of increasing carbonate contents and pH values with increasing soil depth. However, it seems that calcite is the main determinant of this observation, which could be explained by its relatively better solubility in weak acids, such as precipitation and humic acids in soil solution as compared to dolomite; hence, it is preferably translocated into deeper soil horizons in forest soils. This illustrates the importance of considering soil inorganic carbon as a combination of different forms of carbonate, influencing soil chemical processes, and utilizing quantitative spectroscopic approaches, and examples from previous research in marine sediments (Ji et al. 2009) underline this proclamation.

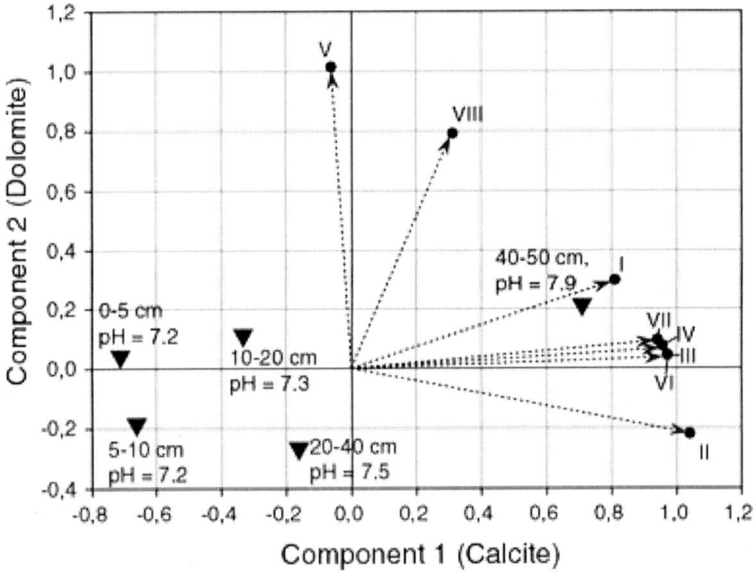

Figure 4. Results of a principal component analysis (PCA) with oblique rotation at KMO = 0.877 and Bartlett's test of sphericity χ^2 (28) = 755.84; $p < 0.001$. Component loadings of mid-infrared and XRD bands are shown (*black dots*, see Experimental section for band assignments and wavelength information) together with genetic soil horizons (*triangles*) and their respective pH value (H_2O)

4. CONCLUSION

For the samples in our study from three different soil reference groups, the model shows a good performance in predicting carbonate contents across the study region. The proposed method can be used as an alternative approach for estimating carbonate content in soils. Here, we show that PLSR is a suitable method, even if the sample size is small, and that the model performance is increased by including information on the actual occurring forms of carbonate, for example, bands V and VII as a proxy for dolomite. As compared to other published spectroscopic approaches, based typically on a single method, a combination of methods has the potential to improve accuracy. However, we suggest a larger study, based on more samples from different soils and vegetation types to validate our results. Samples with high carbonate contents (>15 %) should be included to test the stability of the proposed method at high concentration levels.

ACKNOWLEDGMENTS

This study was funded by the Austrian Academy of Sciences (ÖAW), Commission for Interdisciplinary Ecological Studies (KIÖS) (P2007-07), and the Austrian Federal Ministry of Agriculture, Forestry, Environment and Water Management (Project No. 100185). We explicitly want to thank Michael Tatzber who contributed throughout the project with valuable discussions and methodical hints, Yan Shuai and Yoseph Delelegn for their indefatigable and precise laboratory work, Christina Delaney for proofreading, and anonymous reviewers for valuable comments.

REFERENCES

1. Bruckman VJ, Yan S, Hochbichler E, Glatzel G (2011) Carbon pools and temporal dynamics along a rotation period in Quercus dominated high forest and coppice with standards stands. For Ecol Manag 262:1853–1862
2. Cools N, de Vos B (2010) Sampling and analysis of soil. Manual part X. In: Manual on methods and criteria for harmonized sampling, assessment, monitoring and analysis of the effects of air pollution on forests. UNECE, ICP Forests, Hamburg
3. Gunasekaran S, Anbalagan G, Pandi S (2006) Raman and infrared spectra of carbonates of calcite structure. J Raman Spectrosc 37:892–899
4. IUSS Working Group WRB (2006) World reference bases for soil resources 2006. World soil resources reports no. 103. Food and Agricultural Organization of the United Nations FAO, Rome
5. Janik LJ, Skjemstad JO (1995) Characterization and analysis of soils using mid-infrared partial least-squares. II.*Correlations with some laboratory data. Aust J Soil Res 33:637–650
6. Ji JF, Ge Y, Balsam W, Damuth JE, Chen J (2009) Rapid identification of dolomite using a fourier transform infrared spectrophotometer (FTIR): a fast method for identifying Heinrich events in IODP Site U1308. Mar Geol 258:60–68
7. Kamogawa MY, Nogueira ARA, Miyazawa M, Artigas J, Alonso J (2001) Determination of soil calcareous efficiency using flow system with pervaporative separation. Anal Chim Acta 438:273–279
8. Lambkin DC, Canti MG, Piearce TG, Hodson ME (2011) Dissolution rates of earthworm-secreted calcium carbonate. Appl Geochem 26(Suppl):67–69

9. Mevik BH, Wehrens R (2007) The pls package: principal component and partial least squares regression in R. J Stat Softw 18:1–24
10. ON L 1084-99 (1999) Chemical analyses of soils—determination of carbonate. In: Austrian Standards Institute (ed) ÖNORM L 1084. Austrian Standards Institute, Vienna
11. R Development Core Team (2011) R: a language and environment for statistical computing. R Foundation for Statistical Computing, Vienna
12. Rossel RAV, Walvoort DJJ, McBratney AB, Janik LJ, Skjemstad JO (2006) Visible, near infrared, mid infrared or combined diffuse reflectance spectroscopy for simultaneous assessment of various soil properties. Geoderma 131:59–75
13. Schlichting E, Blume HP, Stahr K (1995) Bodenkundliches Praktikum—eine Einführung in pedologisches Arbeiten für Ökologen, insbesondere Land- und Forstwirte, und für Geowissenschaftler. Blackwell Wiss.-Verl., Berlin; Wien u.a.
14. Tatzber M, Stemmer M, Spiegel H, Katzlberger C, Haberhauer G, Gerzabek M (2007) An alternative method to measure carbonate in soils by FT-IR spectroscopy. Environ Chem Lett 5:9–12
15. Tatzber M, Mutsch F, Mentler A, Leitgeb E, Englisch M, Gerzabek MH (2010) Determination of organic and inorganic carbon in forest soil samples by mid-infrared spectroscopy and partial least squares regression. Appl Spectrosc 64:1167–1175
16. Tatzber M, Mutsch F, Leitgeb E, Englisch M, Mentler A, Gerzabek M (2011) Mid-infrared spectroscopy for determinations of carbonate in soils-what is important? Am Lab 43:13–17

CHAPTER 16

Mobility of toxic elements in carbonate sediments from a mining area in Poland

Natalia Ospina-Alvarez1, Łukasz Głaz 1, Krzysztof Dmowski, Beata Krasnode2¸bska-Ostre¸ga1

1 Faculty of Chemistry, University of Warsaw, Pasteura 1, 02-093 Warsaw, Poland

2 Faculty of Biology, University of Warsaw, Z_ wirki i Wigury 101, 02-097 Warsaw, Poland

ABSTRACT

The Bolesław Bukowno mining area in Poland is highly polluted by elements such as Zn, Pb, Cd and As. The reactivity and mobility of toxic elements such as Tl are poorly known. Here, we studied by sequential extraction the mobility of As, Cd, Co, Cr, Cu, Mn, Mo, Pb, Tl and Zn in sediments from two water reservoirs near Bukowno. Results show that available As, Co, Mn, Pb and Zn are found in carbonate minerals. Available Cd, Cu and Tl are found in sulphides and organic matter. The extractability of As, Cr, Mo and Tl was rather poor. By contrast, 85 % of total Cd, Pb and Zn was mobile. We discuss Tl and Mo association in carbonate sediments from ore deposits.

Keywords: Sequential extraction, BCR, Mining–tailing area, Thallium, Molybdenum Poland

1. INTRODUCTION

Bolesław–Bukowno mining area (Upper Silesia) is one of the most polluted areas in Poland. Approximately 2.2 and 0.1 million tons of Zn and Pb, respectively, have been smelted there since 1952 (Verner et al. 1996). Extensive sulphide ore mining, flotation treatment and smelting are the major sources of anthropogenic dispersion of metals in the vicinity of the zinc–lead smelters. The semi-liquid post-flotation wastes are deposited in large ponds located at the top of 25–30 m high heap. The heap is considered to be a main source of trace metals pollution in this area. The dried wastes can be easily spread by winds in form of dust in a range of several kilometres, due to vast surface and open character of the heap (Trafas 1996). Several elements such as As, Cd, Co, Cr, Cu, Mn, Mo, Pb, Tl and Zn are of high environmental interest due to their toxicity and therefore should be monitored, especially in polluted areas.

The post-flotation wastes contain dolomite, calcite, barite, sulphide and oxide minerals of iron, zinc and lead ores but also tailing metals such as Cd, Cu and Tl (Lis et al. 2003). These minerals are main components of the studied water reservoir bottom sediment. Trace metals exist mostly as carbonates or sulphides. Even small changes in water properties (pH and/or oxidation potential) can influence the leaching of these metals associated with carbonate and sulphide minerals. In the neighbourhood of the post-flotation waste heap (about 100 and 500 m far from its base), in the ground depression filled with seeping water, two ponds were formed.

Krasnodębska-Ostręga et al. (2005a) determined dissolved element fraction in surface water and total content in sediments for these water reservoirs, and their results pointed out that As, Cd, Cu, Mn, Pb, Zn, Se, Sb and Tl were available in bottom sediments in much higher concentrations than in water. Similar relation between metal content in water and sediments was found in comparable areas in Spain (Marqués et al. 2001) and China (Zhang et al. 2004).

After dissolved total quantification of elements in water and sediments, the next step of monitoring studies in this kind of reservoirs should be an estimation of mobility and fractionation of the studied elements. To estimate the mobile metal fraction available during natural processes in bottom sediments (for example, organic matter decomposition and oxidation of sulphide minerals and/or dissolution of carbonate minerals resulting in pH lowering of water being in equilibrium with the sediments), various reagents have been proposed. Considering the diversity of procedures and the lack of uniformity in different protocols, a sequential extraction scheme of the Standards, Measurement and Testing programme (SM&T, formerly BCR) was developed (Ure et al. 1993). This procedure have been accepted by other specialists and directly applied (Yuan et al. 2004; Zhang et al. 2004; Kowalska et al. 2002) or modified (Sutherland and Tack 2007; Prego et al. 2014; Kaplan et al. 2011). The acetic acid-available metal fraction is widely accepted as mobile fraction of trace metals in solid samples. The hydrogen peroxide-available fraction estimated basing on SM&T sequential extraction scheme is accepted as a metal fraction associated with organic matter and sulphide minerals.

The preliminary study of mineral components of the examined bottom sediments indicated that the sediments contain mostly carbonate minerals. Therefore, the aim of this study was to define the fractionation of As, Cd, Co, Cr, Cu, Mn, Mo, Pb, Tl and Zn in sediments rich in carbonates and poor in reducible minerals. The SM&T procedure of sequential extraction was modified taking in account the properties of investigated sediments.

2. EXPERIMENTAL

2.1 Sampling site

The sampling area is located in southern Poland, near zinc–lead mining area of Bolesław and smelting works in Bukowno (50.29°N, 19.49°W). The region belongs to the Silesian–Cracowian ore deposits of the 'Mississippi Valley' type, and the region is relatively rich in thallium (Lis et al. 2003). The sediment samples were collected from water reservoirs situated 100 m (Buk 2) and 500 m (Buk 3) away from the base on the heap in Bukowno. The ponds were formed in ground depressions filled with seeping waters. There are two main sources of trace metals: sediment–water exchange processes and the emission of metal-rich dust from the heap. The Buk 2 pond is regularly used during spring by frogs and toads as mating and breeding area, and Buk 3 pond is used by local inhabitants for fishing and recreation activities.

Bukowno–Bolesław district is characterised for the presence of minerals related to mining waste and ore smelting. The area is rich in meteoric minerals such as cohenite and graphite, sulphides of Fe, Cu, Pb, Zn (chalcopyrite, pyrrhotite, sphalerite, wüstite), spinels of Cr, Mg, Fe, Mn and Zn (chromite, magnetite), monoxides of Fe, Mg, Mn and Zn (wüstite, periclase) and silicate minerals (augite, omphacite, kirschsteinite). Mineral dressing waste contains mostly dolomite, limonite, Zn carbonates, Pb carbonates, sulphates, pyrite, and zinc and lead sulphides (Lis et al. 2012). Physical properties of the soil in the sampling area are detailed in Gruszecka and Wdowin (2013) (profiles I-A, II-A and III-0). In the first 10 cm of the soil, pH ranged from 6.6 to 8.0, organic matter from 30 to 80 % and reduction potencial from −21 to 21 mV.

2.2 Reagents

Nitric acid (68 %, *d* = 1.42 g mL^{-1}), hydrochloric acid (30 %, *d* = 1.15 g mL^{-1}), hydrofluoric acid (40 %, *d* = 1.13 g mL^{-1}), boric acid (4 %, *d* = 1.44 g mL^{-1}) and acetic acid (96 %, *d* = 1.06 g mL^{-1}) Suprapur® (Merck, Germany) were used. Ammonium acetate and hydrogen peroxide were analytical grade (POCh Gliwice, Poland). Diethylene triamine pentaacetic acid, sodium acetate and sodium chloride were pro analysis grade (Merck). Multi-element standard solutions used for ICP MS calibration were prepared before measurements by dilution from a stock standard solution (SPEX CertiPrep®, Metuchen, NJ, USA). All solutions were prepared with Milli-Q water (Millipore, USA).

2.3 Sediment digestion and sequential extraction

The sediment samples (three mixed sub-samples) were collected using PE vessels from a layer at a depth of 0–10 cm. The sediments with natural water content were open-air dried at the laboratory during 24 h and then oven-dried at 50 °C for 5 h. Dried materials were sieved through a 1-mm sieve. The material was homogenised by shaking. PE vessels with sediments were stored in darkness and ambient temperature until analysis.

Sediments were completely decomposed before determination by inductively coupled plasma mass spectrometry (ICP MS), following a three-step digestion using a microwave-system (Anton Paar, Austria) with high-pressure PTFE vessels. The chosen decomposition method was validated using certified reference materials (Schramel et al. 1996). Sediment samples intake was about 110 mg. Sediment samples were weighed into PTFE vessels, 2 mL of concentrated HNO_3 and 1 mL of concentrated HCl were added to each vessel and screwed up tightly. The HPR rotor with vessels was placed on turntable of microwave oven, and microwave energy was applied in cycles: 5 min: 100 up to 500 W, 10 min: 500 W, 5 min: 500 up to 1,000 W and 45 min: 1,000 W. After 15 min of cooling down, the vessels were opened and 0.5 mL HF was added. The energy was applied again with the same cycles before described. After cooling, 5 ml of saturated H_3BO_3 solution was added and energy was applied as follows: 5 min from 100 up to 500 W and then 60 min at 600 W. After 15 min of cooling, the samples were dissolved in distilled water, quantitatively transferred into volumetric flasks of 30 mL and diluted to the mark.

For sequential extraction, one gram of sediment was transferred to an Erlenmeyer flask and 50 mL of CH_3COOH (0.43 mol L^{-1}) were added. The obtained mixture was shaken in horizontal position (16 h, 200 rpm) by means of a rotary shaker (Elpan 357). After shaking, extracts were immediately filtered through a filter paper previously rinsed with diluted nitric acid, followed by distilled water. The filtrates were collected in polyethylene tubes, and 0.1 mL of concentrated HNO_3 was added.

Extraction with hydrogen peroxide, alone as well as in sequence with acetic acid, was done in quartz tubes. One-gram sample intake was leached with 10 mL of 30 % H_2O_2. The mixture was kept in room temperature for 1 h, until reaction nearly stopped. Then, the mixture was warmed up and kept for 1 h at 80–90 °C. Finally, 40 mL of CH_3COONH_4 (1 mol L^{-1}) were added, and the solution was directly filtered. The extracts were stored at 4 °C until analysis. Blank extractions were carried out for each set of analysis using the same reagents as described above. All the blanks were handled and prepared following the same procedure used for the samples.

2.4 Analytical procedure

Trace elements were determinate by means of ICP MS (SCIEX Elan 6000, Perkin Elmer, USA), according to the following parameters: sweep: 5; replicates: 5; dwell time: 0.1 s; ICP RF power: 1375 W; lens voltage: 11.3 W; nebulizer gas flow: 1.06 L min^{-1}; plasma gas flow: 15.5 L min^{-1}. The isotopes monitored were as follows: As 75, Cd 114, Co 59, Cr 52, Cu 63, Mn 55, Mo 95, Pb 208, Tl 205 and Zn 64. Total Quant™ software was used to automatically correct intensities for interference due to isobaric and molecular ions. Calibration curve method was applied as quantitative analysis approach.

Certified reference material (CRM: Trace elements in water, NIST-SRM 1643d,) was used to assess ICP MS measurements. Limit of detection (LOD) of the analytical procedure was calculated as a mean value plus 3 times the standard deviation of the blank ($\bar{x}\bar{x}$ ± 3 SD, n = 8). Reported values were always above detection limits.

In order to verify ICP MS determinations, Tl and Pb were analysed by means of anodic stripping voltammetry in some solutions after digestion and extraction according to Krasnodębska-Ostręga et al. (2005b) (data not shown). Tl was determined by means of differential pulse anodic stripping voltammetry (DP-ASV) using a μAutolab type II (ECO-Chemie, BV, The Netherlands) coupled with a 663VA Stand electrode system (3 mol L^{-1}Ag/AgCl reference electrode) (Metrohm, Switzerland), according to the following parameters, amplitude: 50 mV; scan rate: 10.5 MV s^{-1}; deposition potential: −0.75 V vs 3 mol L^{-1} Ag/AgCl; deposition time: 2–5 min. 0.015 mol L^{-1} DTPA in 0.06 mol L^{-1} acetate solution (pH = 6.2) was used as supporting electrolyte. Determination of Pb was performed using subtractive anodic stripping voltammetry (S-ASV) with square wave mode (amplitude: 10 mV; step potential: 2.5 mV; frequency: 25 Hz; deposition potential: −0.80 V vs AgQRE; deposition time: 2–5 min). The solution containing 10 mmol L^{-1} NaCl and 10 mmol L^{-1} HNO_3 was not deoxygenated. Silver rods (area: 1 cm^2) dipped directly into solutions were used as working electrode and counter/reference electrode. S-ASV curves were obtained by automatic subtraction of oxidation currents recorded with and without accumulation.

SEM–EDS, scanning electron microscope (Zeiss LEO 435VP) equipped with EDS analyser (Roentec), was used for elementary analyses of main components in bottom sediments. Results of SEM–EDS analysis are presented as measured value ± uncertainty (Smith 2002).

3. RESULTS AND DISCUSSION

The first step of the study was the determination of the main components of the collected sediments (Table 1). The obtained results show that Mn content was lower than 0.1 %, and content of Fe did not exceed 1.9 %. The ratio of Ca:Mg was close to 1, and sulphuric total content was not higher than 0.5 %. These results confirm that the samples were rich in dolomite and poor in sulphide and Fe–Mn minerals.

Mean element content in sediment samples after single and sequential extraction is presented in Table 2. Two-tailed t test were applied to (1) results by ICP MS and ASV (data not shown) and (2) obtained mean values and known values (CRM) (data not shown). Significant differences were not observed between both data sets ($p < 0.05$).

According to the results obtained by SEM–EDS (Table 1), the fractionation study also indicated that bottom sediments contained rather dolomite. Acetic acid is commonly accepted reagent for estimation of easily exchangeable metal in sediments and soils (0.11 and 0.43 mol $L^{-1}CH_3COOH$ for sediments and soils, respectively) (Fletcher 1981; Relić et al. 2013). The almost complete dissolution of the sediment matrix was the first visible effect of the leaching with acetic acid solution. The residue contained mainly silicates and some brown matter. The large amount of Mn extracted with acetic acid could come from Mg/Ca carbonates; Mn^{2+} can substitute Mg^{2+} as well as Ca^{2+} in carbonate minerals, rather Mg in dolomite than Ca in calcite (Arunachalam et al. 1996). The extractability (relative mobility) of Zn and Co was large and reached even 90 % of their total content (Figure 1). These elements are known to be associated with carbonates (Arunachalam et al. 1996; Zhang 2008). While calcite could be dissolved in relatively short time and at high pH, dolomite could be dissolved during longer time and lower pH (Zhang 2008). Therefore use of 0.43 mol L $^{-1}$ acetic acid was suitable for evaluation of mobile fraction in particular type of bottom sediment (Ure et al. 1993).

Table 1. Matrix of elements determination using SEM–EDS

Element/line		C (%) ± UNC	
		Buk 2 pond	Buk 3 pond
C	K-ser	8.05 ± 0.77	12.6 ± 1.1
Ca	K-alpha	1.72 ± 0.12	1.89 ± 0.19
O	K-ser	51.9 ± 4.3	57.8 ± 4.5
Fe	K-alpha	1.45 ± 0.18	1.58 ± 0.29
Mg	K-ser	1.48 ± 0.11	1.62 ± 0.16
Al	K-ser	2.85 ± 0.17	3.12 ± 0.14
Si	K-ser	29.8 ± 1.3	14.9 ± 0.6
S	K-ser	<0.3	0.52 ± 0.06

Data are presented as mean ± UNC

SEM–EDS scanning electron microscope with energy-dispersive detector, *UNC* uncertainty (calculated according to Smith 2002)

Unexpectedly high concentration of Pb, even 80 % of total content in acetic acid extracts (Figure 1) indicated that the oxides minerals are not important for immobilisation of the metal. The published data pointed out that lead in presence of oxides minerals is mainly bound to this phase (Jakubowska et al. 2006; Krasnodębska-Ostręga et al. 2001; Sutherland and Tack 2007). The extractability discussed above and the results obtained using SEM–EDS revealed that metal fraction associated with reducible minerals is not important for immobilisation of lead in the studied bottom sediments.

Table 2. Mean element concentration in sediment samples and extracts

	Buk 2-pond				Buk 3-pond			
	HAc	H_2O_2 – [Sq]	H_2O_2 – [S]	Total	HAc	H_2O_2 – [Sq]	H_2O_2 – [S]	Total
As	3.0 ± 0.2	0.21 ± 0.03	0.22 ± 0.08	22 ± 2	48 ± 2	0.84 ± 0.09	1.2 ± 0.03	320 ± 8
Cd	1.2 ± 0.3	0.17 ± 0.04	0.83 ± 0.07	1.5 ± 0.3	8.4 ± 0.7	90 ± 3	65 ± 2	110 ± 4
Co	1.3 ± 0.3	0.24 ± 0.09	0.82 ± 0.21	5.0 ± 3	2.4 ± 0.8	1.0 ± 0.4	3.1 ± 1.8	3.0 ± 2.1
Cr	1.3 ± 0.5	0.7 ± 0.2	0.7 ± 0.2	12.5 ± 0.5	1.5 ± 0.3	7.0 ± 3.0	8.4 ± 0.5	48 ± 1
Cu	64 ± 4	29 ± 1	15 ± 1	155 ± 43	2.6 ± 0.3	25 ± 1	28 ± 2	45 ± 5
Mn	129 ± 10	25 ± 3	113 ± 21	320 ± 26	760 ± 31	91 ± 4	608 ± 28	900 ± 32
Mo	0.13 ± 0.04	0.19 ± 0.06	0.20 ± 0.04	5.0 ± 0.2	0.17 ± 0.03	1.0 ± 0.2	0.63 ± 0.02	5.0 ± 0.4
Pb	357 ± 19	39 ± 2	37 ± 3	440 ± 37	1,491 ± 67	684 ± 30	48 ± 6	5,070 ± 260
Tl	0.14 ± 0.03	0.05 ± 0.02	0.12 ± 0.02	0.33 ± 0.05	1.2 ± 0.1	3.1 ± 0.1	4.0 ± 0.2	14 ± 3
Zn	1,106 ± 95	267 ± 10	593 ± 22	1,110 ± 110	15,110 ± 917	7,387 ± 188	10,332 ± 230	29,700 ± 210

Buk 2 and Buk 3 ponds correspond to the water reservoirs sampled in the study area of Bukowno–Bolesław (Southern Poland). Extractions were carried out with 30 % H_2O_2, alone as well as in sequence with 0.43 mol L^{-1} CH_3COOH

[Sq]: extraction in sequence scheme; [S]: single extraction. Data are presented in mg kg^{-1} dry weight, as mean ($n = 4$) and standard deviations in parentheses

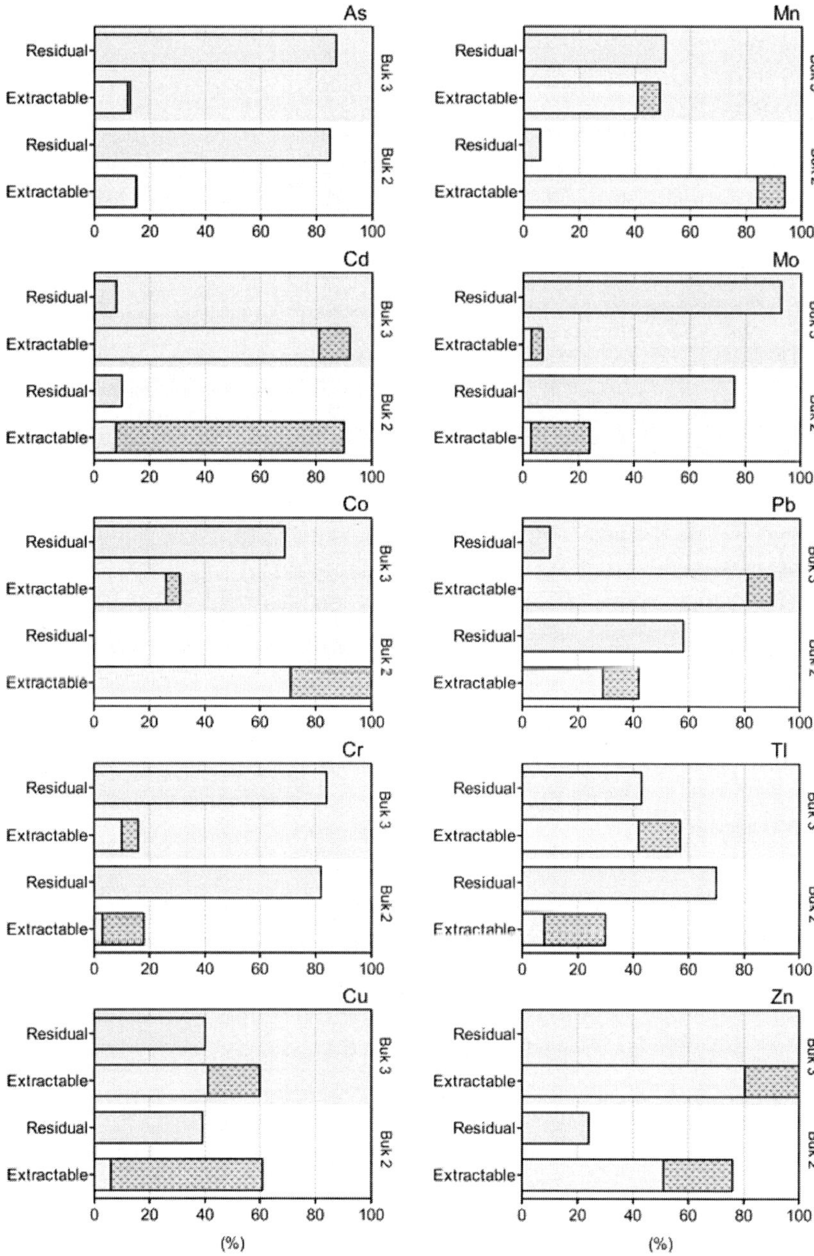

Figure 1. Relative proportion of extractability from bottom sediments of Bukowno–Bolesław mining/smelting area. *Light grey* area corresponds to the sampling site Buk 2 pond, and *dark grey area* to the Buk 3 pond. Extractable fraction corresponds to the sequential extraction carried out

using hydrogen peroxide (*dark blue, dashed pattern*) and acetic acid (*light blue, without pattern*). Values are given in per cent (%) of total content. High extractability could be observed for Cd, Pb and Zn in contrast with the low extractability reached for As, Cr, and Mo, indicating different mechanism of mobilisation for these elements. The data were plotted using GraphPad Prism 5.0 for Mac OS-X (GraphPad Software Inc., 2007). (Color figure online)

The extraction efficiency for 0.1 mol L^{-1} hydroxylamine hydrochloride ($NH_2OH{\cdot}HCl$, pH 2) was also tested, and the results were close to the detection limits for most of studied elements. The relatively high extraction efficiency for Cr (6–15 %), Zn (25 %), Tl (15–22 %), Cu (20–55 %) and Cd (even 80 % for Buk2) obtained using hydrogen peroxide indicated that these elements were associated with the oxidable phase (Figure 1). Studies of fractionation of solid samples by Kowalska et al. (2002) and Yuan et al. (2004), suggest that As, Cd, Cu, Cr, Ni and Pb are strongly bound to organic matter or sulphide minerals in the absence of Mn–Fe oxides phase (Zhang 2008); therefore, taking into account the results achieved with other extractants, the extraction with solution of hydroxylamine hydrochloride should not be considered in sediments with high content of carbonates.

It was noted a low extractability of As (Figure 1), organic and sulphide fraction of As was less then 1 % of total arsenic content. On the other hand, the acetic acid-available arsenic fraction was relatively high—about 15 %. The extractability of As, Cr and Mo estimated basing on two steps stepwise extraction was lower than 20 % (Table 2). Besides Cd, Cu and Tl, the carbonate fraction was greater than the oxidable fraction in the proposed sequential extraction scheme (Table 2). Cd, Pb and Zn were in general the most mobile elements; even 90 % of total content of Cd was extracted in two steps of the proposed sequential extraction (Table 2, Buk 3 in contrast to Buk 2). Comparing the obtained results with the published ones about fractionation of elements based on BCR sequential extraction, (Guevara-Riba et al. 2004; Kowalska et al. 2002; Yuan et al. 2004; Zhang 2008), one might conclude that even small changes in pH and/or oxidising condition in soil environment could cause mobilisation of Cd, Pb and Zn. Apart from the stepwise extraction (acetic acid and hydrogen peroxide), single step extraction with hydrogen peroxide (pH 2) was also carried out. The reagent alone should be capable of simultaneous dissolution of carbonates and oxidising organic mater and sulphides.

In both sediment samples, it was noted that the sum of acetic acid and hydrogen peroxide fractions was always greater than release of the elements in single extraction with hydrogen peroxide solution (Table 2). The difference in leaching is especially noticeable for Pb and Zn. These results indicated that acetic acid solution did affect the trace element bound to organic phase. The hydrogen peroxide solution acidified with nitric acid up to pH 2 was not able to completely leach element fraction associated with acetic acid dissoluble phase of these bottom sediments. It was found for all studied elements that the total content and the mobile fraction (the fraction leached under SM&T extraction conditions) were higher in sediments collected from Buk 3 pond, than from Buk 2 pond. The extractability highly depends on composition of the sediment, and particle-size distribution and organic matter are relevant to the potential metal mobility (de Santiago-Martín et al. 2013); therefore, even small changes in the characteristics of the sediment could affect the extractability of the metals.

Previous studies indicated that the dissolved element fraction in surface water from Buk 3 pond was threefold to fivefold higher than in Buk 2 pond (Krasnodębska-Ostręga et al. 2005a). The extraction study showed that Co, Cd, Cu, Mn, Pb, Tl and Zn mainly bound to carbonates and sulphides (main components in the bottom sediments), are quite mobile. Basing on the fractionation study, it could be concluded that molybdenum extractability was rather poor and similar to As and Cr extractability from sediment matrix rich in carbonates and sulphides, in the absence of oxides and organic matter. The results of Cd and Tl leaching from these two sampling points differ in distance from post-flotation wastes heap, and the activity of live organisms could give some indirect evidence of their association with the sediment matrix. The immobilisation of trace metals depends on the presence of Mn–Fe oxides phase in solid matrix (Fletcher 1981; Krasnodębska-Ostręga et al. 2001), which was reflected by the poor immobilisation of Cd, Cu, Pb, Tl and Zn observed in the bottom sediment with low organic content. In these cases, the reducible and oxidable phase content is too low to effectively take part in the immobilisation of those metals (Jakubowska et al. 2006). It could suggest that the mechanism of Tl association with the sediment matrix rich in carbonates and sulphides in the absence of oxides and organic matter was similar to the mechanism of Pb. The results show that mobile fraction of Tl was associated with carbonate minerals.

The extractability of studied elements in sediment collected in Buk 3 pond beside higher total content was mostly equal (As, Cd, Cr and Cu) or even lower (Pb, Tl and Zn) than the extractability from Buk 2 sediment. It was probably correlated with the organic matter content, since if the content of total organic carbon (TOC) in sediment samples is high, the organic matter is mainly responsible for immobilisation of these elements (Kowalska et al. 2002).

The high mobility for Cd, Pb and Zn (≈85 %) here evaluated, opens up further avenues of research to be undertaken in this highly polluted area of Upper Silesia, such as evaluation of metal uptake by plants (phytoextraction, phytostabilisation, rhizofiltration, phytovolatilisation), and specific mobility studies of trace elements in order to assess its effect on the local wildlife.

4. CONCLUSIONS

This study was conducted to evaluate the fractionation of trace elements in sediments rich in carbonates from mining and smelting areas. The results here obtained showed that carbonate minerals were responsible for immobilisation of As, Co, Mn, Pb and Zn, but the mobile fractions of Cd, Cu and Tl were associated with sulphides and organic matter phase. Fractions of As, Cr and Mo were strongly associated with the bottom sediments, and these elements would not be leached into water in relatively short time. It was found that Tl association mechanisms with sediments rich in carbonates and sulphides in the absence of oxides and organic matter tended to be similar to the Pb mechanism. Molybdenum extractability was poor and similar to the presented by As and Cr in the same kind of matrix. These findings provide new insight into Tl and Mo behaviour in complex matrices and could be specifically applied for understanding the role of essential (Mo) and non-essential (Tl) elements in plant metabolism.

ACKNOWLEDGMENTS

Dr. Ospina-Alvarez gratefully acknowledges support by a post-doctoral grant from the AXA Research Fund (2nd wave—2012).

REFERENCES

1. Arunachalam J, Emons H, Krasnodebska B, Mohl C (1996) Sequential extraction studies on homogenized forest soil samples. Sci Total Environ 181(2):147–159.
2. de Santiago-Martín A, Valverde-Asenjo I, Quintana JR, González-Huecas C, Lafuente AL (2013) Soil properties affecting metal extractability patterns in periurban calcareous agricultural soils in the Mediterranean area. IJER 7(4):831–840
3. Fletcher WK (1981) Analytical methods in geochemical prospecting. In: Govett GJS (ed) Handbook of exploration and environmental geochemistry, vol 1. Elsevier Science B.V, Amsterdam
4. GraphPad Software Inc (2007) GraphPad Prism 5.0. San Diego, California, USA
5. Gruszecka AM, Wdowin M (2013) Characteristics and distribution of analyzed metals in soil profiles in the vicinity of a postflotation waste site in the Bukowno region, Poland. Environ Monit Assess 185(10):8157–8168.

6. Guevara-Riba A, Sahuquillo A, Rubio R, Rauret G (2004) Assessment of metal mobility in dredged harbour sediments from Barcelona, Spain. Sci Total Environ 321(1–3):241–255.
7. Jakubowska M, Zembrzuski W, Lukaszewski Z (2006) Oxidative extraction versus total decomposition of soil in the determination of thallium. Talanta 68(5):1736–1739.
8. Kaplan O, Ince M, Yaman M (2011) Sequential extraction of cadmium in different soil phases and plant parts from a former industrialized area. Environ Chem Lett 9(3):397–404. doi:10.1007/s10311-010-0292-0CrossRefGoogle Scholar
9. Kowalska J, Krasnodębska-Ostręga B, Golimowski J (2002) Electroanalytical methods for determination of the metal content and acetic-acid-available metal fractions in soils. Anal Bioanal Chem 373(1–2):116–118.
10. Krasnodębska-Ostręga B, Emons H, Golimowski J (2001) Selective leaching of elements associated with Mn–Fe oxides in forest soil, and comparison of two sequential extraction methods. Fresenius J Anal Chem 371(3):385–390.
11. Krasnodębska-Ostręga B, Dmowski K, Stryjewska E, Golimowski J (2005a) Determination of Thallium and other elements (As, Cd, Cu, Mn, Pb, Se, Sb, and Zn) in water and sediment samples from the vicinity of the zinc–lead smelter in Poland. J Soils Sediments 5(2):71–73.
12. Krasnodębska-Ostręga B, Stryjewska E, Golimowski J (2005b) Voltammetric determination of thallium and lead in sediment samples. Chem Anal (Warsaw) 50(4):807–810
13. Lis J, Pasieczna A, Karbowska B, Zembrzuski W, Lukaszewski Z (2003) Thallium in soils and stream sediments of a Zn–Pb mining and smelting area. Environ Sci Technol 37(20):4569–4572.
14. Lis J, Pasieczna A, Mojski JE, Przeniosło S, Sylwestrzak H, Strzelecki R, Wołkowicz S (2012) Geochemical atlas of Poland (Atlas Geochemiczny Polski). (trans: Leszczyński K). Polish Geological Institute, Warsaw
15. Marqués MJ, Martínez-Conde E, Rovira JV, Ordóñez S (2001) Heavy metals pollution of aquatic ecosystems in the vicinity of a recently closed underground lead–zinc mine (Basque Country, Spain). Environ Geol 40(9):1125–1137.
16. Prego R, Caetano M, Ospina-Alvarez N, Raimundo J, Vale C (2014) Basin-scale contributions of Cr, Ni and Co from ortegal complex to the surrounding coastal environment (SW Europe). Sci Total Environ 468–469:1–10.
17. Relić D, Đorđević D, Sakan S, Anđelković I, Pantelić A, Stanković R, Popović A (2013) Conventional, microwave, and ultrasound sequential extractions for the fractionation of metals in sediments within the petrochemical industry, Serbia. Environ Monit Assess 185(9):7627–7645.
18. Schramel P, Wendler I, Knapp G (1996) Total digestion of silicate containing matrices (plants, soil, sludges) using a pressure ashing device with PFA-vessels. Anal Bioanal Chem 356(8):512–514.
19. Smith E (2002) Uncertainty analysis. In: El-Shaarawi AH, Piegorsch WW (eds) Encyclopedia of environmetrics, vol 4. Wiley, Chichester, pp 2283–2297
20. Sutherland RA, Tack FMG (2007) Sequential extraction of lead from grain size fractionated river sediments using the optimized BCR procedure. Water Air Soil Pollut 184(1–4):269–284.
21. Trafas M (1996) Changes in the properties of post-flotation wastes due to vegetation introduced during process of reclamation. Appl Geochem 11(1–2):181–185.
22. Ure AM, Quevauviller P, Muntau H, Griepink B (1993) Speciation of heavy metals in soils and sediments. an account of the improvement and harmonization of extraction techniques undertaken Under the auspices of the BCR of the Commission of the European Communities. Int J Environ Anal Chem 51(1–4):135–151.
23. Verner JF, Ramsey MH, Helios-Rybicka E, Jêdrzejczyk B (1996) Heavy metal contamination of soils around a Pb–Zn smelter in Bukowno, Poland. Appl Geochem 11(1–2):11–16.
24. Yuan CG, Shi JB, He B, Liu JF, Liang LN, Jiang GB (2004) Speciation of heavy metals in marine sediments from the East China Sea by ICP-MS with sequential extraction. Environ Int 30(6):769–783.

25. Zhang H (2008) A comparison between heavy metals released from soil and its efficient speciation extracted by sequential extraction procedure. Chin J Geochem 27(1):36–40.
26. Zhang G, Liu CQ, Yang Y, Wu P (2004) Characterization of heavy metals and sulphur isotope in water and sediments of a mine-tailing area rich in carbonate. Water Air Soil Pollut 155(1–4):51–62.

CHAPTER 17

Early murine immune responses from endotracheal exposures to biotechnology-related *Bacillus* strains

Azam F. Tayabali , Kathy C. Nguyen & Verner L. Seligy

ABSTRACT

An immunology-based *in vivo* screening regime was used to assess the potential pathogenicity of biotechnology-related microbes. Strains of *Bacillus cereus* (Bc), *Bacillus subtilis* (Bs), *Bacillus thuringiensis* (Bt), and Bt commercial products (CPs) were tested. Balb/c mice were endotracheally instilled with purified spores, diluted CP, or vegetative cells (VC) (live or dead). Exposed mice were evaluated for changes in behavioral and physical symptoms, bacterial clearance, pulmonary granulocytes, and pulmonary and circulatory pyrogenic cytokines (interleukins (IL)-1β, IL-6 and tumor necrosis factor (TNF)-α), as well as acute phase biomarkers (fibrinogen and serum amyloid A). Except for some differences in clearance rates, no marked effects were observed in mice exposed to any spore at 10^6 or 10^7 colony forming units (cfu). In contrast, live Bc or Bt VCs (10^5 or 10^6 cfu) produced shock-like symptoms (lethargy, hunched appearance, ruffled fur, and respiratory distress), and 11–200-fold elevations in pyrogenic cytokines at 2-h post-exposure. In the study, 4-h effects included increased lethargy, ocular discharge, and 1.5–4-fold rise in circulatory acute phase markers, but no indications of recovery. Bs VC did not produce any changes in symptoms or biomarkers. After 2 or 4 h of exposure to dead VC, increases of only plasma IL-1β and TNF-α (4.6- and 12.4-fold, respectively) were observed. These findings demonstrate that purified spores produced no marked effects in mice compared to that of metabolically active bacteria. This early screening regime was successful in distinguishing the pathogenicity of the different *Bacillus* species, and might be useful for assessing the relative hazard potential of other biotechnology-related candidate strains.

Keywords: acute phase, gram-positive, inflammation, pulmonary, spore, vegetative cell

1. INTRODUCTION

The Bacillus cereus (Bc) group encompasses genetically related species of clinical and industrial interest. Within this group, a number of strains typed as B. cereus (Bc) sensu stricto were implicated in poisoning from contaminated foods, bacteraemia, endocarditis, endophthalmitis, neonatal meningoencephalitis, and soft tissue infections (Abusin, Bhimaraj, and Khadra 2008; Castedo et al. 1999; David, Kirkby, and Noble 1994; McIntyre et al. 2008; Stenfors Arnesen, Fagerlund, and Granum 2008). In contrast, many strains of the Bc group member Bacillus thuringiensis (Bt) are known to be entomopathic, and some of them have been developed for large-scale commercial use in forestry (e.g., against spruce budworm, gypsy moth), agriculture (e.g., against corn borer, tobacco budworm) and prevention of disease dissemination (e.g., against mosquitos, blackflies). However, there are reports that some Bt strains are infectious to humans, and similar in properties to Bc strains examined (Damgaard et al. 1997; Hernandez et al. 1998, 1999, 2000; Jackson et al. 1995; McIntyre et al. 2008; Samples and Buettner 1983).

Genomic and proteomic comparisons showed that members of the Bc group are related and possess similar virulence factors, including enterotoxins, exotoxins, hemolysins, phospholipases, and proteases, many of which are controlled by a common regulator, PlcR (Gohar et al. 2005; Rasko et al. 2005; Sergeev et al. 2006, Stenfors Arnesen, Fagerlund, and Granum 2008; Vilas-Boˆas, Peruca, and Arantes 2007). The molecular similarity (sequence and organization) of Bc group chromosomes strongly supports claims that existing species classification need to be re-evaluated (Helgason et al. 2000). Given the close relationship between Bc and Bt, there is a necessity for strain-specific data on their functional properties, especially as they relate to harmful attributes.

There are only a few reports of Bt infections in humans and animals (Hernandez et al. 1998, 1999, Salamitou et al. 2000; Siegel 2001; Siegel and

Shadduck 1990; Siegel, Shadduck, and Szabo 1987). This lack of data creates a need for clarification about specific strain differences in relation to molecular identity, pathogenic capacity, and overlapping environmental niches. Furthermore, experimental exposures demonstrate conflicting information about the capacity of Bc and Bt to infect non-target animals. Hernandez et al. (1998, 1999) found that mice instilled intranasally with high doses (107 –108 cfu per mouse) of either dormant spores or actively growing vegetative cells (VC) of Bc or Bt strains, showed hemorrhagic symptoms, neutrophil infiltration and death. Siegel and colleagues argued that death of mice does not occur with inhalation exposure, and that only intracerebral or intraperitoneal routes result in murine death (Siegel 2001; Siegel and Shadduck 1990; Siegel, Shadduck, and Szabo 1987).

Another Bacillus species, whose strains are often used in industrial applications is Bacillus subtilis (Bs). This species is generally recognized as safe (GRAS). However, like Bt, it is taxonomically complex (Rooney et al. 2009), and there is no published information on its inhalational safety. However, in the analysis of probiotic activity of various Bacillus strains and products, mice exposed intragastrically to spores of a Bs lab strain (PY79) showed increased tissue (spleen, liver, mesenteric lymph nodes, and submandibular glands) cytokine levels (TNF- and IFN-) and plasma immunoglobulin G (Duc et al. 2004). In addition, it was suggested that Bs persists and possibly replicates in the gut, but is not disseminated beyond the gastrointestinal tract (Hoa et al. 2001).

Further successful safe exploitation of Bacillus species would benefit considerably from research which clarifies the species/strain differences of the types and potencies of various virulence factors. Towards this goal, the present study describes a screening methodology using in vivo experiments with indicators of early murine immune responses to evaluate the pathogenicity potential of Bacillus biotechnology strains relevant to the Canadian Environmental Protection Act (CEPA 1999). The methodology exploits immune markers indicative of acute lung inflammation, fever induction and acute phase response (APR) (Conti et al. 2004; Cray, Zaias, and Altman 2009; Kabir et al. 2002). Comparisons were made between strains of Bc and Bt, commercially used Bt subspecies kurstaki and israelensis biopesticides (referred to as CP1 and CP2, respectively), and a biotechnology relevant Bs strain. The effects of purified spores, as well as spore-derived live and dead VC were also compared.

2. MATERIALS AND METHODS

2.1 Preparation of bacterial spores

Bacillus spp. were either obtained from the American Type Culture Collection (ATCC) or as commercial products (CP) from Valent Biosciences, Libertyville, IL (formerly Abbot Laboratories Inc.). The ATCC strains were B. cereus (Bc14579TM).), B. subtilis (Bs6051aTM), and B. thuringiensis (Bt13367TM). The spore-containing CP, referred to as CP1 (Foray 48B, Bt spp. kurstaki) and CP2 (Vectobac 12AS, Bt spp. israelensis) were diluted in physiological saline, based on spore content calculated as colony forming units (cfu). To prepare pure ATCC Bt or Bs(a) spores, approximately 107 cfu were incubated at 28C for 10 days in 50 mL of either Luria Bertani (LB) broth (10 g L^{-1} Tryptone, 5 g L^{-1} Yeast Extract, 10 g L^{-1} NaCl) or 2X SG medium (16 g L^{-1} Difco Nutrient broth, 0.5 g L^{-1} $MgSO_4 \cdot 7H_2O$, 2 g L^{-1} KCl, pH 7) supplemented with 1 mmol $Ca(NO_3)_2 \cdot 4H_2O$, 0.1 mmol $MnCl_2 \cdot 4H_2O$, 1 μmol $FeSO_4 \cdot 7H_2O$, 0.1% (w/v) glucose. Bc spores were prepared according to the procedure of Gaillard et al. (2005). Sporulation plates (40 mg L1 MnSO4, 100 mg L1 CaCl2, 100 mg L1 KCL absorbed into solidified nutrient broth-agar) were inoculated with 200 mL of Bc liquid culture, and incubated at 37C for 1 week. Bacterial cells were scraped from the plates and suspended in ddH2O. The amount of spores, VC and other culture components was monitored daily using a modification of the Schaeffer–Fulton method (Conn 1977). Briefly, preparations were stained with 5% (w/v) malachite green oxalate and counterstained with 3% (w/v) safranin (Sigma, Oakville, ON), and slide smears were visualized by light microscopy. When spore concentrations were 490%, the cell fraction was washed with saline, and any live VC were killed by heating at 60C for 30 min. Spores were further purified from killed VC and other culture components with an aqueous-polymer two phase separation procedure (Sacks and Alderton 1961). This procedure yielded 99–100% viable Bacillus spores, which were concentrated to 108 cfu mL1 and frozen at 80C in PBS with 10% (v/v) glycerol until use. Spore viability and concentration were validated regularly. Preparation of li

2.2 Preparation of live and dead VC

Washed VC were generated by incubating spores for 72 h at 37C in LB to yield cultures containing primarily live and dead VC, but also spores (51%) and insoluble culture debris. These cultures were washed of culture supernatant and debris with three centrifugations (5000 g, 10 min) and re-suspension steps using 5–10 volumes of sterile PBS. Both CP were efficient sporulators, and 72-h cultures contained 99% spores. Therefore, VC for CP-1 and CP-2 were derived from 18-h cultures grown at 37C. For some experiments, VC were killed by heating at 100C for 30 min or incubating with 50 mg mL1 gentamicin for 2 h, followed by three washes with PBS. To track dose delivery, killed Bc VC were stained with Alexa-Fluor555TM Succinimydl Ester (Molecular Probes, Eugene, OR) using a modification of the supplier's protocol. Briefly, bacteria were centrifuged at 5000 g for 10 min, resuspended in 100 mmol sodium bicarbonate (pH 8.3) and mixed with Alexa dye. After constant agitation for 1 h, the stained bacteria were washed 3 10 min at 5000 g to remove unbound dye and resuspended in fresh saline. 31

2.3 Animals and exposures

All procedures involving animals were approved by the Health Canada Animal Care Committee. Female Balb/c mice were purchased from Charles River Laboratories Inc. (Saint-Constant, Quebec) or Taconic Inc. (Germantown, NY), and were maintained and treated in a pathogen-free biosafety level 2 containment facility. Only animals between 18 and 23 g (8–10 weeks) that had been acclimated for at least 2 weeks were used for experiments. Mice had access to food and water ad libitum. For endotracheal instillation, mice were anesthetized with isoflurane and the 25 mL dose of bacteria or saline alone was aerosolized through the trachea into the lungs with a syringe-style nebulizer (MicrosprayerTM, Penn Century, Philadelphia, PA). Mice typically recovered from anaesthesia and dosing within 2 min. Either 3 or 6 mice were used for each treatment group.

The selection of the first time point for routine screening was done following pilot experiments that tracked tissue bacterial content and cytokine levels at 20 min, 2 and 4 h. These experiments demonstrated that the 2-h time point allowed for measurements of both cytokine and bacterial content of target organs. Even though the numbers of bacteria in tissue were lower compared to 20 min exposures, the 2-h interval permitted confirmation of dose delivery. Furthermore, the 2-h point was the earliest time that might accommodate dosing, blood collection and necropsy. The 4-h point was most useful for detecting protein expression changes in the lungs, blood, and liver, but bacterial counts were substantially reduced, preventing reliable estimates of target organ delivery. Some experiments were extended to 4 h to determine if recovery was possible, but the resultant severe shock-like symptoms observed at 4 h after Bc VC exposure deterred routine use of this time point.

2.4 Blood and tissue analysis

Treated mice were anesthetized and blood (approximately 500–1000 mL) was immediately collected by cardiac puncture and transferred to blood collection tubes containing ethylenediaminetetraacetic acid (EDTA). After mixing to prevent clots, the blood was centrifuged at 1000 g for 10 min to collect plasma. The latter was stored at 80C until use. Tissues (liver, lung, and trachea) were collected and also stored at 80C. Plasma samples were analyzed for levels of circulating, pyrogenic cytokines, as well as Th1 and Th2 cytokines. These were measured with pre-conjugated bead kits (BioRad, Mississauga, ON) using a high throughput multiplexed bead array system (Bioplex, BioRad). Cytokines are listed here with their limit of detection in parentheses (pg mL1): IL-1 (9.4), IL-2 (0.6), IL-4 (2.1), IL-5 (0.3), IL-6 (0.2), IL-12(p70) (2.3), TNF- (1.4), granulocyte macrophagecolony stimulating factor (GM-CSF; 2.3), interferon-gamma (IFN- ; 1.2). ELISA kits were used for validation of cytokine levels (Peprotech, Rocky Hill, NJ; R&D Systems, Minneapolis, MN), and also for measurement of fibrinogen (Innovative Research Inc., Southfield, MI; detection limit 30 ng mL1) and serum amyloid A (SAA; Biosource International Inc., Camarillo, CA; detection limit 3 ng mL1). The multiplex and ELISA assays were also used for measurement of cytokine and acute phase protein levels from lung and liver homogenized (PowerGen 125, Fisher Scientific) in the presence of a protease inhibitor cocktail (Roche, Laval, QC). In orde

In order to measure the clearance of bacteria from tissues, animals were sacrificed by cervical dislocation at various times following exposure. The lung, trachea, and esophagus were excised and homogenized in 1 mL saline with a hand-held homogenizer (PowerGen 125, Fisher Scientific). Serial dilutions were plated on LB-agar plates, and cfu content was determined after incubation at 28C for 18 h.

Granulocyte infiltration into tissues was monitored by immunofluorescence microscopy using antibodies to the surface glycoprotein Ly-6G of granulocytes. Tissue pieces were placed into 4% (w/v) paraformaldehyde for 18–24 h, followed by storage in 70% ethanol at 4C until further processing. Tissues were sectioned with a vibrating microtome (Vibratome Inc., St. Louis, MO). Sections (25–60 mm) were incubated for 60 min in blocking/incubation buffer (0.05% Tween-20), 90 min at 37C in 1 : 50 rat anti-LY-6G (eBioscience, San Diego, CA) and 60 min with 1 : 200 TRITC-conjugated anti-rat IgG. Two PBS rinses were conducted between each step. Sections were mounted with ProlongTM antifade solution (Molecular Probes, Eugene, OR) and viewed with a Nikon TE2000 inverted microscope with epifluorescence optics, connected to a Nikon C1 Laser Scanning Confocal unit with three lasers (488, 543, and 633 nm). Three random fields for each mouse (nine fields total for each group) were enumerated for LY-6G-positive cells.

2.5 Statistical analysis

The significance of the differences in Ly-6G, cytokines, and APR markers between bacteria-exposed mice and saline-treated mice was done by conducting an analysis of variance (ANOVA) followed by a post-hoc Dunnett's Multiple Comparison Test (Dunnett 1980). Statistical analyses were done in "R" version 2.11.1 (R Development Core Team 2010).

3. RESULTS

3.1 Bacteria used for murine exposures

Purity and consistency of the bacterial preparations was ensured with quality control monitoring using a staining protocol in which spores, VC and debris were differentially stained. The micrographs in Figure 1 show smears of spores, washed VC and CP stained with Malachite Green (blue-green spores) and counterstained with safranin (red-orange VC and debris). Staining showed that the preparations of purified spores and CP contained no VC, but did contain some pinkish-red stained, amorphous, extracellular debris. Washed VC were predominantly composed of VC (51% spores). Based on cfu counts and hemocytometer measurements of total VC, the ratio of live to dead cells was approximately 1 : 10 for Bt VC, and 1 : 1 for VC derived from Bs(a), Bt and both CP. As expected from previous studies (Seligy et al. 1997, 2000) CP contained large numbers of spores, as well as relatively large safranin-stained amorphous material and smaller regularly shaped structures, resembling parasporal inclusion bodies.

3.2 Tissue clearance and persistence of Bacillus cells

To localize dose delivery within the airway, washed VC of Bc were labeled with a fluorescent dye before exposure. Spores were not used because they did not stain with sufficient intensity for microscopic detection. At 2-h after exposure with 106 cfu per mouse, tissue was excised, fixed, and sectioned. Micrographs in Figure 2 show that fluorescentlylabeled VC (orange) were localized around the bronchioles and alveoli (Figure 2a, inset) Other tissues had trace numbers of bacteria in the trachea (arrows in Figure 2b) and esophagus (not shown).

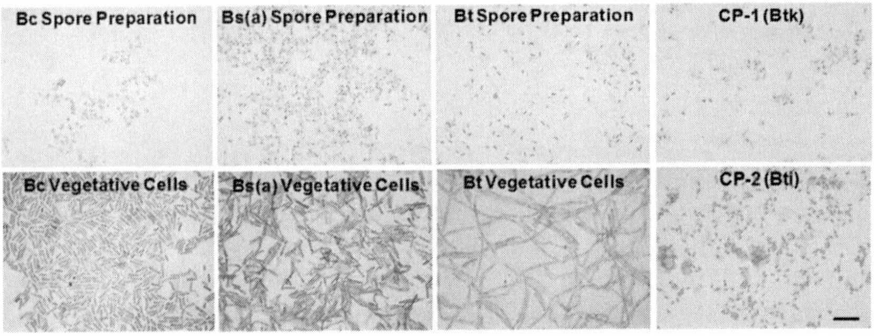

Figure 1. (Color online). Preparations of Bacillus strains used for murine exposures. Micrographs of Bacillus spores and washed VC were prepared as described in the section "Materials and methods". Samples were dried onto slides and stained using a modification of the Schaeffer–Fulton Method, which results in blue-green spores (Malachite Green) and red-orange VC debris (Safranin). Bc, B. cereus 14579; Bs(a), B. subtilis 6051a; Bt, B. thuringiensis 13367; CP-1 (Btk), commercial product from Bt subsp. kurstaki; CP-2 (Bti), commercial product from Bt subsp. israelensis. Bar in lower right of CP-2 represents 5 mm, which is the same scale for all micrographs.

To determine the dose distribution within lungs following delivery by the endotracheal method, lung lobes were excised and homogenized, and serial dilutions were plated on LB-agar. Colonies were counted after incubation at 37C for 18 h. Measurements in different areas of the lungs (Figure 2c, d) indicated bacteria were equally distributed between left and right lungs. For the right lung, the greatest proportion of bacteria was found in the median and caudal lobes, and fewest bacteria were in the cranial lobe.

Figure 3 summarizes bacterial counts in the lungs and trachea following exposure. At 2 h for all strains, 1.3–341 cfu mg1 (total 260–68,000 cfu lung1) were recovered, which equates to 0.03–6.8% of the applied dose. Since 2 h was the earliest that could be done to accommodate all testing activities, including dosing and blood collection, an experiment for 20 min following exposure was done. Data revealed that almost the entire original dose could be retrieved from the trachea and lungs, which indicates that as much as 99% of pulmonary clearance occurs between 20 and 120 min. To determine if there was a difference in the rate of elimination of bacteria from the lung depending on the species, bacterial content was measured over a 1-week period. Bs(a) was cleared from the lungs within 48 h, at least 2 days before the other bacteria. The commercial Bt preparations showed delayed clearance from both the trachea and lungs since bacteria could still be recovered 3–7 days post-exposure. A difference between CP-1 and CP-2 was that mice exposed to CP-2 showed an apparent transitory fall in pulmonary and tracheal bacteria at 24 h post-exposure, followed by an elevation in bacterial counts in both tissues, which may represent transit from another tissue.

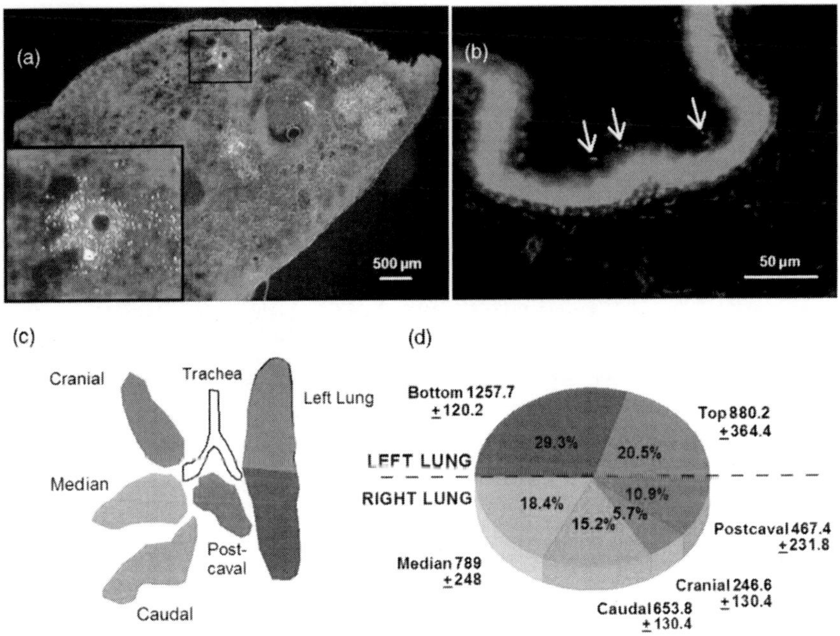

Figure 2. (Color online). Bacillus dose distribution in tissues. VC of Bc (106 cfu per mouse) were labeled with Alexa Fluor 555 succinimydl ester and used for endotracheal exposures. Two hours after exposure, airway tissues were excised and either fixed and sectioned for microscopy, or homogenized and plated on LB-agar for bacterial enumerations. Micrographs show immunofluorescence of (a) lung sections with bacteria (orange) around the bronchioles and within alveoli, and (b) traces of bacteria in the trachea as indicated with arrows. SytoxTM Green (nuclei) was used as a counterstain for both (a) and (b). The pulmonary bacterial distribution at 2 h is shown in (c) and (d). The colors of lung segments in (c) correspond to the colors in the pie chart in (d). Numbers adjacent to pie segments in d are cfu mg1 of tissue obtained from homogenized lung segments. Percentages in d represent the proportion of total cfu's recovered. Data are mean values from three mice   standard deviation.

3.3 Physical/behavioral symptoms of Bacillus exposure

The behavior and appearance of mice were monitored throughout every exposure for symptoms of discomfort or stress. Mice treated with 106 purified spores or CP showed no apparent symptoms over a 1-week period. Mice treated with 107 spores had noticeable ruffled fur at 24 h, but otherwise resembled saline-treated controls. In contrast, experiments involving washed Bc and Bt VC were prematurely terminated at 2 h in accordance with animal care guidelines, because animals showed obvious signs of stress, including hypoactivity, hunched-over appearance, ruffled fur and respiratory distress. The proportion of treated animals that displayed such symptoms differed for Bc (all symptomatic), Bt (two of four symptomatic) and Bs(a) (asymptomatic).

3.4 Immune effects of purified spores, CP, and washed VC

Several biomarkers were monitored following endotracheal instillaton with 106 cfu of either purified spores or diluted CPs. Enumeration of LY-6G-positive granulocytes, and pulmonary and blood inflammatory cytokines IL-1, IL-6, and TNF-), and Th1 and Th2 cytokines (IL-2, IL-4, IL-5, IL-12(p70), GM-CSF, IFN-) showed that there were no significant changes compared to saline-treated controls when monitored over a 1-week period, or even during weekly Bc spore exposures for 1 month. In contrast, confocal micrographs of exposures to washed Bc, Bt, and CP-1 VC showed a significant elevation in the number of pulmonary granulocytes (Figure 4b) compared to those treated with saline (Figure 4a). The box plots in Figure 4c illustrate that individual responses were variable, as indicated by the large area within the boxes. Yet the means were statistically different compared to controls. In contrast, the Bs(a) or CP-2 strains did not induce a significant elevation in granulocytes.

The increased numbers of pulmonary granulocytes suggested that several cytokines may have been elicited by lung cells. To test this hypothesis, excised lung tissues and plasma of mice treated for 2 h with washed VC were screened for the pyrogenic cytokines IL-1, IL-6, and TNF-. None of the pyrogenic cytokines levels for any of the strains were significantly altered in plasma. In lung tissue, Bc showed increases in the mean levels of IL-1, IL-6, and TNF- by 4.5, 63, and 3.5-fold, respectively (Figure 4d) compared to controls.

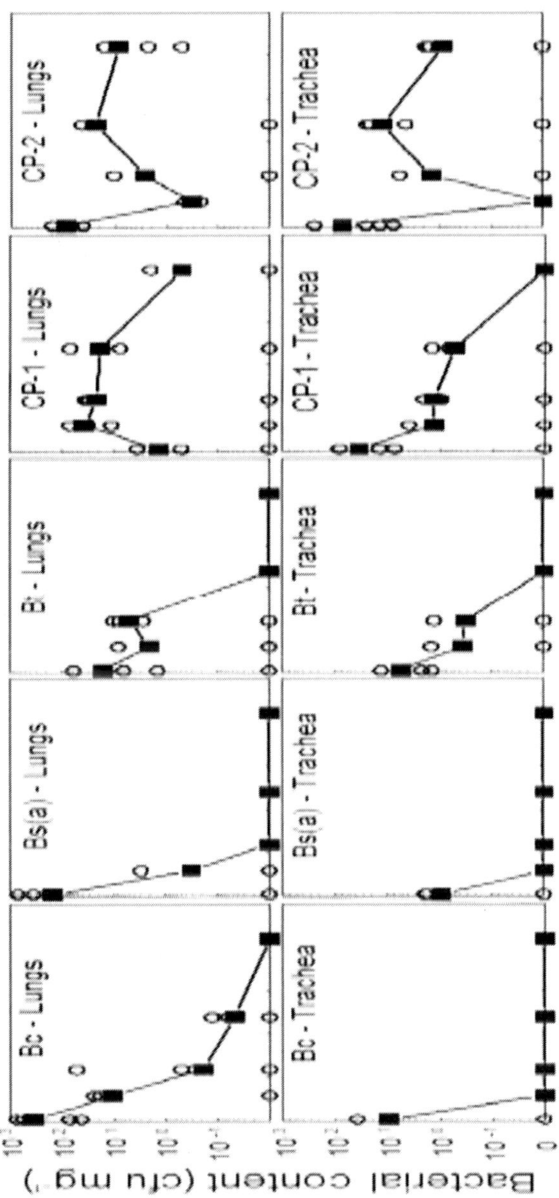

Figure 3. Tissue clearance and persistence of Bacillus cells. Mice inoculated endotracheally with 106 cfu of spores or VC per mouse were sacrificed at various time intervals, and the lungs and trachea were homogenized and spread-plated for estimation of bacterial content. Means (filled squares) of 4 control and 3 exposed mice (empty circles) are shown.

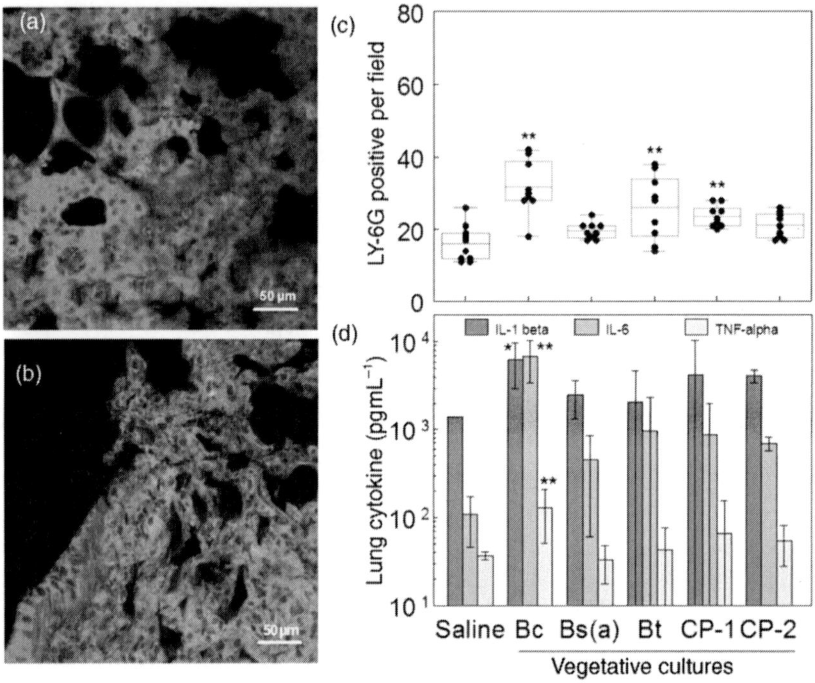

Figure 4. (Color online). Immune effects of washed VC. Three mice were instilled endotracheally with either saline or 106 cfu per mouse of VC, and were sacrificed at 2-h post-exposure and lung sections were stained with rat anti-mouse-LY-6G and anti-rat-TRITC antibodies. Examples of confocal micrographs shown in (a) (saline) and (b) (Bc VC) were used to enumerate three random fields from each lung section of three mice (nine fields total per group) of LY-6G-positive cells (red– orange), and are summarized with box-whisker plots in (c). Box limits represent the 25th and 75th percentiles and whiskers represent the 5th and 95th percentiles. Means (horizontal line within boxes) are for 10 individual counts (solid circles). Pulmonary pyrogenic cytokine levels (d) were derived from homogenized lung tissue using a multiplex bead array system against antigenic standards supplied by the manufacturer. Error bars represent the standard deviation. Asterisks indicates significantly different values from pooled saline exposures, as determined using ANOVA and Dunnett's Multiple Comparison Test (*p5 0.1; **p5 0.05).

Table 1. Cytokine levels (pg mL[1]) 2 h after Bc14579 vegetative cell exposure.

Dose (cfu per 25 μL)	IL-1β	IL-6	TNF-α
Lung cytokines			
0 (vehicle)	614.6±545.5	47.6±26.8	15.0±2.4
10^2	520.5±74.6	60.8±24.5	20.4±1.8
10^3	322.07±91.0	42.4±26.5	12.2±3.0
10^4	592.6±387.6	43.0±24.1	9.8±3.8
10^5	6515.4±4571.1	6336.3±5226.7	181.5±169.8
10^6	6873.0±6321.9	9520.2±9478.3	271.6±379.6
Plasma cytokines			
0 (vehicle)	Not detected	4.6±1.4	0.0±0.0
10^2	Not detected	5.5±4.5	0.7±0.0
10^3	Not detected	3.5±0.9	0.7±0.0
10^4	Not detected	5.9±0.9	0.7±0.4
10^5	Not detected	133.6±149.6	3.3±3.6
10^6	Not detected	122.2±137.6	4.3±0.0

Notes: Mice (three per group) instilled endotracheally with either saline or various doses of VCs were sacrificed at 2-h post-exposure. Pyrogenic cytokines were measured in homogenized portions of lung tissue and plasma using a multiplex bead array system. The high standard deviation in some exposures resulted from usually one non-responsive animal per group.

3.5 Inactivated vegetative cell exposures

Washed VC contain variable amounts of dead, inactive VC. To test whether they could contribute to immunostimulation, mice were dosed with VC of each bacterial strain killed with either heat or gentamicin. Both inactivation methods resulted in no bacterial colonies after plating or bacterial metabolic activity after staining with MTT, which is a viability indicator that detects the presence of even a single metabolically active bacterium (Seligy et al. 1997). At a dose of 106 or 108 Bc inactivated VC (cfu-equivalents), exposed mice showed no symptoms at 2 h. There was also no significant elevation of granulocytes compared to controls. The only effects observed were an increase in the means of plasma IL-1 (4.6-fold) and TNF- (12.4-fold). Mean levels of IL-6 did not change during any exposure to dead VC. Data demonstrate that live bacteria are necessary to generate the majority of effects observed during exposures to washed VC.

3.6 Dose effects of washed VC

To further investigate the effects of washed VC and determine the no observed effect level (NOEL) for immune responses, exposures were conducted with log10 serial dilutions of Bc, Bs(a), and Bt washed VC (106 – 102 cfu per mouse). These experiments showed that some animals (usually one of three) did not respond to the exposures and produced high standard deviations (Table 1). For those that did respond, pyrogenic cytokines were significantly elevated in blood and lungs for 105 and 106 cfu doses of washed Bc VC but not significantly for the same doses of Bs(a) or Bt. The NOEL for cytokine changes was 104 cfu for both lung and blood.

3.7 Extended exposures with washed VC

To investigate whether longer exposures may result in recovery or worsened effects, 106 cfu of washed VC of each Bacillus strain was administered to mice, and observations were made 4-h post-exposure. All Bc-exposed animals had ruffled fur at 2 h, and 4 of 6 showed additional symptoms beginning at 2.5 h, which included hypoactivity and ocular discharge resulting in a ring of damp fur around the eyes. Effects did not diminish at 4 h when the experiments were terminated. Almost equivalent effects were observed with mice treated with washed VC of Bt (2 of 3 positive), CP-1 (1 of 3 positive) or CP-2 (2 of 3 positive). However, still no marked effects were observed with Bs(a) exposure.

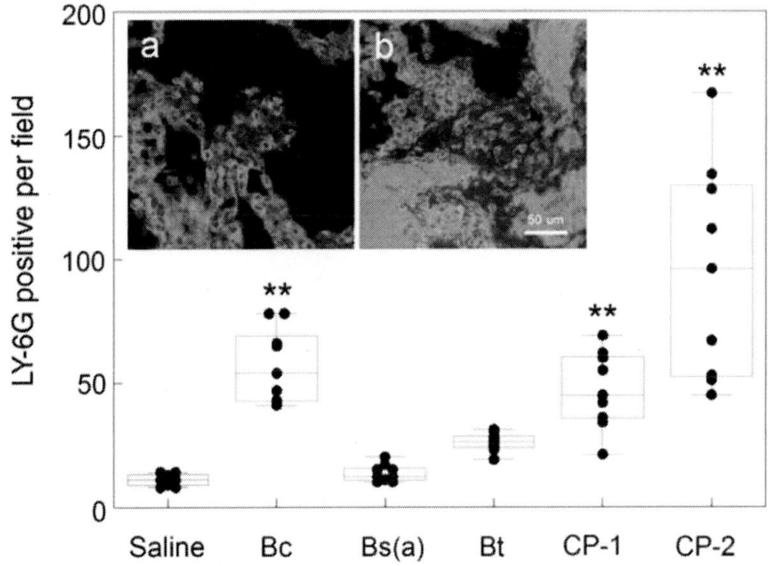

Figure 5. (Color online). Granulocyte infiltration 4 h after exposure. Three mice were instilled endotracheally with either saline or 106 cfu per mouse of VC, and were sacrificed at 4-h post-exposure. LY-6G-positive granulocytes were enumerated from three random fields of each lung section from three mice (nine fields total per group). Insets show lung sections from mice treated with either saline (a) or Bc (b). Asterisks indicates significantly different values from pooled saline exposures, as determined using ANOVA and Dunnett's Multiple Comparison Test (*p5 0.1; **p5 0.05).

Plating of homogenized lung, trachea, and esophagus showed that the number of bacteria recovered was essentially equivalent at 2 and 4 h post-exposure. At 4 h, the mean number of pulmonary granulocytes was 6–9-fold greater compared to controls (Figure 5) and up to 2-fold greater than those at 2-h post-exposure. Levels of plasma and pulmonary cytokine markers, IL-1, IL-6, and TNF-, were also elevated (Figure 6a, b), but not substantially greater than those at 2 h. Compared to saline controls, Bc exposures produced mean levels of IL-1, IL-6, and TNF- to increase by 25, 15, and 16-fold, respectively in lung tissue, and 4.7, 11, and 27-fold, respectively in plasma. No statistical significance was observed with these data, suggesting that the elevation in cytokines levels previously observed at 2 h post-exposure was transient and probably diminishing by 4 h post-exposure.

The presence of pyrogenic cytokines in the circulation suggested that APR proteins may be expressed by the liver, since hepatic cells are known to possess receptors to these cytokines (Gruys et al. 2005). Therefore, two APR proteins, fibrinogen and serum amyloid A (SAA), known to be elevated in the liver and blood during APR response (Gruys et al. 2005) were tested. Measurement 2-h post-exposure did not show increases in either of these markers, but at 4-h post-Bc exposure both markers were elevated in the blood, and post-Bt exposure SAA was significantly elevated (Figure 7).

4. DISCUSSION

This study describes an animal screening methodology which builds on in vitro virulence tests in order to assist in differentiating between pathogenic and non-pathogenic Bacillus organisms proposed for use in biotechnology. One of the tenets of the ethical use of experimental animals is the reduction in the numbers used, so experiments were intentionally designed with a minimal number of animals per exposure group. The confidence in observations was enhanced with preliminary testing using in vitro cytotoxicity assays (Tayabali and Seligy 2000), employing toxicity indicators from various sampling sites (lung, liver, and blood) of the same animal, presentation of data in formats such as box plots to highlight where inter-animal variation is significant, as well as dose–response experiments and time courses to demonstrate reproducibility and trends. In some exposure groups, a proportion of animals were unresponsive, while the majority showed clear and large changes compared to saline-treated animals. Individual animals and their tissues were tracked throughout the analysis, and no significant correlation could be made with the cfu's recovered from the lungs, and extent of response observed. The reason for the inter-animal variability is currently unknown, but was observed to the same extent when the sample size was increased to 6 instead of 3 for 4-h post-exposure measurement of cytokines and APR markers. The larger sample sizes elevated the confidence in the data (lower p values), but did not alter the overall result, which still clearly demonstrated the potential effects of the exposure in affected animals.

The murine exposure data presented here show that for all strains, washed VC containing live VC are most immunoreactive and toxic (physical symptoms, local and systemic effects), and that the potency of spp. can be ranked as Bc4Bt 4 Bs(a). In comparison, dilute commercial Bt products (mainly CP-spores), purified spores, or dead VC elicited little or no inflammatory response. Since the spores were known to be viable (99%) up to the point of instillation, and there was low recovery of cfu from the lung at 2 h, this lack of inflammatory response may have been due to early clearance, inhibition of spore germination, or that early germinants were inactivated prior to replication. However, for both dead and live VC, there was an increase in expression of IL-1 and TNF-, which is similar to the production of these cytokines by monocytes during lipopolysaccharide exposure (Schindler et al. 1990; Werling et al. 1996). The lack of IL-6 expression with killed VC may be due to the early stage in the cytokine expression cascade, since IL-6 is known to be a secondary inflammatory mediator (Uhlar and Whitehead 1999).

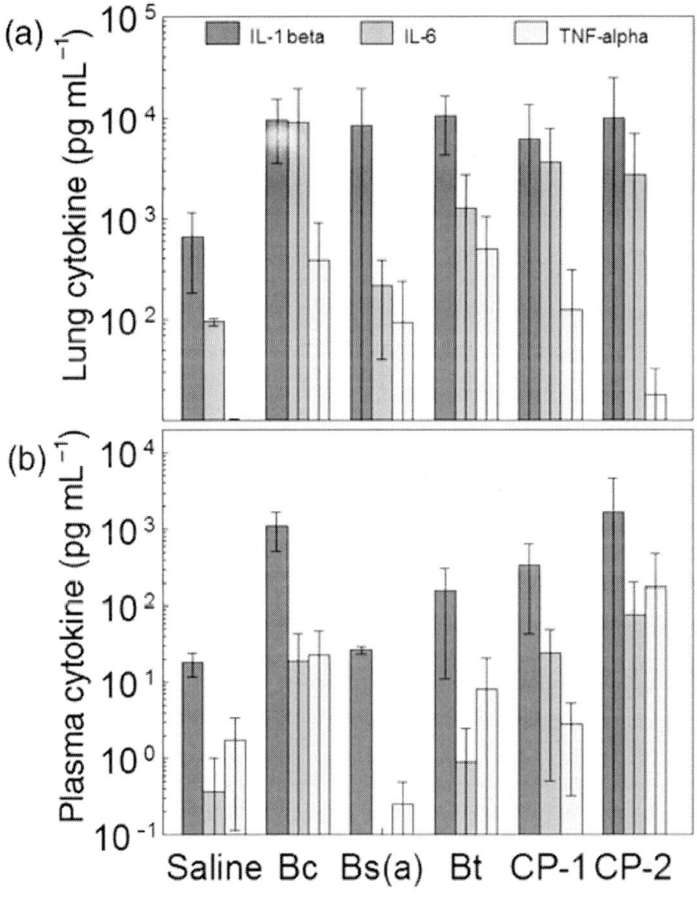

Figure 6. Pyrogenic cytokine levels 4 h after exposure. Mice instilled endotracheally with either saline or 106 cfu per mouse of VC were sacrificed at 4-h post-exposure. Pyrogenic cytokine levels were measured using a multiplex bead array system in portions of lung tissue (a) and plasma (b). Data are mean values from three mice standard deviation.

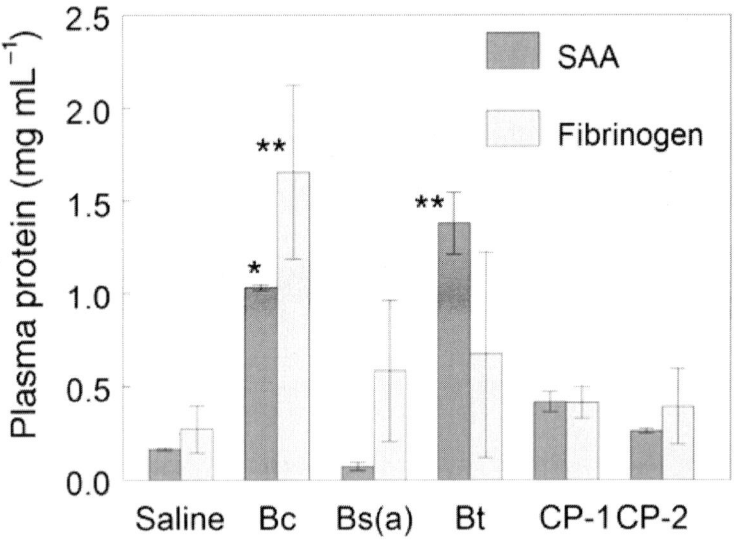

Figure 7. Levels of APR Indicators. Mice instilled endotracheally with either saline or 106 cfu per mouse of VC were sacrificed at 4-h post-exposure. APR markers, fibrinogen and SAA were measured by ELISA. The graph summarizes plasma fibrinogen and SAA levels from 4-h after exposure to all sources of washed VC. Data are mean values from three mice standard deviation. Asterisks indicates significantly different values from pooled saline exposures, as determined using ANOVA and Dunnett's Multiple Comparison Test (*p5 0.1; **p 50.05).

The immunostimulatory capacity of the three different Bacillus species was not the same, and may relate to findings recently reported. Ryu et al. (2009) showed that different species of gram-positive bacteria have different TNF- and nitric oxide inducing capacities in RAW 264.7 macrophage, which were related to their respective lipotechoic acid (LTA) content. Further, Ryu et al. (2009) suggested that the greater D-alanine content of LTA from pathogenic species results in enhanced immunological properties. Elevated LTA D-alanine levels may function to increase the net negative charge of the bacterial cell wall, influencing interactions with mammalian cellular components.

The findings in the present study support investigations previously done in vitro with Bacillus spp. for rapid throughput screening of strains from public repositories and CP (Tayabali and Seligy 1995, 1997, 2000). These studies demonstrated that a number of animal cell lines were useful for predicting toxicity and the potential for eliciting innate immune responses. Spores from commercial sources, and type-strains of Bc and Bt, but not

Bs(a), were not toxic if inactivated with antibiotic. However, if spores were permitted to germinate and replicate in the presence of HT29 epithelial cells, rapid and severe necrotic cell death was observed, which resulted from exoproteins produced during vegetative growth. The present experiments showed that VC exerted their effects beginning at about 1.5 h, and that equivalent numbers of killed VC did not produce similar pathogenesis as live VC. Our previous in vitro findings for effects of washed VC also reflected these observations and showed that the cytolytic effects resulting from toxin accumulation occurred at about 1.5 h, which is approximately equivalent in duration to one doubling of VC number (Tayabali and Seligy 2000). Data presented suggest that VC toxin production and replication probably also occurs in vivo.

Washed VC produced shock-like symptoms that corresponded to sensitive local and systemic cellular responses, and elevations in molecular indicators such as local inflammation, pyrogenic cytokines, and APR proteins. Biomarker elevations were dependent on the bacterial species: animals exposed to live Bc elicited the most inflammation, pyrogenicity, and early APR, whereas VC of Bt or Bs(a) resulted in relatively attenuated immune stimulations. One possible explanation for lowered APR from commercial Bt strains is their propensity to rapidly re-sporulate. Significant elevations of the APR proteins were detected (means were 2–3-fold compared to controls). These markers have been recommended for the diagnosis of bacterial infections (Gruys et al. 2005; Urbach et al. 2004). SAA levels increase 5-1000-fold in the blood beginning at 4–6 h (Uhlar and Whitehead 1999), making it a sensitive indicator of APR. The multiple roles of SAA in recovery from bacterial infections is well-documented, and was reviewed in depth by Uhlar and Whitehead (1999). During the APR, fibrinogen levels rise 2–3-fold, as observed in our study, and its role in coagulation is well known.

Some of our findings are consistent to those of Hernandez et al. (1999) and Salamitou et al. (2000) who reported that intranasal instillation of Bt subsp. Konkukian or Bc14579 (107 –108 spores) and 108 Bt serotype 407, 3a3b or H34 induced shock-like symptoms such as nasal bleeding pulmonary hemorrhaging, neutrophil infiltration, and lesions from acute bronchitis, followed by mortality at 24 h. Their control soil Bt serotype H12 only produced limited pulmonary inflammation. These effects were observed with either spores or VC. However, our 24 h exposures with 107 spores of Bc (same strain they used), or spores of our other test strains, failed to induce any immune effects, but rather showed that bacteria were rapidly cleared from the lungs. Differences in methodologies could account for the difference in observations. First, our pilot experiments with saline alone showed that a 50 mL volume, such as that used in studies by Hernandez et al. (1999) and Salamitou et al. (2000) was too large for the small Balb/c mice, and often resulted in death from asphyxiation, not produced by bacteria. Furthermore, significant precautions were taken to minimize the amount of VC material in our spore preparations (Materials and methods). Data presented demonstrate that killing VC by heat, which was used by Hernandez et al. (1999) in preparation of spores, results in dead VC that still produce low level inflammation. Another difference between experimental observations is that their spores persisted for at least 10 days, whereas our spore preparations were cleared by 99.9% within 4 days. This discrepancy may be due to the heat-killed VC debris present in their spore preparations, adding to the particulate burden in the lungs. This is reminiscent to the slow clearance observed with Bt CPs which contain large amounts of acellular material.

Siegel (2001) argued that the pathogenic effects seen by Hernandez et al. (1999) and Salamitou et al. (2000) are due to doses that were unrealistic by 41000-fold based on human exposures from Bt spray operations (Noble, Riben, and Cook 1992). According to Burges (1980), a microbial control agent can be considered toxic if a dose of 106 cfu per mouse produces mortality or clinical/pathological changes similar to the symptoms observed here. Our routine exposures were at, or below this level and all were sublethal for the duration of the study.

A recent study examined the lethal dose of VC and spores of clinical Bt strains using Balb/c mice exposed intratracheally by surgery (Ghelardi et al. 2007). The study revealed that VC produced 50% lethality with doses of 1.8 106 cfu at 48 h after inoculation. Similar to our study, no marked effects were observed at or below 7.2 104 cfu, and all animals showed effects with 1.7 107 cfu. Although our study was not designed to intentionally measure lethal effects, similar death rates with VC-treated animals may have been observed based on our observations at 4 h. Moribund animals in Ghelardi et al. (2007) study noted identical physical symptoms to those described here. Interestingly, spores from the same strains could also kill the mice, albeit with 20-fold greater cfu. The exaggerated effects observed by Ghelardi et al. (2007) may be invariably due to the virulence of their strains, and the relatively more invasive intratracheal route of exposure.

In our findings, none of the Bs(a)-treated mice showed physical symptoms, but high VC doses still resulted in increased pulmonary granulocytes. However, there were no increases in any cytokine, which is in contrast to Duc et al. (2004), who showed that Bs strains PY79 and probiotic BiosubtylTM were able to induce IFN- and TNF- from spleens, livers, submandibular glands, and mesentery lymph nodes of intragastrically infected Balb/c mice. The discrepancy in observations may be accounted for by considering their intragastric route of exposure, Bs strain-specific characteristics, and high dose level (109 cfu).

Given the findings reported here, a general model for early immune responses resulting from exposure to these Bacillus strains can be summarized as follows. Spores at concentrations up to 107 cfu per mouse and live VC up to 104 cfu per mouse, were cleared by professional macrophages without intervention of granulocytes. No marked changes were observed in the levels of granulocyte infiltration (LY-6G) or tissue and blood cytokines, and yet almost all test bacteria were completely cleared from the lungs 1-week after exposure. In exposures containing 4104 live VC, at least Bc was able to produce toxins, as observed in vitro (Tayabali and Seligy 2000). Macrophage-induced clearance was likely inadequate and necessitated augmentation with granulocyte action. This same mechanism of macrophage-assisted bacterial elimination at low doses, and impaired clearance and granulocyte recruitment at higher doses, was reported previously (Delclaux and Azoulay 2003; Marriott and Dockrell 2007). The recruitment of granulocytes is brought on by expression of pulmonary and possibly circulatory chemokines, which are expressed by macrophage and lung epithelial cells (Delclaux and Azoulay 2003). These granulocytes have potent microbicidal properties against Bacillus, as demonstrated by their ability to kill B. anthracis (Mayer-Scholl et al. 2005) and their involvement in the defence against Bt (Ghelardi et al. 2007). The more severe effects that were observed at 4 h after exposure to Bc VC resemble those of early APR, including elevated systemic levels of liver APR markers, fibrinogen and SAA (Cray, Zaias, and Altman 2009). In this model of Bacillus infection, it is not known whether our VC-exposed animals are capable of recovery without therapeutic intervention.

The testing regime used here was successful in comparing the toxicity of prioritized biotechnology-related Bacillus species, and may be applied to any microbe whose pathogenicity potential is uncertain. This study revealed that for clearance and early immune effects, pure spores, as well as other substances and additives in diluted CP, exhibit no observable effects compared to those elicited by live, metabolically active bacteria (VC). As such, the commercial preparations tested here are not expected to be toxic following inhalational exposure to non-target mammals, including humans, and should be safe when used as intended. Furthermore, data underscore the need for scrutiny of organisms beyond the genus and species levels, especially if proposed for large scale use. Specific strains and their purity need to be evaluated through their life cycle in product preparations, as well as various phases of proposed product use.

ACKNOWLEDGMENTS

We thank Drs Phil Shwed, Hari Vijay and Ivan Curran of Health Canada for reviewing the manuscript and Dr Martha Navarro of the Health Canada Animal Resources Division (HC-ARD) for veterinary advice, technical help, and hematology analysis. We also thank Kevin Kittle, Douglas Parks, and Svetlana Popovic of HC-ARD for technical help, animal maintenance, dose administration and blood collection, as well as Remi Gagne and Andrew Williams for help with the "R"-based statistical analysis. This research was supported by the Canadian Regulatory System for Biotechnology (CRSB HC3.3/4.4) to VLS.

REFERENCES

1. Abusin, S, Bhimaraj, A and Khadra, S. 2008. Bacillus cereus endocarditis in a permanent pacemaker: A case report. Cases Journal, 1: 95
2. Burges, H D. 1980. "Safety, testing and quality control of microbial pesticides". In Microbial control of pests and plant diseases 1970–1980, Edited by: Burges, H D. 738–767. New York: Academic Press Inc.
3. Canadian Environmental Protection Act (CEPA). 1999. Available at: http://laws.justice.gc.ca/en/C-15.31/text.html (accessed March 31, 2010)
4. Castedo, E, Castro, A, Martin, P, Roda, J and Montero, C G. 1999. Bacillus cereus prosthetic valve endocarditis. Annals of Thoracic Surgery, 68: 2351–2. doi:10.1016/S0003-4975(99)01163-7
5. Conn, H J. 1977. Biological stains, , 9th, Baltimore: The Williams & Wilkins Co.
6. Conti, B, Tabarean, I, Andrei, C and Bartfai, T. 2004. Cytokines and fever. Frontiers in Bioscience., 9: 1433–49.
7. Cray, C, Zaias, J and Altman, N H. 2009. Acute phase response in animals: A review. Comparative Medicine, 59: 517–26.
8. Damgaard, P H, Granum, P E, Bresciani, J, Torregrossa, M V, Eilenberg, J and Valentino, L. 1997. Characterization of Bacillus thuringiensis isolated from infections in burn wounds. FEMS Immunoology and Medical Microbiology, 18: 47–53.
9. David, D B, Kirkby, G R and Noble, B A. 1994. Bacillus cereus endophthalmitis. British Journal of Ophthalmology, 78: 577–80.
10. Delclaux, C and Azoulay, E. 2003. Inflammatory response to infectious pulmonary injury. European Respiratory Journal, 22(Suppl. 42): 10s–4s.
11. Duc, L H, Hong, H A, Barbosa, T M, Henriques, A O and Cutting, S M. 2004. Characterization of Bacillus probiotics available for human use. Applied and Environmental Microbiology, 70: 2161–71.
12. Dunnett, C W. 1980. Pairwise multiple comparisons in the unequal variance case. Journal of the American Statistical Association, 75: 796–800.
13. Gaillard, S, Leguerinel, I, Savy, N and Mafart, P. 2005. Quantifying the combined effects of the heating time, the temperature and the recovery medium pH on the regrowth lag time of Bacillus cereus spores after a heat treatment. International Journal of Food Microbiology, 105: 53–8.
14. Ghelardi, E, Celandroni, F, Salvetti, S, Fiscarelli, E and Senesi, S. 2007. Bacillus thuringiensis pulmonary infection: Critical role for bacterial membrane-damaging toxins and host neutrophils. Microbes and Infection, 9: 591–8.
15. Gohar, M, Gilois, N, Graveline, R, Garreau, C, Sanchis, V and Lereclus, D. 2005. A comparative study of Bacillus cereus, Bacillus thuringiensis and Bacillus anthracis extracellular proteomes. Proteomics, 5: 3696–711.
16. Gruys, E, Toussanint, M JM, Niewold, T A and Koopmans, S J. 2005. Acute phase reaction and acute phase proteins. Journal of Zhejiang. University Science, 6B: 1045–56.
17. Helgason, E, Okstad, O A, Caugant, D A, Johansen, H A, Fouet, A, Mock, M, Hegna, I and Kolsto, A. 2000. Bacillus anthracis, Bacillus cereus, and Bacillus thuringiensis: One species on the basis of genetic evidence. Applied and Environmental Microbiology, 66: 2627–30.

18. Hernandez, E, Ramisse, F, Cruel, T, Vangueresse, R le and Cavallo, J -D. 1999. Bacillus thuringiensis serotype H34 isolated from human and insecticidal strains serotypes 3a3b and H14 can lead to death of immunocompetent mice after pulmonary infection. FEMS Immunology and Medical Microbiology, 24: 43–7.
19. Hernandez, E, Ramisse, F, Ducoureau, J P, Cruel, T and Cavallo, J -D. 1998. Bacillus thuringiensis subsp. konkukian (serotype H34) superinfection: Case report and experimental evidence of pathogenicity in immunosuppressed mice. Journal of Clinical Microbiology, 36: 2138–9.
20. Hernandez, E, Ramisse, F, Gros, P and Cavallo, J. 2000. Super-infection by Bacillus thuringiensis H34 or 3a3b can lead to death in mice infected with the influenza A virus. FEMS Immunology and Medical Microbiology, 29: 177–81.
21. Hoa, T T, Duc, L H, Isticato, R, Baccigalupi, L, Ricca, E, Van, P H and Cutting, S M. 2001. Fate and dissemination of Bacillus subtilis spores in a murine model. Appllied and Environmental Microbiology, 67: 3819–23.
22. Jackson, S G, Goodbrand, R B, Ahmed, R and Kasatiya, S. 1995. Bacillus cereus and Bacillus thuringiensis isolated in a gastroenteritis outbreak investigation. Letters in Applied Microbiology, 21: 103–5.
23. Kabir, K, Gelinas, J P, Chen, M, Chen, D, Zhang, D, Luo, X, Yang, J H, Carter, D and Rabinovici, R. 2002. Characterization of a murine model of endotoxin-induced acute lung injury. Shock, 17: 300–3.
24. Marriott, H M and Dockrell, D H. 2007. The role of the macrophage in lung disease mediated by bacteria. Experimental Lung Research, 33: 493–505. doi:10.1080/01902140701756562
25. Mayer-Scholl, A, Hurwitz, R, Brinkmann, V, Schmid, M, Jungblut, P, Weinrauch, Y and Zychlinsky, A. 2005. Human neutrophils kill. Bacillus anthracis PloS Pathogens, 1: 179–86.
26. McIntyre, L, Bernard, K, Beniac, D, Isaac-Renton, J L and Naseby, D C. 2008. Identification of Bacillus cereus group species associated with food poisoning outbreaks in British Columbia, Canada. Applied and Environmental Microbiology, 74: 7451–3.
27. Noble, M A, Riben, P D and Cook, G J. 1992. Microbiological and epidemiological surveillance programme to monitor the health effects of Foray 48B Btk spray. Report to the British Columbia Ministry of Forests, September 30 1992
28. R Development Core Team. 2010. R: A language and environment for statistical computing. R Foundation for Statistical Computing, Vienna, Austria. ISBN 3-900051-07-0, URL http://www.R-project.org
29. Rasko, D A, Altherr, M R, Han, C S and Ravel, J. 2005. Genomics of the Bacillus cereus group of organisms. FEMS Microbiology Reviews, 29: 303–29.
30. Rooney, A P, Price, N P, Ehrhardt, C, Swezey, J L and Bannan, J D. 2009. Phylogeny and molecular taxonomy of the Bacillus subtilis species complex and description of Bacillus subtilis subsp. inaquosorum subsp. nov. International Journal of Systematic Evolutionary Microbiology, 59: 2429–36.
31. Ryu, Y E, Baik, J E, Yang, J S, Kang, S, Im, J, Yun, CKim, D W. 2009. Differential immunostimulatory effects of gram-positive bacteria due to their lipotechoic acids. International Immunophamacology, 9: 127–33.
32. Sacks, L E and Alderton, G. 1961. Behavior of bacterial spores in aqueous polymer two-phase systems. Journal of Bacteriology, 82: 331–41.
33. Salamitou, S, Ramisse, F, Brehélin, M, Bourguet, D, Gilois, N, Gominet, M, Hernandez, E and Lereclus, D. 2000. The plcR regulon is involved in the opportunistic properties of Bacillus thuringiensis and Bacillus cereus in mice and insects. Microbiology, 146: 2825–32.
34. Samples, J R and Buettner, H. 1983. Ocular infection caused by a biological insecticide. Journal of Infectious Diseases, 148: 614
35. Schindler, R, Macilla, J, Endres, S, Ghorbani, R, Clark, S C and Dinarello, C A. 1990. Correlations and interactions in the production of interleukin-6 (IL-6), IL-1, and tumor necrosis factor (TNF) in human blood mononuclear cells: IL-6 suppresses IL-1 and TNF. Blood, 75: 40–7.

36. Seligy, V L, Beggs, R W, Rancourt, J M and Tayabali, A F. 1997. Quantitative bioreduction assays for calibrating spore content and viability of commercial Bacillus thuringiensis insecticides. Journal of Industrial Microbiology and Biotechnology, 18: 370–8.
37. Seligy, V L, Douglas, G R, Rancourt, J M, Tayabali, A F, Otvos, I, van Frankenhuyzen, K, Dugal, J, Rousseau, G and Szabo, A G. 2000. "Comparative performance of conventional and molecular dosimetry methods in environmental biomonitoring: Assessment using Bacillus-based commercial biopesticides as models". In Rapid methods for the analysis of biological materials in the environment, NATO ASI Series, Edited by: Stopa, P J. 279–97. Netherlands: Kluwer Academic Publishers.
38. Sergeev, N, Distler, M, Vargas, M, Chizhikov, V, Herold, K E and Rasooly, A. 2006. Microarray analysis of Bacillus cereus group virulence factors. Journal of Microbiology Methods, 65: 488–502.
39. Siegel, J P. 2001. The mammalian safety of Bacillus thuringiensis-based insecticides. Journal of Invertebrate Pathology, 77: 13–21.
40. Siegel, J P and Shadduck, J A. 1990. Clearance of Bacillus sphaericus and Bacillus thuringiensis ssp. israelensis from mammals. Journal of Economic Entomology, 83: 347–55.
41. Siegel, J P, Shadduck, J A and Szabo, J. 1987. Safety of the entomopathogen Bacillus thuringiensis var. israelensis for mammals. Journal of Economic Entomology, 80: 717–23.
42. Stenfors Arnesen, L P, Fagerlund, A and Granum, P E. 2008. From soil to gut: Bacillus cereus and its food poisoning toxins. FEMS Microbiology Review, 32: 579–606.
43. Tayabali, A F and Seligy, V L. 1995. Semiautomated quantification of cytotoxic damage in cultured insect cells exposed to commercial Bacillus thuringiensis biopesticides. Journal of Applied Toxicology, 15: 365–73.
44. Tayabali, A F and Seligy, V L. 1997. Cell integrity biomarkers and evaluation of cytotoxic responses of Bacillus thuringiensis-containing commercial bioinsecticides. Ecotoxicology and Environmental Safety, 37: 152–62.
45. Tayabali, A F and Seligy, V L. 2000. Human cell exposure assays of Bacillus thuringiensis commercial insecticides: Production of Bacillus cereus-like cytolytic effects from outgrowth of spores. Environmental Health Perspectives, 108: 919–30.
46. Uhlar, C M and Whitehead, A S. 1999. Serum amyloid A, the major vertebrate acute-phase reactant. European Journal of Biochemistry, 265: 501–23.
47. Urbach, J, Shapira, I, Branski, D and Berliner, S. 2004. Acute phase response in the diagnosis of bacterial infections in children. Pediatric. Infectious Disease Journal, 23: 159–60.
48. Vilas-Bôas, G T, Peruca, A P and Arantes, O M. 2007. Biology and taxonomy of Bacillus cereus, Bacillus anthracis, and Bacillus thuringiensis. Canadian Journal of Microbiology, 53: 673–87.
49. Werling, D, Sutter, F, Arnold, M, Kun, G, Tooten, P CJ and Gruys, E. 1996. Characterization of the acute phase response of heifers to a prolonged low dose infusion of lipopolysaccharide. Research in Veterinary Science, 61: 252–7.

CHAPTER 18

Identification and quantification of phthalates in water and sediment of Ori Stream, Iwo, Southwestern Nigeria using high performance liquid chromatography

Godwin O. Olutona* Modupe O. Dawodu

Department of Chemistry and Industrial Chemistry, Bowen University, Iwo, Nigeria.

ABSTRACT

This study provides data on distribution of phthalate esters (PAEs) in the water and sediments in the Ori stream, Iwo, Nigeria. The liquid-liquid extraction method was employed, followed by cleaned up in an alumina column using hexane. Chromatographic separation and quantification were done using Zorbax Eclipse AAA C18 analytical column. The concentrations of DMP, DBP, DEP, and DPP in water (μg/L) ranged from 1.29±1.3 in DBP to 938±780 in DMP while in sediment (μg/kg) ranged from 12.7±15 in DEP to 152±140 μg/kg in DBP. These concentrations were above the USEPA limits of 3 μg/L as recommended for phthalates in water.

Keywords: Phthalates, liquid-liquid extraction, high performance liquid chromatography (HPLC), endocrine disruptors, stream.

1. INTRODUCTION

Phthalate esters (PAEs) are well known polymer additives and widely used in industries as plasticizers for polymeric products especially cellulose esters and vinyl chloride copolymers (Adeniyi et al., 2011; EU-RAR, 2008); and non-plasticizers such as paints, varnishes, adhesive, cosmetics, insect repellants, insecticides carriers, propellants, building materials, automobile parts and food packaging (EU-RAR, 2008; Staples et al.1997; Gobas et al., 2004; Zheng et al., 2010) with the aim of improving their flexibility, transparency, durability, longevity and altering physical properties like malleability and flame resistant of synthetic products (Arago et al., 2010; Kanchanamayoon et al., 2012). Chemically, phthalates are stable, colour-, odour- and flavorless, soluble in water liquids over a long temperature range (Clara et al., 2010). Estimate of 8 billion tones of phthalate esters are reported being used globally each year particularly as additives to plastics products, industrial chemicals and parts of several consumer products (Blount et al, 2000).

Due to the additives nature of these contaminants, phthalates exist freely and are not chemically bound to the polymer chain in a freely mobile phase and drain away; hence they can drain away, migrate or evaporate into the environment via domestic and industrial effluents, sewage sludge, storm water, runoffs and indiscriminate dumping of phthalate products (Adeniyi et al., 2011; Fujii et al., 2003; Gomez-Hens et al., 2003; Afshari et al., 2004). These contaminants have been found in air, food, water and soil, organism, tissues and fluid of wild life, treated and untreated waste water (EU-RAR, 2004, 2008; Tan et al., 2007; Dargnat et al., 2009; Adewuyi, 2012),sewage sludge ((EU-RAR, 2004/2008), sediment (EU-RAR, 2008; Yuwatini et al., 2006); storm water (Björklund et al., 2009) and have found their way into the human body in various ways.

Phthalates have been classified as carcinogenic agent and endocrine disruptor for humans. Many health issues associated with these contaminants are well documented in literatures, for instance early puberty in females (Colon et al., 2000); genital defects and low testosterone production (Fisher et al., 2003; Swan et al., 2005); testicular cancer lessened with sperm quality and sperm damage in men (Swan et al., 2005; Rozati et al., 2002; Duty et al., 2003); pre- mature delivery (Duty et al., 2003; Latini, 2003) breathing related health issues like airway obstruction, lung malfunction (Jaakkola et al., 1999) and asthma (Bornehag et al., 2004; Hoppin et al., 2004). These health problems are capable of increasing abnormal development in animals, such as cleft palate, skeletal malformations and fetal death (Arago´et al., 2012); abnormal development in amphibians and crustaceans and to induce genetic lapses (EU-RAR, 2004).

Health risks assessment of PAEs are subject of global concern due to the fact that human beings are unavoidably exposed to PAEs from drinking contaminated water and aquatic animals (Adeniyi et al., 2011). The rural dwellers are highly prone to associated health risk problems due to phthalates than people in semi-urban areas because they depend on stream water for their daily domestic and agricultural usage. The construction of reservoirs all over the world in the past century has constituted to negative and complex trace metals and trace organics contaminant burdens which were adsorbed in the sediment deposited behind these reservoirs (Rayne and Friesen, 2009; Thi et al., 2010). River bottom-sediments are significant origins and act as sinks for the distribution and bioaccumulation of PAEs in river and have been greatly used as an environmental indicator for evaluation of PAEs contamination (Gobas et al., 2004). Thus, stream bottom sediments are of great concern because of accumulation effect of these persistent organic pollutants under the modified hydrodynamic conditions (Bendz et al., 2005; Zhou et al., 2006).

It has been reported that the environmental condition of the Ori stream continues to degrade with the ever expanding population of Iwo (Akindele et al., 2015). This study is the first to be conducted on the occurrence of PAEs in water and sediments in this area. The focal thrust of this study was to assess the levels and distributions of PAEs in water and sediment of Ori stream.

2. MATERIALS AND METHODS

2.1 Study area

Ori stream is one of the main streams that pass through the ancient semi-urban city of Iwo (Figure 1). Unlike Aiba stream, the stream is situated along the outskirt part of the town but various developmental projects have extended to the area. Due to the rising human needs, a lot of activities such as open markets, motor parks, abattoir, irrigation farming activities, car wash, filling stations, technical school and spiritual bathing taking place along the course of the stream has led the stream to be subjected to various eutrophic, disruptive and thermal stress as it flows through the city and empty itself into Oba river (Akindele et al., 2015).

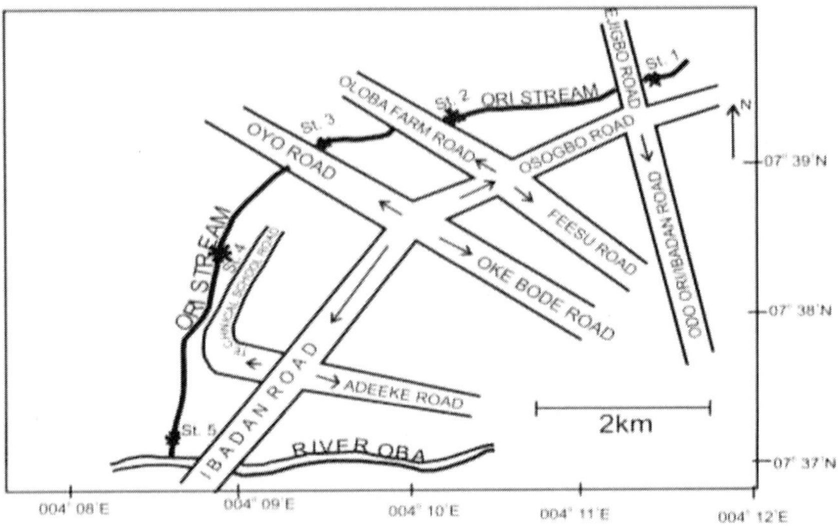

Figure 1. Map of Ori stream showing the sampling sites (Adapted from Akindele et al., 2015).

2.2 Sampling

Water and sediment samples were collected from five different sites (St 1-5) as shown in the study area map (Figure 1).Water samples were collected in amber whinchester bottle (2.5 L) to prevent bacteria growth. The water and sediment samples were collected with stainless steel auger and grab samplers. Sediment samples were scooped using a pre-cleaned stainless steel scoop into solvent-rinsed aluminium containers. The samples were stored cold in the refridgerator prior to analysis.

2.3 Apparatus and reagent used

All bottles and glassware used in the study were earlier steeped in 10% nitric acid for 48 h, washed with detergent, rinsed three times with tap water before finally rinsed with distill water then with acetone (99.5%) and n-hexane (99.0%) mixture (1:1 v/v) and then heated in an oven at a temperature of 105^{0}C for 1 h prior to use. The entire reagent used, viz. dichloromethane and n-hexane (GFS chemicals, Inc Colombus); acetone, acetonitrile, ethyl acetate (Park Scientific Ltd. Northamptom, United Kingdom); Aluminia from Lab Tech Chemicals; Sodium sulphate and sodium chloride (BDH, England). Standard mixture of phthalate ester and n-butyl benzoate were purchased from Sigma Aldrich, South Africa.

2.4 Extraction procedures

Liquid–liquid chromatographic extraction method was employed for this experiment. 500 mL of the sample were measured into a separatory funnel and this was spiked with 0.1 mL (100 ppm) of the internal standard. Ten (10 g) of NaCl was used to saturate the sample so as to destabilize the formation of emulsions from the solvent. Three portions of 25 mL dichloromethane were then employed to extract the sample. Three portion extracts from each sample were put together in another separatory funnel and further extracted with three times of 5 mL 0.1 M Sodium carbonate. This was done to remove the free-fatty acid (FFA) from the organic phase that had been extracted by dichloromethane (Fatoki and Ogunfowokan, 1993; Ogunfowokan et al., 2006). The extracts were then dried over anhydrous sodium sulphate. The extracts were evaporated over steam bath and the residue re-dissolved in 1 mL of dichloromethane. Sediment samples were extracted in soxhlet extractor with dichloromethane. This was then clean-up in alumina column.

2.5 Clean up procedure

A column of about 15 cm × 1 cm (i.d) was packed with about 10 g activated aluminia prepared in a slurry form in n-hexane. About 0.5 cm^3 of anhydrous sodium sulphate was placed at the top of the column to absorb any water in the sample or the solvent. The column was pre-eluted with 15 mL of n-hexane and 30 mL ethyl acetate. The eluate was collected, was then concentrated to about 1 mL by evaporation to dryness under a stream of analytical grade nitrogen gas. 1 ml of acetonitrile was added to the reduced extract for HPLC analysis (Fatoki and Ogunfowokan, 1993; Ogunfowokan et al., 2006). The extract was refrigerated prior to instrumental analysis was conducted.

2.6 Recovery experiment

In this study, no standard phthalate esters reference materials were available; hence, recovery analysis was conducted in order to establish the efficiency of the analytical procedures used. Recovery study was carried out by spiking 500 mL organic free water with 0.1 mL (1000 ppm) of phthalate standard mixture and was extracted and treated the same way as the sample.

2.7 Stock solution preparation and estimation of response factor

One hundred ppm stock solution of the mixture of the dibutyl (DBP), dimetyl (DMP), diethyi (DEP) and diphenyl (DPP) and n-butyl benzoate (internal standards) was prepared by weighing 1.0 mg each of the DMP, DEP, DBP and DPP and internal standards using weighing boat into a beaker, and make up to mark in 10 mL standard flask with acetonitrile. The stock solution with internal standard was run on high performance liquid chromatography for the determination of response factor which was calculated using the equation below (Fatoki and Noma, 2001).

Response Factor = Peak Area of Phthalate/Peak area of Internal Standard.

2.8 Instrumental analysis

High-performance liquid chromatography (HPLC) was done with Agilent HPLC 1260 infinity series available at the Central Science Laboratory of Bowen University, Iwo, which was used in determining the presence and level of phthalate ester in the water and sediment samples. The HPLC conditions were as follows: Zorbax Eclipse AAA C18 column 150 mm × 4.6 mm i.d., particle size of 3.5 μm, 228 nm wavelength, temperature 40^0C, injection volume 20 μl and gradient elution condition using acetonitrile and water (90:10 v/v) as mobile phase. Under this condition separation lasted for about 4 min with flow rate of 1 mL/min for the first four minutes then 2 ml/ min. Identification of individual phthalate ester was established based on its retention time and quantification was performed by combination of internal standardization and response factor.

3. RESULTS AND DISCUSSION

The efficiency and reliability of the analytical method employed was determined by the analysis of the response factor, retention time and percentage recovery. The phthalates were eluted from the column in the order of dimethyl- (DMP), diethyl- (DEP), dibutyl- (DBP), and diphenyl phthalates (DPP). The response factor and retention time were indicated in Table 1. The recovery analysis was done to proof the reliability of the analytical method employed. The percentage recoveries ranged from 78.0% in diethyl- and 94.3% in diphenyl-.

Table 1. Values of response factor, retention time and percent recoveries.

Compound	Response factor	Retention time (min)	% Recovery
DEP	0.27	1.86	78.0
DBP	0.85	2.96	85.5
DMP	0.21	1.62	88.7
DPP	0.47	2.25	94.3
n-Butyl benzoate: IS		1.28	

The quantitative results of phthalates in water and sediment at the designating sampling locations along the Ori stream are presented in Tables 2 and 3. The concentrations of phthalates obtained in this study showed a number of occurrence of phthalate in both water and sediment of Ori stream.

Table 2. Levels of phthalate esters (µg/L) in water samples from Ori stream.

Station	DEP	DBP	DMP	DPP	Σ_4PEs
St1 Ori	ND	ND	74.8±45	ND	75
St 2 Agogo	ND	ND	1980±62	ND	1980
St 3 Osadep	505±74	ND	1440±570	ND	1945
St 4 Technical	ND	6.43±3.10	81.4±48	21.3±13	109
St 5 Oba	ND	ND	1110±560	6.35±9.0	1116
Mean	101±16	1.29±1.3	938±780	5.53±6.5	

Table 3. Levels of phthalate esters (µg/kg) in sediment samples from Ori stream.

Station	DEP	DBP	DMP	DPP	Σ_4PEs
St 1 Ori	21.2±19	110±59	12.9±8.3	18.0±1.1	162
St 2 Agogo	10.8±5.3	213±130	11.7±6.5	27.2±16	360
St 3 Osadep	2.45±3.5	86.1±9.9	28.5±20	24.1±13	141
St 4 Technical	25.5±7.5	54.3±10	118±13	51.1±3.1	249
St 5 Oba	3.50±2.9	295±160	ND	55.2±46	354
Mean	12.7±15	152±140	34.2±28	35.1±26	

Table 2 presents the levels of phthalate esters in water samples. The phthalates esters determined in the water > DBP. A quick glance at the concentrations of phthalate esters showed the presence of dimethyl phthalate in all sites .Generally, the levels of DMP are higher than all other phthalate compounds. Dibutyl phthalate was detected only in St 4 (Technical) while diphenyl phthalate were detected in St 4 and 5 (Technical and Oba). The mean levels of phthalate in the stream water ranged from

1.29 μg/L for dibutyl phthalate to 938 μg/L for dimetyl phthalate. The level of dilution between DBP and DMP was above 450 fold. This could be attributed to the weak and van der waal force interaction of phthalate esters which contributes to the slow decomplexation of phthalate-solute complexes observed. The Σ_4PEs concentrations ranged from 75 μg/L at St1 (Ori) to 1980 μg/L at St 2 (Agogo). The highest Σ_4PEs concentrations were found at St2, St 3 (Osadep) followed by St 5(Oba). It could also be observed that the levels of phthalate ester decreases from the Upstream St 1(Ori) downstream. The presence of phthalate ester in this water body could be due to local contamination from domestic wastes including disposable plastic materials (such as tiles, wire coatings, synthetic leathers, wall papers, wrapping materials and plastic soft drink bottles) and agricultural wastes, from motor park and market area thrown into the water body at St 1(Ori), farming as well as recreational activities occurring along the stream bank. Burning of tyres at abattoir located at St 1(Ori) and refuse incineration could also be a good source of phthalates in this water body. The values reported in this study for the stream are above the USEPA water criteria of 3 μg/L for the survival of fish and other aquatic life in river (USEPA, 1980).

Our results for the stream, are above those reported elsewhere for river, e.g New England Rivers, United States (1-30 μg/L) (Hites, 1973); the Delaware River, United States (0.3 – 50 μg/L for DBP and DEHP) (Sheldon and Hites, 1979); River Irwell and Etherow in the Manchester area of United kingdom (0.2 – 33.5 μg/L) (Fatoki and Vernon, 1990); River Ronneybyan (0.32-3.10 μg/l and River Svartan (0.39 to 1.98 μg/l), Sweeden (Thuren 1986), University Teaching Hospital Sewer waste and receiving stream 2.05 to 62.81 μg/L and 0.18 to 9.17 μg/L , respectively (Adewuyi, 2012).

Table 3 present the levels of phthalate esters in sediment samples of Ori stream. All the four phthalate esters analyzed were detected at all the sites with exception of DMP which was not detected at St 5 (Oba). The mean values of phthalate esters in the sediment samples ranged from 12.7 μg/l for DEP to 152 μg/l for DBP. The order of prominence of phthalate esters detected were DBP > DPP > DMP > DEP. The levels of DBP and DPP were extremely low in water sample while DEP were extremely low in sediment sample. The Σ_4PAEs concentrations ranged from 141 μg/kg at St 3 (Osadep) to 354 μg/kg at St 5 (Oba). It could also be observed that the sediment record display relatively diversified results which could be an indication of photolysis and biodegradation conditions. Comparative study of PAEs concentration in sediment of Ori stream with those measured in other locations over the world revealed a medium concentrations magnitude (Table 4).

Table 4. Comparison of PAEs in sediment measured in this study with those reported from other locations (μg/g)

Location	No. of target PAEs	No of sample sites	Range of ΣPAEs	DMP	DEP	DBP	DPP	Sample year	Reference
Ori stream	4	6	0.141-0.36	0.034	0.013	0.152	0.035	2015	This study
Qiantang River, China	16	23	0.585-6.74	0.053	0.018	-	-	2011	Sun et al. (2013)
Gomti River, India	5	30	0-0.364	0.011	0.005	-	-	2008	Mathur et al. (2010)
False Creek Harbour, Canada	13	4	0.004-2.1	0.038	0.021	-	-	2004-06	Mackintosh et al. (2006)
Ogun River, Nigeria	4	6	0.325-2.88	0.214	0.200	-	-	2005	Adeniyi et al. (2011)
Lake of Guangzhou, China	15	16	2.27-74.94	0.039	0.13	-	-	2005	Zeng et al. (2008)
Yellow River, China	5	7	30.52-85.16	2.594	5.049	-	-	2004	Xia et al. (2007)
Yongding River, China	16	12	0.679-2.119	0.009	0.022	-	-	2003	Wang et al. (2006)
Rivers of Taiwan	8	6	2.9-27.1	-	0.200	-	-	2000	Chang et al. (2002)

The occurrence of DMP and DEP PAEs in water and sediment of Ori stream might be as a result of their widely use as additives in products, such as perfumes, lotions, cosmetics, toys, adhesives, inks, varnishes (Sun et al., 2013). Similarly, the occurrence of DBP usually used in epoxy resin and cellulose ester containing products such as disposable medical tubes, blood storage bags, toys that found their ways into the water body (Heise and Litz, 2004). The Σ_4PEs levels in surface water were higher than that of the sediments. This may suggest PAEs fresh inputs into the water body and partitioning of organic contaminant in the aquatic ecosystem depends on the physical and chemical parameters of the individual compounds such as sorption coefficient, vapour pressure, solubility, suspended organic matter and sediment characteristics (Zeng et al., 2008). Other factors that may responsible for different levels of PAEs at different sites and aqueous matrices include adsorption, deposition of suspended particles, microbial or photo-chemical degradation and dilution in the partitioning systems of dissolved and sediment-associated phase (Tao et al., 2005). The hydrophobic nature couple with high octanol-water ratio and low aqueous solubility of PAEs tend to be bounded to particulate and dissolved matter as a result of low flow velocity (Chen et al., 1999). The low flow velocity of the stream enhances the deposition of suspended particles which are easily bonded by organic contaminants than coarse particles (Chen et al., 2006).

4. CONCLUSION

This study provides first hand information on the contamination levels of four PAEs in the water and sediments of the Ori stream, Iwo, south-western, Nigeria. The levels of phthalate esters recorded in this study were at levels that raise concern. The state environmental protection agency needs to adopt a constant monitoring plan so that the concentrations of this persistent organic pollutant discharged into the water body are in conformity with international standard.

CONFLICT OF INTERESTS

The authors have not declared any conflict of interests.

ACKNOWLEDGEMENTS

The authors are very grateful for the contribution of Miss Folasade Olanrewaju to the success of this work and the constructive comments and suggestions of the anonymous reviewers of this manuscript.

REFERENCES

1. Adeniyi AA, Okedeyi OO, Yusuf KA (2011). Flame ionization gas chromatographic determination of phthalate esters in water surface sediments and fish species in the Ogun river catchments, Ketu, Lagos, Nigeria. Environ. Monit. Assess. 172:561-569.
2. Adewuyi GO (2012). High performance liquid chromatographic identification and estimation of phthalates in sewer waste and a receiving river in Ibadan city, southwestern Nigeria. J. Water Resour. Prot. 4:851-859.
3. Afshari AL, Gunnarsen PA, Clausen HV (2004). Emission of phthalates from PVC and other materials, Indoor Air 14:120-128.
4. Akindele EO, Adedapo TA, Olawoye OO, Olutona GO, Adu BW (2015). Preliminary Limnological Survey of Ori Stream, Iwo, Osun State, Nigeria. Int. J. Biol. Chem. Sci. 9(1):329-341.
5. Arago'n M, RMarce'n RM, Borrull F (2012). Determination of phthalates and organophosphate esters in particulated material from harbour air samples by pressurised liquid extraction and gas chromatography–mass spectrometry. Talanta 101:473-478.
6. Bendz D, Paxeus NA, Ginn TR, Loge FJ 2005 Occurrence and fate of pharmaceutically active compounds in the environment, a case study: Hoje River in Sweden. J. Hazard Mater. 122:195-204.
7. Björklund K, Cousins AP, Strömvall AM, Malmqvist PA 2009. Phthalates and nonylphenols in urban runoff: occurrence, distribution and area emission factors. Sci. Total Environ. 407:4665-4672.
8. Blount BC, Milgram KE, Silva MJ, Malek NA, Reidy JA, Needham LL 2000 Quantitative Detection of Eight Phthalate Metabolites in Human Urine Using HPLC-APCI-MS/M. Analytical Chemistry 72(17):4127-4134.
9. Bornehag CG, Sundell J, Weschler CJI 2004 The Association between Asthma and Allergic Symptoms in Children and Phthalates in House Dust: A Nested Case- Control Study. Environ. Health Perspect. 112(14):1393-1397.
10. Chang BV, Yuan SY, Liu C, Liao CS (2002). Occurrence and microbial degradation of phthalate esters in Taiwan river sediments. Chemosphere 40:1295-1299.
11. Chen W, Kan AT, Pu G, Vignona LC, Tomson MB (1999). Adsorption-desorption behaviours of hydrophobic organic compounds in sediments of Lake Charlse Louisiana, USA. Environ. Toxicol. Chem. 18:1610-1616.
12. Chen SJ, Gao XJ, Mai BX, Chen ZM, Luo XJ, Sheng GY, Fu JM, Zeng EY (2006). Polybrominated diphenyl ethers in surface sediments of the Yangtze river Delta: Levels, distribution and potential hydrodynamic influence. Environ. Pollut. 144:951-957.
13. Clara M, Windhofer G, Hartl W, Braun K, Simon M, Gans O, Scheffknecht C, Chovanec A (2010). Occurrence of phthalates in surface runoff, untreated and treated wastewater and fate during wastewater treatment. Chemosphere 78:1078-1084.
14. Colon I, Caro D, Bourdony CJ, Rasario O (2000). Identification of Phthalate Esters in the Serum of Young Puerto Rican Girls with Premature Breast Development, Environ. Health Perspect. 108(9):895-900.

15. Dargnat C, Teil MJ, Chevreuil M, Blanchard M (2009). Phthalate removal throughout wastewater treatment plant: case study of Marne Aval station (France). Sci. Total Environ. 407:1235-1244.
16. Duty SM, Silva MJ, Barr DB, Brock J, Ryan WL, Chen Z, Herrick RF, Christiani DC, Hauser R (2003). Phthalate Exposure and Human Semen Parameters. Epidemiology, 14(3):269-277.
17. EU-RAR (2004). European Union Risk Assessment Report on Dibutyl Phthalate. Institute of Health and Consumer Protection (IHCP), European Chemicals Bureau, 1st Priority List P 29.
18. EU-RAR (2008). European Union Risk Assessment Report on Bis (2-ethylhexyl) Phthalate (DEHP). Institute of Health and Consumer Protection (IHCP), European Chemicals Bureau, 2nd Priority List 80.
19. Fatoki OS, Ogunfowokan AO (1993). Procedural clean- up technique for determination of phthalate esters in an aquatic environment. Int. J. Environ. Stud. 44(4):237-243.
20. Fatoki OS, Noma A (2001). Determination of phthalates esters in the aquatic environment. S. Afr. J. Chem. 54:69-83.
21. Fatoki OS, Vernon F (1990). Phthalate esters in river of the Greater Manchester area, UK. Sci. Total Environ. 95:227-232.
22. Fisher JS (2003) Human Testicular Dysgenesis Syndrome: A Possible Model Using In-Utero Exposure of the Rat to Dibutyl Phthalate, Human Reproduction 18(7):1383-1394.
23. Fujii M, Shinohara N, Lim A, Otake T, Kumagai K, Yanagisawa Y (2003). A study on emission of phthalate esters from plastic materials using a passive flux sampler. Atmos. Environ. 37:5495-5504.
24. Gobas FAPC, Mackintosh, CE, Maldonado J, Jing HW, Hoover N, Chong A, Ikonomou MG (2004). Distribution of phthalate esters in a marine aquatic food web comparison to polychlorinated biphenyls, Environ. Sci. Technol. 38:2011-2020.
25. Gomez-Hens A, Aguilar-Caballos MP (2003). Social and economic interest in the control of phthalic acid esters, Trends Anal. Chem. 22:847-857.
26. Heise S, Litz N (2004). Deskstudy- Phthalates. German federal Environmental Agency, Berlin, Germany. 41 p.
27. Hites A (1973). Analysis of trace organic compounds in New England River. J. Chromatogr. Sci. 11:570-574.
28. Hoppin JA, Ulmer R, London SJ (2004). Phthalate Exposure and Pulmonary Function. Environ. Health Perspect. 112(5):571-574. Doi:10.1289/ehp.6564
29. Jaakkola JJK, Oie L, Nafstad P (1999). Interior Surface Materials in the Home and the Development of Bronchial Obstruction in Young Children in Oslo, Norway. Am. J. Public Health 89(2):188-192.
30. Kanchanamayoon W, Prapatpong P, Chumwangwapee S, Chaithongrat S (2012) Analysis of phthalate esters contamination in drinking water samples. Afr. J. Biotechnol. 11(96)16263-16269.
31. Latini G (2003). In-Utero Exposure to Di-(2-ethylhexyl) Phtha late and Human Pregnancy Duration. Environ. Health Perspect. 111(4):1783-1785.
32. Mackintosh CE, Maldonado JA, Ikonomou MG, Gobas FAPC (2006). Sorption of phthalate esters and PCBs in a marine ecosystem. Environ. Sci. Technol. 40:3481-3488.
33. Mathur PK, Srivastava A, Sharma VP, Tripathi R, Kumar R, Patel DK (2010). Occurence of phthalate esters in Gomti river sediment, India. Environ. Monit. Assess. 169:397-406.
34. Ogunfowokan AO, Torto N, Adenuga AA, Okoh EK (2006). Survey of level of phthalate ester plasticizers in a sewage lagoon effluent and a receiving stream. Environ. Monit. Assess. 118(1-3):457-480.
35. Rayne S, Friesen KJ (2009). Contaminant trapping behind large dams: a mini-review, NaturalProtection.hdl:
36. Rozati R, Reddy PP, Reddanna P, Mujtaba R (2002). Role of Environmental Estrogens in the Deterioration of Male Factor Fertility, Fertility and Sterility 78(6):1187-1194.
37. Sheldon SC, Hites RA (1979). Organic compounds in the Delaware River. Environ. Sci. Technol. 12:1188-1194.

38. Staples CA, Peterson DR, Parkerton TF, Adams WJ (1997). The environmental fate of phthalate esters: A literature review. Chemosphere 35:667-749.
39. Sun J, Huang J, Zhang A, Liu W, Cheng W (2013). Occurrence of phthalate esters in sediments in Qiantang River, China and inference with urbanization and river flow regime. J. Hazard Mater. pp. 248-249, 142-149.
40. Swan SH, Main KM, Liu FSL, Stewart SL (2005). De-crease in Anogenital Distance among Male Infants with Prenatal Phthalate Exposure. Environ. Health Perspect. 113(8):1056-1061.
41. Tan BLL, Hawker DW, Müller JF, Leush FDL, Tremblay LA, Chapman HF (2007). Modelling the fate of selected endocrine disruptors in a municipal wastewater treatment plant in South East Queensland, Australia. Chemosphere 69:644-654.
42. Tao S, Shi Z, Pan B, Fan W, He XC, Zuo Q, Wu SP, Li BG, Cao J, Liu WX, Xu FL, Wang XJ, Shen WR, Wong PK (2005). Contamination of rivers in Tainjin, China by polycylic aromatic hydrocarbons. Environ. Monit. Pollut. 134:97-111.
43. Thi HD, Coyne A, Orange D, Blanc G, Etcheber H, Lan AL (2010). Long-term monitoring (1960-2008) of the river-sediment transport in the Red River Watershed (Vietnam): Temporal variability and dam-reservoir impact. Sci. Total Environ. 408:4654-4664.
44. Thuren A (1986). Determination of phthalates in aquatic environments. Bull. Environ. Contam. Hydrol. 36:33-40.
45. USEPA (US Environmental Protection Agency) (1980). Ambient water quality criteria for phthalate esters. Washington, DC. Office of water
46. Wang XT, Ma LI, Sun YZ, Xu XB (2006). Phthalate esters in sediments from Guanting reservoir and the Yongding River, Beijing People's Republic of China. Bull. Environ. Contam. Toxicol. 76:799-806.
47. Xia XH, Sha YJ, Yang GH (2007). Distribution of PAEs in the middle and lower reaches of the Yellow River, China Environ. Monit. Assess. 124:277-287.
48. Yuwatini E, Hata N, Taguchi S (2006). Behavior of di(2-ethylhexyl) phthalate discharged from domestic waste water into aquatic environment. J. Environ. Monit. 8:191-196.
49. Zeng F, Cui K, Xie Z, Liu M, Li Y, Zeng Z, Li F (2008). Occurence of phthalate esters in water and sediment of urban lakes in a subtropical city, Guengzhou, south China. Environ. Intl. 34:372-380.
50. Zeng F, Lin YJ, Cui KY, Wen JX, Ma YQ, Chen HL, Zhu F, Ma ZL, Zeng ZX (2010). Atmospheric deposition of phthalate esters in a subtropical city. Atmos. Environ. 44:834-840.
51. Zhou RB, Zhu LZ, Yang K, Chen YY (2006). Distribution of organochlorine pesticides in surface water and sediments from Qiantang River, East China. J. Hazard. Mater. 137:68-75.

CHAPTER 19

Forecasting natural hazards, performance of scientists, ethics, and the need for transparency

Fausto Guzzetti

ABSTRACT

Landslides are one of several natural hazards. As other natural hazards, landslides are difficult to predict, and their forecasts are uncertain. The uncertainty depends on the poor understanding of the phenomena that control the slope failures, and on the inherent complexity and chaotic nature of the landslides. This is similar to other natural hazards, including hurricanes, earthquakes, volcanic eruptions, floods, and droughts. Due to the severe impact of landslides on the population, the environment, and the economy, forecasting landslides is of scientific interest and of societal relevance, and scientists attempting to forecast landslides face known and new problems intrinsic to the multifaceted interactions between science, decision-making, and the society. The problems include deciding on the authority and reliability of individual scientists and groups of scientists, and evaluating the performances of individual scientists, research teams, and their institutions. Related problems lay in the increasing subordination of research scientists to politics and decision-makers, and in the conceptual and operational models currently used to organize and pay for research, based on apparently objective criteria and metrics, considering science as any other human endeavor, and favoring science that produces results of direct and immediate application. The paper argues that the consequences of these problems have not been considered fully.

Keywords: natural hazard, risk, landslide, forecast, predictability, accountability, performance, ethics, transparency

1. INTRODUCTION

I am a geologist, and I work as a research scientist for the Italian Consiglio Nazionale delle Ricerche, the single largest research organization in Italy (CNR www.cnr.it). Our mission is to design, promote, and execute research in all fields of science, and to transfer and disseminate knowledge to foster scientific, technological, economic, and social development in Italy. My personal research interest is on natural hazards, chiefly landslides, a slippery field in all senses when it comes to predictions

In this paper, I present ideas and considerations on problems that scientists face when attempting to predict natural hazards: landslides in my case. First, I examine the approaches commonly used by scientists to predict natural hazards (including landslides), outlying the conceptual limitations of the approaches. Next, I discuss the role of scientists as advisors of decision-makers, the supposed contrast between basic and applied science, and the possible subordination of science to politics and decision-making. This is followed by considerations on the lack of understanding of the societal role of scientists, and the need for specific training on ethics in scientific research. Finally, I criticize current methods to evaluate the performance of scientists based on infometrics, and I call for improved transparency of the science evaluation systems.

The ideas expressed in the paper are personal and subjective. They result from 30 years of work spent detecting, mapping, and predicting landslides, and attempting to ascertain landslide risk. I shaped my ideas executing research to improve the ability of the Italian civil protection system to cope with landslides, a widespread and frequent hazard with significant human and economic consequences in Italy (Guzzetti 2000; Guzzetti and Tonelli 2004; Guzzetti et al. 2014; Salvati et al. 2010, 2014). The Italian (local) perspective has conditioned my thoughts, but I hope that my other activities, including being president of the Natural Hazards division of the European Geosciences Union (EGU www.egu.eu) between 2002 and 2006, and an editor of the journal Natural Hazards and Earth System Sciences, have helped me retaining a broader perspective. Finally, my interest in the philosophy of science and in the societal role of scientists has influenced the ideas expressed in the paper, which were first written in a shorter paper published in the Italian magazine Ecoscienza (Guzzetti 2013).

2. LANDSLIDES AND THEIR PREDICTION

The paper deals with the prediction of landslides and their consequences. But what is a landslide? A landslide is the movement of a mass of earth, debris, or rock down a slope, under the influence of gravity (Cruden and Varnes 1996; Hungr, Leroueil, and Picarelli 2013) (Figure 1). The definition is clear and simple, apparently. Reality is different, and landslides are complex and diverse phenomena (Guzzetti et al. 2012). Landslides can fall, topple, slide, and flow, and many landslides exhibit a combination of these simple types of movements (Hungr, Leroueil, and Picarelli 2013), at the same time or through the lifetime of a landslide that spans the range from a few seconds (e.g., in the case of a rock fall) to several thousand years. The area and volume of most terrestrial landslides are in the ranges 100 < AL < 108 m2 and 10¡4 < VL < 1011 m3 , respectively (Brunetti, Guzzetti, and Rossi 2009; Guzzetti et al. 2009). The Naschitti landslide, in New Mexico, USA, extends for AL D 4 £ 108 m2 , with an estimated volume VL D of 2.2 £ 1011 m3 (Guzzetti et al. 2012), and the Saidmarech or Kabir Kuh landslide in Iran possibly the largest known sub-aerial terrestrial landslide extends for AL D 2 £ 1011 m2 . Submarine landslides are generally larger than the terrestrial failures, in the ranges 104 < AL < 1011 m2 and 102 < VL < 1012 m3 (Haflidason et al. 2004; Guzzetti et al. 2009; Brunetti, Guzzetti and Rossi 2009). Different natural triggers and human actions cause landslides (Guzzetti et al. 2009, 2012). The main natural triggers are intense or prolonged rainfall, rapid snowmelt, earthquakes, volcanic activity, and freeze thaw cycles. Human causes of landslides include excavations, overloading, construction works, leakage from water or sewage lines, irrigation, deforestation, and traffic. On Earth, a natural trigger can result in a single landslide (or no landslides at all), or in tens or even hundreds of thousands of landslides in periods ranging from a few seconds to a few days. Landslides are not limited to the Earth, and were found on planets and moons (Lucchitta 1978; Lucchitta 1981; McEwen 1989; Quantin et al. 2004; Brunetti et al. 2014). Despite the simple definition, landslides are diversified and complex phenomena. The complexity and variability of the landslides make it difficult to predict landslides, their single and collective properties, and most important for their societal consequences the impact of landslides on vulnerable elements, including the population.

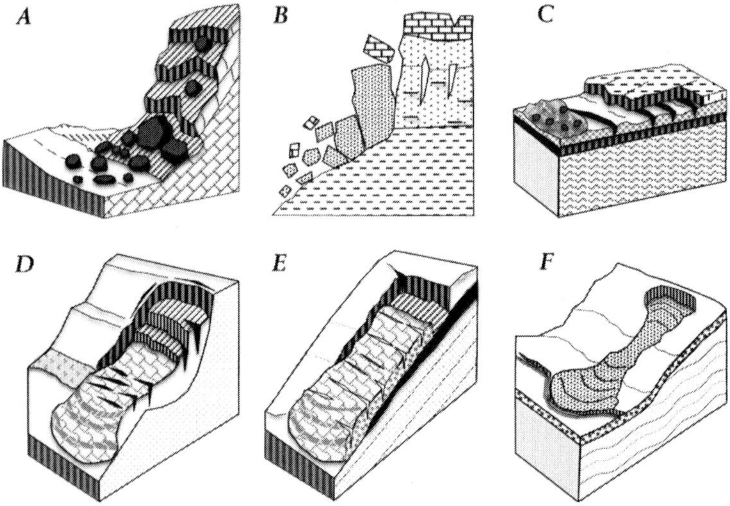

Figure 1. Landslide types: (A) rock fall, (B) topple, (C) lateral spread, (D) rotational slide, (E) translational slide, and (F) flow. Modified after Varnes (1978) and Cruden and Varnes (1996).

3. PREDICTING NATURAL HAZARDS

Scientists attempt to predict natural hazards (including landslides) and their consequences using a variety of approaches, methods, techniques, and tools, which to some extent depend on the type of the hazard and the information available for the prediction. Despite the differences, all the approaches can be loosely grouped into two broad categories: theoretic (mechanistic, deterministic, physically based, "hard") and empirical (statistical, functional, "soft") approaches. In very general terms, all natural phenomena, including natural hazards, can be classified based on their level of predictability (Prigogine and Stengers 1979; Prigogine 1996; Taleb 2007; Stein and Stein 2014). The motion of a bullet, pendulum, or planet exhibits comparatively little uncertainty, and can be predicted accurately in space and time using models expressed by simple mathematical equations (Figure 2). These natural phenomena share another important characteristic. Their physical behavior can be inferred from repeated observations. For these phenomena, time series and records of past events are important to determine the physical laws that control the phenomena, and for predicting future occurrences.

Other natural phenomena, including geophysical (e.g., earthquakes), meteorological (e.g., hurricanes, rainfall), and geomorphological (e.g., landslides, floods, erosion) phenomena that constantly shape the surface of the Earth (but also biological, ecological, and economic phenomena), are characterized by low predictability (which is different from unpredictability) (Stein and Stein 2014) (Figure 2). These phenomena are typically chaotic, and their low predictability originates in their unstable nature that can evolve catastrophically (Bak, Tang, and Wiesenfeld 1987, 1988; Rundle, Turcotte, and Klein 1996; Turcotte 1997). For these phenomena, repeated observations may not help much in the prediction of future events (Taleb 2009; Bl€oschl and Montanari 2010), particularly when the future events are outside the range covered by the past observations. Many geophysical, meteorological, and geomorphological phenomena exhibit distinct nonlinear behaviors revealed by their "heavy-tailed" statistics (Rundle, Turcotte, and Klein 1996; Turcotte 1995, 1997). This complicates further the possibility of using past occurrences to predict future events.

It must be stressed that attempting to forecast a natural phenomenon (including landslides) with limited or incomplete information or knowledge is not only inherently diffi- cult, but it may prove dangerous (Prigogine and Stengers 1979; Prigogine 1996; Fenton 2011; Nosengo 2012).

4. PREDICTING LANDSLIDES

If addressed in mechanistic terms, the problem of the prediction of a landslide is relatively simple, or so it may appear. A block that slides along an inclined plane controlled by friction (Skempton 1948; Taylor 1948; Janbu 1954; Hoek and Bray 1977), or a point mass that falls and bounces along parabolic trajectories until it rests when all the energy is lost in impacts (Guzzetti et al. 2002), are reasonable representations of landslides (Figure 3). Both representations are relatively simple to model using basic physics principles, and a number of numerical models and software exist that can be used to predict the behavior of single landslides, or of unstable slopes (Guzzetti et al. 2002; Wyllie and Mah 2005).

However, a pure mechanistic approach does not take us far, particularly when attempting to predict populations of landslides or the behavior of multiple unstable slopes over large areas (Alvioli et al. 2014; Raia et al. 2014; Mergili et al. 2014). This is because real landslides are more complex than their representation in the adopted physically based models, and because for most landslides, we do not know with sufficient accuracy (or we ignore completely) the geometrical and mechanical parameters that control the instabilities, from their points of failure to those of arrest. The behavior of a boulder falling from a cliff is impossible to predict "exactly" after just a few bounces, and rock falls are an example of a simple mechanical system whose behavior cannot be predicted even if the initial conditions and the driving force (i.e., gravity) are known (Guzzetti et al. 2002). To a reader unfamiliar with landslides, it may seem odd, but for most of the very many landslides in a typical landscape (Trigila, Iadanza, and Spizzichino [2010] have mapped more than 450,000 landslides in Italy, an average density of 1.5 landslides per square kilometer), we do not know the geometry, i.e., the extent, depth, and shape of the sliding surface, the volume of the instability, and the time or period of failure. For most landslides caused by rainfall or rapid snowmelt, we ignore the amount of water that has infiltrated into the ground causing the slope to fail (Guzzetti et al. 2007). This limits greatly our ability to predict landslides using the deterministic approach.

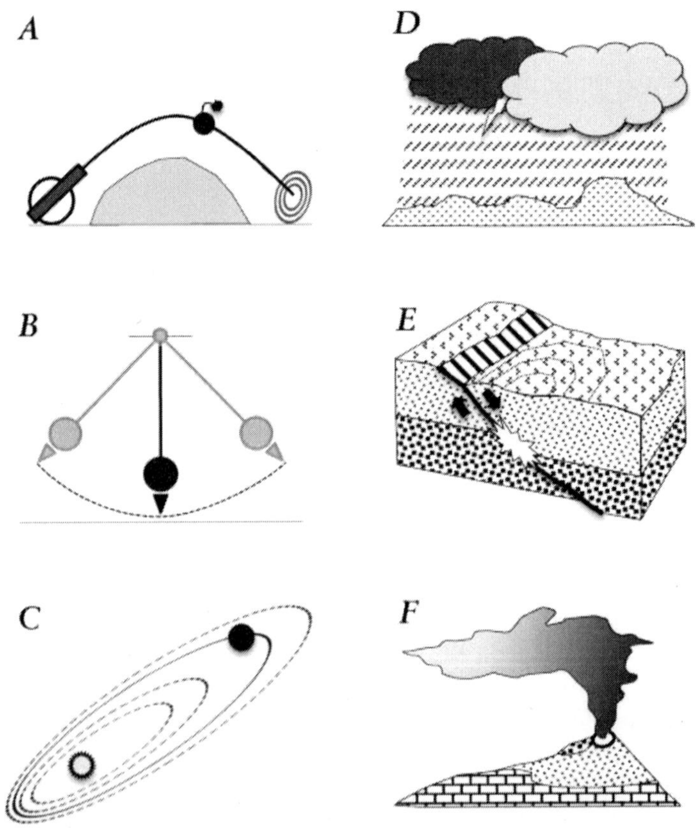

Figure 2. Natural phenomena and their predictability: (A) trajectory of a bullet, (B) pendulum, (C) orbit of a planet, (D) intense rainfall, (E) earthquake, and (F) volcanic activity. Left column depicts phenomena characterized by high predictability, and right column depicts phenomena characterized by low predictability.

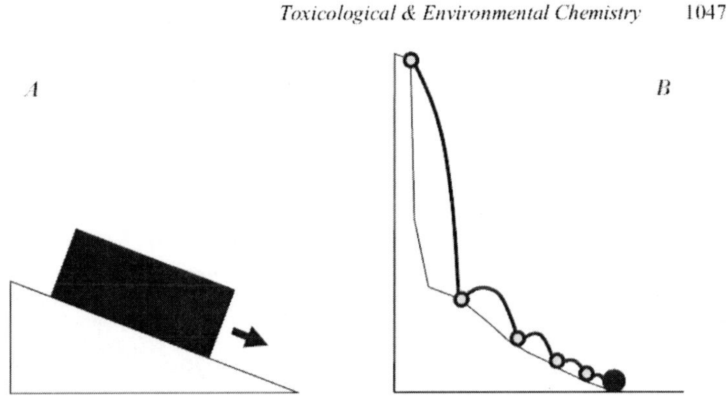

Figure 3. Simple but realistic representations of landslides: (A) a block that slides along an inclined plane is a good representation of a slide; (B) a point mass that falls and bounces along parabolic trajectories until all the energy is lost in impacts is a good representation of a rock fall, a particularly harmful type of landslide.

As for other natural phenomena, when the mechanistic approach fails to provide good results (for whatever reason, including the lack of a theory, of adequate knowledge, or of sufficient information) scientists revert to "soft" approaches that typically analyze the existing empirical observations using numerical (statistical) correlation analysis methods. Through the analysis of empirical observations, statistics is used in the attempt to identify patterns or relationships that can then be used to make predictions. However, as pointed out, for example, by Dondi and Moser (2014) in this volume, correlation does not prove cause effect evidence, necessarily. In a recent paper, Guangmeng and Jie (2013) have used satellite images showing "cloud anomalies" observed hours before an earthquake to "predict" three earthquakes in Bulgaria, Iran, and Italy. Unfortunately, the authors have failed to consider events for which similar "anomalies" were observed and earthquakes did not occur, and to demonstrate the cause effect evidence needed for predictions, something that was attempted by Harrison, Aplin, and Rycroft (2014). Lack of evidence of a cause effect relationship obtained through statistical correlation of empirical data is a potentially significant limitation of "soft" approaches, and it should always be considered when attempting to predict future events based on the results of "soft" modeling approaches. Further, "soft" methods require proper understanding of statistics and probability, and a correct interpretation of the mathematical results (D'Agostini 2011; Fenton 2011).

A benefit of statistical ("soft") models lays in the fact that they can measure the uncertainty associated with a predictive model, allowing to determining how much or how frequently a prediction is expected to be right or wrong. This can also be obtained with deterministic models, perturbing the initial conditions (e.g., like in ensemble forecasting, Leutbecher and Palmer [2008]) or by sampling randomly from known or inferred distributions of the parameters controlling the deterministic models (e.g., Raia et al. 2014; Mergili et al. 2014). The model uncertainty has different sources, including deficiencies in the model formulation, the approximate numerical methods used to solve the equations, and errors introduced using imperfect initial and other modeling conditions. Overall, this part of the uncertainty measures the lack of knowledge necessary for an accurate prediction. The uncertainty also measures the unpredictability of the phenomenon. Unpredictability and lack of knowledge are problematic to disentangle, and hard to communicate. The latter has many reasons, including the fact that scientists have to admit their lack of knowledge (i.e., ignorance) and cognitive limitations. Recognition of a (current) limit of science may be seen in contrast with the expectation that science can be used to predict all aspects of nature (Prigogine and Stengers 1979), including those related to natural hazards and their consequences. This may result in a loss of credibility by the general public. However, I maintain that there is an ethical duty to communicate the uncertainty associated to a prediction (Pozzati 2004). I further maintain that much work is needed to improve the ability of scientists (and science) to communicate their individual and collective ability (or inability, or the impossibility) to predict future hazardous events, and their consequences.

This is not an easy task, and it is complicated by the assortment and the sophistication of modern media, the increased demand of scientific information, and the inherent diffi- culty in communicating uncertain information. It is also unclear who should communicate, i.e., government (official) institutions, academic institutions, individual scientists, or independent organizations. The idea that only official (government) sources communicate reliable information may prove ineffective, particularly when the information is not delivered timely, and where other (independent) sources of information exist and are effective.

5. SCIENTISTS, DECISION-MAKERS, AND SOCIETY

This brings me to the relationship between scientists, decision-makers, and the society; an old problem. Decision-makers, once represented by emperors, kings, and other monarchs, and today by presidents, prime ministers, ministers, legislators, governors, and policy and decision-makers, who decide and act on behalf of all of us, regardless of their rank and role have always sought the advice of scientists, and many scientists have advised decision-makers. Ludovico Sforza ("il Moro"), duke of Milano, asked Leonardo da Vinci to build for him machines, innovative weapons, and defensive structures. Isaac Newton served the British government as Warden and Master of the Royal Mint. The US President Franklin D. Roosevelt asked scientists of the caliber of Enrico Fermi, Richard P. Feynman, and J. Robert Oppenheimer to help the USA and its allies win World War II.

He also asked Venner Bush, director of the wartime Office of Scientific Research and Development, "how the lessons that had been learned (...) could be applied in the days of peace". Bush (1945) responded with the famous report "Science, the endless frontier". It is the rationale behind the saying "when the going gets tough, the toughs have to get going".

As a civil servant working for the largest research organization in Italy, I do not have a problem (moral, ethical, or else) with scientists working for, or advising decision-makers. I am convinced that science (and scientists) can be useful to decision-makers, and benefi- cial to the society. Otherwise, I would not have accepted to be part of the "Commissione Nazionale per la previsione e la prevenzione dei Grandi Rischi", a group of experts that collectively advise the Prime Minister and the Italian National Department for Civil Protection on natural and human-induced risks. However, I perceive a number of problems in the relationship between scientists, decision-makers, and the society. Some have always existed; others are new or were exacerbated recently (Sidle et al. 2013; Bilotta, Milner, and Boyd 2014; Nursey-Bray et al. 2014).

A first problem lays in the selection of who should advice the decision-makers. Ideally, decision-makers should seek for the "best" scientists in the pool. But who decides who is the "best", i.e., who is authoritative (influential) and who is not? The question is not trivial, particularly if one considers that often the problems for which scientists are asked for their advice are outside the fuzzy boundary of consolidated science, and require the scientists' "educated" or "informed" opinion. In the past, the difficult and uncertain selection was based (largely, if not entirely) on the cursus honorus of the scientist, i.e., on the results he/she had obtained in his/her career, demonstrated by the scientific work accomplished and the papers published, summarized in the CV. The opinion of the peers (i.e., the other scientists) was important to determine who was influential (authoritative) in a specific field, and who was not. The system was far from perfection, and established scientists could (and did) abuse of their authority. An example for all: Isaac Newton considered impossible, in face of contrary evidence, that the problem of determining the longitude at sea could be solved with (then) innovative mechanical clocks, a breakthrough in navigation (Sobel

1995). Here, I maintain that the track record of a scientist is important (mandatory) but that his/her future record is unpredictable. The fact that a scientist has obtained significant results or has made successful predictions in the past is no guarantee that the same scientist will be right on the next prediction

I feel that in the recent years the relevance of the cursus honorus (the track record) and the opinion of the peers have diminished, significantly. Others contribute to deciding who is authoritative (influential) in a specific field. Further, when a scientist is (or is considered to be) an authority in a field, it is easy to consider her/him an authority in (apparently) related fields. Thus, an (true) expert in climate can be considered an expert in meteorology, and even in the various impacts of intense rainfall including floods, flash floods, landslides, and erosion. This, in spite of the fact that climate and, say, landslides are different phenomena, investigated using rather different methods and techniques. But who are the "others" who decide on authoritative scientists? The media and chiefly TV and the Web play a significant role. With an increased number of laws regulating medical, environmental, ecological, civil protection, and other personal and collective safety issues, the legal responsibilities of individuals and organizations have increased, or have materialized where they did not existed before. The judiciary system (e.g., courts of laws, judges, prosecutors, lawyers) exploits science (and scientists) more than ever before (Fenton 2011), and it is also (implicitly, and possibly unwillingly) contributing to the selection of influential scientists.

Politicians and policy and decision-makers use the advice of scientists, and contribute to decide who is authoritative and, implicitly, who is not. The later can generate a (potentially dangerous) short circuit, with decision-makers selecting scientists that fit their needs and expectations, and scientists fulfilling the needs and the expectations of the decisionmakers. In Italy, many regional and even local administrations rely on experts working for local or nearby universities and research centers. However, it is unrealistic to think that in all the local universities and research centers, one can find leading experts and influential scientists. It is not sufficient to be employed by a university or a research center, or to publish in a scientific journal, to be an expert or an influential scientist. Further, there are excellent scientists employed by public, non-academic administrations and by private enterprises and businesses, who do not publish regularly in international journals. I maintain that universities and research centers, and the society as a whole, should acknowledge this evidence.

A second problem consists in the potential (or factual) increased subordination of scientists to policy and decision-makers. In my broad field of research (i.e., understanding and predicting natural hazards), the relationship between scientific knowledge and its exploitation by policy and decision-makers is changing, rapidly. On the one hand, more than ever before scientists are becoming advisors (or consultants) to decision-makers. A consultant serves the interests of the client, and when the "consultant" is a scientist and the "client" is a decision-maker, there is a subtle, but not negligible, risk that the scientist can lose the status of an independent and "free" scholar. On the other hand, decision-makers, and particularly elected officials, tend to lay the burden of hard and often unpleasant decisions on their advisors (i.e., the scientists). This is unfair and inadequate, because the roles and responsibilities of decision-makers and scientists are different. It is of paramount importance that scientists always perform their job of "advisors" to their "clients" adopting the highest possible standards and the best available knowledge. This will contribute to mitigate the problem, but will not resolve it, totally. Also, we should acknowledge that the highest standards and the best knowledge might not be accessible to all scientists, always and everywhere. This is known, for example, in the medical field.

I acknowledge that the different roles and responsibilities are often difficult (or even impossible) to separate. However, I maintain that a role of scientists in modern societies is also that of "whistle blowers" (Shrader-Frechette 2007; Benchekroun and Pierlot 2011). In sports, referees and umpires blow the whistle or raise a flag when a fault is committed. Similarly, scientists should blow their whistles and raise their flags when policy and decision-makers do not act, or do not act properly, or effectively. This can prove useful to prevent and mitigate risks (Dondi and Moser 2014). In sports, good referees and umpires are the independent referees and umpires. The same is for science, and scientists must be independent to be good and respectable referees who blow the whistle only when it is necessary.

It is worth investigating the reasons for the changing balance between science and decision-making. I maintain that it is simplistic to blame politics and the decision-makers. Certainly, in Italy as in other countries politics has percolated deeply into the structure of the modern society. The niche of the society represented by scientists and their organizations, including universities and research institutes and centers, is no exception and the influence of politics on science has increased, undoubtedly. In Italy, the government appoints the presidents and the administration boards of all the major public research organizations, including the National Research Council (CNR). I am not arguing this is negative, necessarily. I notice that this top-down approach that does not stop at the level of the presidents and the administration boards is not balanced by bottom-up (e.g., "elected") counterweighs, that existed in the past and were eliminated.

I argue that others and more important reasons exist for the increased, and not necessarily all healthy, links between scientists, politics, and decision-makers. The first and possibly the most important is that to execute research one needs adequate resources (in other words "funding"), and politics and decision-makers control the public resources, entirely. Again, I am not arguing that this is negative, necessarily. What I want to point out is that how research is executed should not be decided by politics or the decision-makers, but by the scientists. On the other hand, scientists should not interfere directly on the adoption (or lack of adoption) of scientific findings and innovation by the policy and the decision-makers. I acknowledge that complete separation is difficult, but I maintain that in a healthy society, the roles of science (and scientists) and of politics (and decisionmakers) are and should remain different.

A measure of the changing balance between those who provide the resources (politics, decision-makers, funding agencies) and those who use (consume) the resources (the scientists) is a not so novel anymore paradigm (model) for the organization and the execution of research. Nowadays, (almost) all the funding is given to execute "committed" research, i.e., research (more or less) explicitly requested by a "costumer", e.g., a funding agency or an administration that has requested a specific "result". I call it "hired science", without any negative or diminutive implications. Research is now bounded by legal contracts (see, e.g., the Grant Agreements regulating the projects funded by the European Commission that, together with their annexes, explain in great detail what shall be produced (the "deliverables") and when (the "milestones")). Supposedly, advantages of this model to organize research include (1) an increased competition among scientists, who are stimulated by the need to obtaining resources (a consequence of the dreadful rule of "publish or perish"), (2) an improved evaluation of the results obtained ("accountability"), which are compared to the results expected (e.g., promised), and (3) a straightforward exploitation of the innovations obtained (if any). This organizational method illustrates the outspoken modern interest towards "useful" science, which generates results for immediate application, preferably with commercial implications. The European Horizon 2020 flagship research and innovation program (Council of the European Union 2013) is framed around this paradigm. I maintain that scientists share the responsibility for the adoption of this paradigm. Scientists have often used significant resources without control, and not providing reasonable results and feedbacks. This was bad. Also, scientists have contributed in various ways to conceiving and implementing the paradigm, allegedly in an attempt to improve their accountability. I acknowledge the importance of accountability, but I am not convinced this is a way to obtain it.

I see a number of drawbacks arising from the implementation of the described (and prescribed) method of organizing and funding science. The interest of scientists to collaborate has decreased significantly. Individual scientists, research teams, and even research organizations pool together to compete for (and participate in) projects to obtain resources (i.e., to get funded), and not for a driving (and healthy) interest in the collaboration with their peers. This is exemplified by the increasing number of scientists from the same team or laboratory that participate to competing proposals in response to the same call, with the sole scope of getting the grant. For many scientists, the main scope is getting the grant, and not doing the research. This attitude, justified by the shortage of resources (compared to the number of scientists), has another drawback. An increasing number of research groups and even of individual scientists pretend (or attempt) to be knowledgeable and experienced in fields for which they are not. Every competing group attempts to have all expertizes internally, reducing the needs for collaborations, and contributing to generate a self-referenced system.

The obsessive need to demonstrate the ability to obtain quick results has increased the tendency to produce results which are minor, marginal, and repetitive (Errami and Garner 2008), and of (almost) exclusive interest to the "client" of the specific research. There is also a propensity to overestimate (and overemphasize) the relevance and the applicability of the results obtained. A related problem is a tendency to ignore (unwillingly) or to hide (voluntarily) the problems encountered during a research, the workarounds adopted, the limitations of the results obtained, and the uncertainties inherent in any scientific endeavor. I maintain that the medium- to long-term consequences of these drawbacks have not been considered sufficiently. A consequence that is becoming manifest is the increasing lack of proper scientific discussion. The peer review systems that lays at the base of the self-evaluation of the quality of science is becoming less effective, with an increasing number of papers with wrong or irrelevant results published in peer-reviewed journals.

6. BASIC OR APPLIED SCIENCE?

This brings me to the presumed or concrete contrast between basic (i.e., theoretic, fundamental) and applied (i.e., problem solving, hired) science. I acknowledge that differences exist between basic and applied science. Applied science cannot exist (or does not have much sense) if it does not rely on theoretic principles discovered by fundamental science. Applied science can also serve as a stimulus for basic science (for landslide studies, see, e.g., Picarelli [2009]). Differences exist (or emerge) in the missions of the institutions that perform science. In many countries, the mission of universities is focusing on basic (fundamental) research (in addition to education), and the mission of research councils (including the

Italian CNR) and of national institutes and centers is concentrating on applied science (in addition to transfer and exploitation of knowledge for the common societal and economic benefit). Despite these differences, I maintain that in the modern world the distinction between basic and applied science is irrelevant, and the sole relevant difference is between good and poor science. This is not a popular opinion, today

The common opinion   particularly among decision-makers and funding organizations   is that "good" science is "useful" science. I have nothing against applied science (quite the contrary), and I welcome efforts to exploit scientific results and technological innovations for practical purposes. I also welcome educational activities and outreach efforts that help bring science "out of the labs". As I wrote before, I am firmly convinced that science can be useful and beneficial to the society. What I am noticing is that in my country   and in others too   decision-makers, including the highest levels of governments, are unable to ask scientists difficult questions relevant to the society. They are tailing the new discoveries and innovations, rather than stimulating the scientific community with challenging questions that may (and will) result in new discoveries and advancements. This is different from the approach followed, for example, by the UK government in 1714 when it offered the "Longitude Prize" for anyone who could find a practical solution for the precise determination of longitude at sea (Sobel 1995), a challenging problem for navigation at the time. The approach of challenging scientists with hard problems is well exemplified by the speech given by the American President John F. Kennedy at the Rice University, in Houston, TX, on 12 September 1962. To endorse the new space program, Kennedy said, passionately:

We choose to go to the Moon in this decade and do the other things, not because they are easy, but because they are hard, because that goal will serve to organize and measure the best of our energies and skills, because that challenge is one that we are willing to accept, one we are unwilling to postpone, and one which we intend to win, and the others too.

The driving idea is that research fosters economic growth. I share this idea, deeply, but I maintain that the link between science and economic growth is complex and not linear. Today, science is considered as any other venture. I respect entrepreneurs, for their ability to produce innovation, for their willingness to take chances, and for their vision and wisdom. I am happy when a new product or service makes my life better, or easier. However, I do not think that scientists should be (or attempt to be) entrepreneurs, necessarily. I maintain that there is more to science than venture, and that considering science as any other endeavor may not be such a good idea.

With some exception (e.g., Rees 2014), it is self-evident that today politics and decision-making react to, and do not anticipate the ongoing or the expected societal, economical, and environmental changes (and challenges).

It is perhaps less evident that this is happening in science as well. I will give an Italian example. Italy is exposed to most (if not all) natural hazards, and is the country in Europe with the largest number of human consequences produced by natural hazards. In the 64-year period, 1950 2013, more than 7000 people have died, went missing, or were injured by landslides and floods alone. In the same period, the number of evacuated or homeless people due to geo-hydrological hazards exceeded 700,000 (Guzzetti, Stark, and Salvati 2005; Salvati et al. 2010, 2014). Between 1944 and 2012, the direct economic damage caused by all natural hazards in Italy was estimated to exceed €240 billion (2011), a yearly average of €3.5 billion. Landslides and floods accounted for 25% of the total, i.e., €61.5 billion (2011) (ANCECRESME 2012). With significant uncertainties (Guzzetti et al. 2014), the figures are increasing, and measure the societal and economic problem posed by natural hazards, and geo-hydrological hazards, in particular, in Italy. Despite the problem being manifest, Italy does not have a scientific program on natural hazards, their consequences, and for the design of sustainable mitigation and adaptation strategies. There is no coordination of the research activities on natural hazards. The government National Research Program 2014 2020 ignores natural hazards and their societal and economic consequences. The government has announced an investment of €2 billion to mitigate geo-hydrological hazards. This unique effort will be conducted without any scientific support. This is not a matter of basic vs. applied science, or of good vs. poor science. This is evidence that politics and decision-making in Italy are shortsighted, and not interested in exploiting science and its potential contribution to the society. The general public should be aware of this shortsightedness. I maintain that in Italy a shift in paradigm is necessary to bring science at the core of the solution of the problems posed by natural hazards and their associated risk.

7. A MATTER OF SPEED AND ETHICS

Information is more abundant today than in any other period. Abundant information contributes to democracy, because informed decisions are more conscious and, in general, better decisions. Not only information is more abundant, it also circulates faster. Almost everything is faster today than it was in the past, and scientific research is no exception. I take 1991 as the year to mark a significant change in pace. In 1991, the European Organization for Nuclear Research CERN announced the World Wide Web, and today it is (nearly) impossible to execute relevant research disconnected from the Internet. The Internet gives individual scientists and research groups fast and constant access to the existing knowledge and new information (e.g., through publications, conference proceedings, web seminars). Scientists receive through the net new data to analyze, and they process vast amount of data using innovative network technologies (e.g., grid and cloud technologies, search engines). Once obtained, scientists publish their results on the Internet that disseminates the new findings quickly and effectively.

Although research goes fast, it still proceeds slower than desired. The medical and the pharmaceutical fields are good examples. Progress is made virtually every day in these fields, but not fast enough to fulfill the needs and expectations of individuals and the general public, which are growing at a faster pace. Despite the efforts and the resources invested, in many fields of medicine, progress remains slow and an increasing number of people turn to untested and unreliable treatments. Lack of ethics and of full understanding of the societal role and responsibility of science and scientists (Shrader-Frechette 1994), and poor understanding of how science works, chiefly by the media and the general public, favor the sprouting of "miraculous" solutions. There is increasing impatience for "fast" scientific results, not considering that the inability (or the impossibility) to obtain immediate results does not mean that scientists are inadequate or unable, necessarily. Reality is different, and sufficient time is required to design and execute experiments, to verify results, and to falsify the proposed models and theories. The (apparent) slowness of science contrasts with the inclination of the "customers" (including the general public) who are unwilling to wait, or to postpone the adoption of results that were not peer reviewed or validated or, even worse, failed to pass proper scientific validation. The issue is manifest in the medical and the pharmaceutical fields, but it can and will touch other fields of science, including those related to the prediction of natural and environmental hazards and their consequences.

Many geoscientists are unaware of the role, and of the societal and ethical consequences of their work (Shrader-Frechette 1994; Wyss and Peppoloni 2014). This is largely due to a lack of education and specific training, more than the absence of interest. Training in philosophy, and on the ethical problems faced by geoscientists, and scientists in general, should be introduced as part of all university courses. Research institutes and professional organizations should adopt ethical codes. Matteucci et al. (2012) have proposed for geologists an oath similar to the "Hippocratic Oath" that binds medical doctors to a code of ethics. The full content and exact scopes of an ethical code for geoscientists are a matter of discussion (Wyss and Peppoloni 2014), but the design and the acceptance of a shared code of ethics may (1) promote the social responsibility of geologists as scientists and professionals, (2) increase the awareness for the expectations and needs of citizens, decision-makers, and the society, (3) favor the acceptance and use of geoscientific and environmental knowledge and information, and (4) raise awareness for the social mission of geoscientists. Although I have no na€ıve illusions, I argue that the introduction of a code of ethics, associated with proper deontological training, can contribute to the improved use of geoscientific knowledge, and to an ethical approach to the management of natural hazards and their consequences.

8. PERFORMANCE AND TRANSPARENCY

In the recent years, in Italy and elsewhere in the World, universities, research organizations, and funding agencies are increasingly using "infometrics" (or "scientometrics") (Tague-Sutcliffe 1992; Hood and Wilson 2001; Bar-Ilan 2008; Harzing 2011) to measure and rank the performances of individual scientists. I am not against the use of metrics to evaluate performances, including the performances of scientists, groups, and research organizations. However, I maintain that (1) it is illusory to use solely infometrics/scientometrics for the evaluation of scientists, and (2) scientists should be told beforehand what are the rules and metrics used to evaluate their performances. The latter has not occurred in Italy.

It is worth asking why these numerical "metrics" were introduced, and are now used (and abused). They were introduced in an attempt to measure and rank "objectively" the performances, first of scientific journals and next of individual scientists and research organizations. But why was this needed? It became necessary (or deemed necessary) when older evaluation and ranking methods based chiefly on the subjective judgment of peers failed to work. I argue that the failure was largely a result of a lack of ethics among the peers, and not an inherent flaw in the evaluation system. Scientists started to judge and rank their peers partially, with a more benevolent eye toward their academic friends, and a more critical eye for their academic opponents. Partiality and unfair judgment resulted in a loss of credibility in the evaluation system. I argue that the attempt to solving the problem of a reduced credibility by introducing "objective" metrics does not solve the problem. Quite the opposite, as it can foster the growth of lobbies and pressure groups that attempt to bend the system to their benefit. Also, no one that knows how research works can argue that a single metrics, or even a collection of metrics, can capture all the multifaceted aspects of the performance of a scientist, or of a research institute or organization.

What we need to judge science and rank scientists are authoritative, experienced, and independent reviewers, and a more transparent evaluation system. Transparency is a key, but it is difficult to implement. A number of international journals, including most of the journals of the EGU are adopting an "open discussion" system that allows reviewers and other peers to comment publicly on papers before they are published in the journals. The system is far from perfection, but it contributes to fair reviews and to the quality of the publications and the peer review system. I maintain that funding organizations should also embrace transparency in their evaluations. The evaluation system adopted by the European Commission does not shine for transparency, and some of the evaluators have never coordinated a European project. One wonders how these referees can properly judge the proposals they are asked to evaluate.

9. CONCLUDING REMARKS

Like other natural hazards, landslides are difficult to predict. The difficulty arises from the poor understanding of the phenomena that control landslides, and from the inherent low predictability of landslides, a result of their complexity and chaotic nature (Turcotte et al. 2002). However, low predictability is different from unpredictability (Stein and Stein 2014), and efforts should be made to improve our ability to forecast landslides and their consequences. Low predictability results in large uncertainties in the forecasts, and new efforts are needed to determine and to communicate the uncertainties. I argue that the latter is an ethical responsibility of scientists.

Attempting to predict landslides is a problem of scientific interest and of societal relevance, particularly in Italy where landslides are widespread (Trigila, Iadanza, and Spizzichino 2010) and pose a severe threat (Guzzetti et al. 2007; Salvati et al. 2010). Scientists experimenting landslide forecasting are exposed to issues inherent to the complex interactions between science, politics, and the society. A first problem consists in deciding who is authoritative (influential, reliable, trustworthy), and who is not (noninfluential, unreliable, untrustworthy). The difficultly is in the fact that even an excellent track record is no guarantee of the future performances of a forecaster. A second problem is the increasing subordination of science to politics and decision-making. For multiple reasons, scientists are turning into "advisors" or "consultants" to decision-makers, often without a complete understanding of their societal and ethical responsibility. A third problem lays in the way science is organized, executed, and funded, based on apparently objective criteria and metrics, and considering the scientific endeavor as any other human venture. The pressure to obtain quick results of immediate application has increased the tendency to produce results which are minor, marginal, repetitive, and of limited general interest. A related problem is a tendency to ignore or to hide critical issues and limitations of the results obtained.

I argue that the consequences of these problems are serious, and have not been considered sufficiently. Finding a solution to these problems in not trivial, and requires the collaboration of all the parties involved. Training on ethics in scientific research (Shrader-Frechette 1994; Wyss and Peppoloni

2014), the adoption of research and professional ethical codes (Matteucci et al. 2012), and more transparent evaluation and review systems, may contribute to solve or at least to face the problems. Further research on these and related issues is needed.

ACKNOWLEDGMENTS

I would like to acknowledge Stefano Tibaldi for inviting me to the Forum on Scienza, Responsabilità e Governo — conoscere, decidere e comunicare nella società del rischio held on 12 June 2013 at the Accademia delle Scienze di Bologna, Italy; Giancarlo Naldi for giving me the opportunity to contribute to the October 2013 issue of Ecoscienza, where a shorter version of this paper was published; and Francesco Dondi and Hartmut Frank for allowing me to contribute to this thematic issue of Toxicological and Environmental Chemistry. The comments and insights of two anonymous reviewers helped improving the manuscript, considerably.

DISCLOSURE STATEMENT

No potential conflict of interest was reported by the author.

REFERENCES

1. Alvioli, Massimiliano, Fausto Guzzetti, and Mauro Rossi. 2014. "Scaling Properties of Rainfall Induced Landslides Predicted by a Physically Based Model." Geomorphology 213: 38–47.
2. ANCE-CRESME. 2012. Lo stato del territorio italiano. [In Italian.] Rome: ANCE-CRESME.
3. Bak, P., C. Tang, and K. Wiesenfeld. 1987. "Self-organized Criticality: An Explanation of 1/f Noise." Physical Review Letters 59 (4): 381–384.
4. Bak, P., C. Tang, and K. Wiesenfeld. 1988. "Self-organized Criticality." Physical Review A 38 (1): 364–374.
5. Bar-Ilan, J. 2008. "Informetrics at the Beginning of the 21st Century – a Review." Journal of Informetrics 2 (1): 1–52.
6. Benchekroun, T.H., and S. Pierlot. 2011. "Whistleblowers: An Essential Resource for the Sustainable Prevention of Risks in Sociotechnical Systems." Work 41 (Suppl 1): 3051–3061. doi:10.3233/WOR-2012-0563-3051.,
7. Bilotta, G.S., A.M. Milner, and I. Boyd. 2014. "On the Use of Systematic Reviews to Inform Environmental Policies." Environmental Science Policy 42: 67–77.
8. Blöschl, Gunther, and Alberto Montanari. 2010. "Climate Change Impacts – Throwing the Dice?" Hydrological Processes 24, 374–381.
9. Brunetti, M.T., F. Guzzetti, M. Cardinali, F. Fiorucci, M. Santangelo, P. Mancinelli, G. Komatsu, and L. Borselli. 2014. "Analysis of a New Geomorphological Inventory of Landslides in Valles Marineris, Mars." Earth and Planetary Science Letters 405: 156–168.

10. Brunetti, M.T., F. Guzzetti, and M. Rossi. 2009. "Probability Distributions of Landslide Volumes." Nonlinear Processes in Geophysics 16: 179–188.
11. Bush, Venner. 1945. Science, the Endless Frontier: A Report to the President on a Program for Postwar Scientific Research. Washington, DC: National Science Foundation.
12. Council of the European Union. 2013. Council Decision of 3 December 2013 Establishing the Specific Programme Implementing Horizon 2020 – the Framework Programme for Research and Innovation (2014–2020) and Repealing Decisions 2006/971/EC, 2006/972/EC, 2006/973/EC, 2006/974/EC and 2006/975/EC. http://ec.europa.eu/research/participants/data/ref/h2020/legal_basis/sp/h2020-sp_en.pdf.
13. Cruden, D.M., and D.J. Varnes. 1996. "Landslide Types and Processes." In Landslides, Investigation and Mitigation, edited by A.K. Turner and R.L. Schuster, 36–75, Washington, DC: Transportation Research Board.
14. D'Agostini Giulio. 2011. "Probably a Discovery: Bad Mathematics Means Rough Scientific Communication." ArXivorg preprint, arXiv:1112.3620, 21 p.
15. Dondi, F., and F. Moser. 2014. "University and the Risk Society." Toxicological and Environmental Chemistry.
16. Errami, M., and H.R. Garner. 2008. "A Tale of Two Citations." Nature 451 (7177): 397–399.
17. Fenton, N. 2011. "Improve Statistics in Court." Nature 479: 36–37.
18. Guangmeng, G., and Y. Jie. 2013. "Three Attempts of Earthquake Prediction with Satellite Cloud Images." Natural Hazards and Earth System Sciences 13 (1): 91–95.
19. Guzzetti, Fausto. 2000. "Landslide Fatalities and the Evaluation of Landslide Risk in Italy." Engineering Geology 58 (2): 89–107.
20. Guzzetti, Fausto. 2013. "Il dovere di comunicare l'incertezza delle previsioni." [In Italian.] Ecoscienza 4: 26–27.
21. Guzzetti, F., F. Ardizzone, M. Cardinali, M. Rossi, and D. Valigi. 2009. "Landslide Volumes and Landslide Mobilization Rates in Umbria, Central Italy." Earth and Planetary Science Letters 279 (3–4): 222–229.
22. Guzzetti, F., G.B. Crosta, R. Detti, and F. Agliardi. 2002. "STONE: A Computer Program for the Three-Dimensional Simulation of Rock-Falls." Computers and Geosciences 28: 1079–1093.
23. Guzzetti, F., A.C. Mondini, M. Cardinali, F. Fiorucci, M. Santangelo, and K.-T. Chang. 2012. "Landslide Inventory Maps: New Tools for an Old Problem." Earth-Science Reviews 112 (1–2): 42–66.
24. Guzzetti, F., S. Peruccacci, M. Rossi, and C. P. Stark. 2007. "Rainfall Thresholds for the Initiation of Landslides in Central and Southern Europe." Meteorology and Atmospheric Physics 98 (3): 239–267.
25. Guzzetti, F., C.P. Stark, and P. Salvati. 2005. "Evaluation of Flood and Landslide Risk to the Population of Italy." Environmental Management 36 (1): 15–36.
26. Guzzetti, F, and G. Tonelli. 2004. "Information System on Hydrological and Geomorphological Catastrophes in Italy (SICI): A Tool for Managing Landslide and Flood Hazards." Natural Hazards and Earth System Sciences 4: 213–232.
27. Guzzetti, F., R. Zoboli, P. Salvati, C. Bianchi, and M. Mazzanti. 2014. "Quanto sono costate e quanto costano le calamità idrogeologiche in Italia?"[In Italian.] Proceedings XIII Giornata Mondiale dell'Acqua, Accademia Nazionale dei Lincei, Rome, 22 March 2013, Atti dei Convegni Lincei, Vol. 283, 13–31.
28. Haflidason, H., H.P. Sejrup, A. Nygård, J. Mienert, P. Bryn, R. Lien, C.F. Forsberg, K. Berg, and D. Masson. 2004. "The Storegga Slide: Architecture, Geometry and Slide Development." Marine Geology 213 (1–4): 201–234.
29. Harrison, R.G., K.L. Aplin, and M.J. Rycroft. 2014. "Brief Communication: Earthquake–Cloud Coupling Through the Global Atmospheric Electric Circuit." Natural Hazards and Earth System Sciences 14 (4): 773–777.
30. Harzing, Anne-Wil. 2011. The Publish or Perish Book. Melbourne: Tarma Software Research Pty Ltd.
31. Hoek, Evert, and John Bray. 1977. Rock Slope Engineering. 1st ed. London: Institute of Mining and Metallurgy.

32. Hood, William W., and Concepcion S. Wilson. 2001. "The Literature of Bibliometrics, Scientometrics, and Informetrics." Scientometrics 52 (2): 291–314.
33. Hungr, O., S. Leroueil, and L. Picarelli. 2013. "The Varnes Classification of Landslide Types, an Update." Landslides 11 (2): 167–194.
34. Janbu, N. 1954. "Application of Composite Slide Circles for Stability Analysis." Proceedings European Conference on Stability of Earth Slopes 3: 43–49.
35. Leutbecher, M., and T.N. Palmer. 2008. "Ensemble Forecasting." Journal of Computational Physics 227 (7): 3515–3539.
36. Lucchitta, Baerbel K. 1978. "A Large Landslide on Mars." Geological Society of America Bulletin 89: 1601–1609.
37. Lucchitta, Baerbel K. 1981. "Mars and Earth: Comparison of Cold-Climate Features." Icarus 45 (2): 264–303.
38. Matteucci, R., G. Gosso, S. Peppoloni, S. Piacente, and J. Wasowski. 2012. "A Hippocratic Oath for Geologists?" Annals of Geophysics 55 (3): 365–369.
39. McEwen, Alfred S. 1989. "Mobility of Large Rock Avalanches: Evidence from Valles Marineris, Mars." Geology 17: 1111–1114. d
40. Mergili, M., I. Marchesini, M. Alvioli, M. Metz, B. Schneider-Muntau, M. Rossi, and F. Guzzetti. 2014. "A Strategy for GIS-Based 3-D Slope Stability Modelling Over Large Areas." Geoscientific Model Development 7 (6): 2969–2982,
41. Nosengo, Nicola. 2012. "On the Fault Line." Nature 492: 342–343.
42. Nursey-Bray, M.J., J. Vince, M. Scott, M. Haward, K. O'Toole, T. Smith, N. Harvey, and B. Clarke. 2014. "Science into Policy? Discourse, Coastal Management and Knowledge." Environmental Science Policy 38: 107–119.
43. Picarelli, Luciano. 2009. "Understanding to Predict." In Landslides–Disaster Risk Reduction, edited by Kioji Sassa and Paolo Canuti, 63–88. Heidelberg: Springer-Verlag Berlin.,
44. Pozzati, Piero. 2004. "Il Convenzionalismo nel Calcolo Strutturale Sismico." [In Italian.] Inarcos 1–5.
45. Prigogine, I. 1996. The End of Certainty: Time, Chaos and the New Laws of Nature. New York: The Free Press.
46. Prigogine, I., and I. Stengers. 1979. La nouvelle alliance, mètamorphose de la science. [In French.] Paris: Gallimard.
47. Quantin, C., P. Allemand, N. Mangold, and C. Delacourt. 2004. "Ages of Valles Marineris (Mars) Landslides and Implications for Canyon History." Icarus 172: 555–572.
48. Raia, S., M. Alvioli, M. Rossi, R.L. Baum, J.W. Godt, and F. Guzzetti. 2014. "Improving Predictive Power of Physically Based Rainfall-Induced Shallow Landslide Models: A Probabilistic Approach." Geoscientific Model Development 7 (2): 495–514.
49. Rees, Martin. 2014. "A Longitude Prize for the Twenty-First Century." Nature 509: 401.
50. Rundle, J.B., D.L. Turcotte, and W. Klein, eds. 1996. Reduction and Predictability of Natural Disasters. Santa Fe Institute Studies in the Sciences of Complexity. Reading, PA: Addison Wesley Longman.
51. Salvati, P., C. Bianchi, F. Fiorucci, P. Giostrella, I. Marchesini, and F. Guzzetti. 2014. "Perception of Flood and Landslide Risk in Italy: A Preliminary Analysis." Natural Hazards and Earth System Sciences Discussion 2 (5): 3465–3497.
52. Salvati, P., C. Bianchi, M. Rossi, and F. Guzzetti. 2010. "Societal Landslide and Flood Risk in Italy." Natural Hazards and Earth System Sciences 10: 465–483.
53. Shrader-Frechette, K. 1994. Ethics in Scientific Research. Boston, MA: Rowman & Littlefield Publishers.
54. Shrader-Frechette, K. 2007. Taking Action, Saving Lives. New York: Oxford University Press.,
55. Sidle, R.C., W.H. Benson, J.F. Carriger, and T. Kamai. 2013. "Broader Perspective on Ecosystem Sustainability: Consequences for Decision Making." Proceedings of the National Academy of Sciences 110 (23): 9201–9208,
56. Skempton, A.W. 1948. "The ϕ = 0 Analysis for Stability and Its Theoretical Basis." Proceedings 2nd International Conference on Soil Mechanics and Foundation Engineering 1: 72–77.

57. Sobel, Dava. 1995. Longitude: The True Story of a Lone Genius Who Solved the Greatest Scientific Problem of His Time. New York: Walker and Company.
58. Stein, Jerome L., and Seth Stein. 2014. "Gray Swans: Comparison of Natural and Financial Hazard Assessment and Mitigation." Natural Hazards 72 (3): 1279–1297.
59. Tague-Sutcliffe, J. 1992. "An Introduction to Informetrics." Information Processing and Management 28 (1): 1–3.
60. Taleb, Nassim N. 2007. The Black Swan: The Impact of the Highly Improbable. New York: Random House.
61. Taleb, Nassim N. 2009. "Errors, Robustness, and the Fourth Quadrant." International Journal of Forecasting 25 (4): 744–759.
62. Taylor, D. 1948. Fundamentals of Soil Mechanics. New York: Wiley.
63. Trigila, A., C. Iadanza, and D. Spizzichino. 2010. "Quality Assessment of the Italian Landslide Inventory Using GIS Processing." Landslides 7 (4): 455–470.
64. Turcotte, Donald L. 1995. "Chaos, Fractals, Nonlinear Phenomena in Earth Sciences." Reviews in Geophysics 33 (S1): 341–343.
65. Turcotte, Donald L. 1997. Fractals and Chaos in Geology and Geophysics. 2nd ed. Cambridge: Cambridge University Press. ,
66. Turcotte, D.L., B.D. Malamud, F. Guzzetti, and P. Reichenbach. 2002. "Self-organization, the Cascade Model, and Natural Hazards." Proceedings of the National Academy of Sciences of the United States of America 99 (Suppl 1): 2530–2537.
67. Varnes, David J. 1978. "Slope Movements: Types and Processes". In Landslide Analysis and Control, edited by R.L. Schuster and R.J. Krizek, National Academy of Sciences, Special Report 176, 11–33. Washington, DC: Transportation Research Board.
68. Wyllie, D.C., and C.W. Mah. 2005. Rock Slope Engineering: Civil and Mining. 4th ed. London: Spon Press.
69. Wyss, Max, and Silvia Peppoloni. 2014. Geoethics. 1st ed. Amsterdam: Elsevier.

CHAPTER 20

Recycling greenhouse gas fossil fuel emissions into low radiocarbon food products to reduce human genetic damage

Christopher P. Williams

Radiocarb Genetics, Fairlawn, USA

ABSTRACT

Radiocarbon from nuclear fallout is a known health risk. However, corresponding risks from natural background radiocarbon incorporated directly into human genetic material have not been fully appreciated. Here we show that the average person will experience between 3.4×10^{10} and 3.4×10^{11} lifetime chromosomal damage events from natural background radiocarbon incorporated into DNA and histones, potentially leading to cancer, birth defects, or accelerated aging. This human genetic damage can be significantly reduced using low radiocarbon foods produced by growing plants in CO_2 recycled from ordinary industrial greenhouse gas fossil fuel emissions, providing additional incentive for the carbon sequestration.

Keywords: Radiocarbon Mutations, DNA, Histones, Carbon dioxide, Sequestration

1. INTRODUCTION

At first glance, radiocarbon-induced genetic damage in humans and global climate change from fossil fuel greenhouse gas emissions appear to be entirely unrelated problems. But these disparate issues share a surprising environmental interconnection, which until now has been overlooked, and which offers the possibility of addressing them both with a single, novel approach–recycle a seemingly useless and environmentally harmful industrial waste material into heretofore unavailable low radiocarbon food products.

1.1 Natural background radiocarbon

Libby and his coworkers were the first to recognize the significance of radiocarbon in the environment (Anderson 1947; Libby 1960). Approximately, one in 750 billion carbon atoms in the biosphere (that is, the atmosphere, oceans, and living organisms) is radioactive carbon-14, or radiocarbon, with a half-life of 5,730 years. The natural background level of radiocarbon is in general the result of new ^{14}C produced from cosmic ray interaction with ^{14}N in the upper atmosphere achieving equilibrium with the radioactive decay of ^{14}C back into ^{14}N. Libby found that even this trace amount of natural background radiocarbon resulted in a decay rate of 15.367 dpm g^{-1} for ordinary carbon. Because photosynthetic plants and microorganisms use atmospheric and aqueous CO_2 containing this radiocarbon, all living organisms that derive their food directly or indirectly from plants, including man, contain this same natural background level of radiocarbon. Libby realized that once an organism died and failed to take in fresh carbon containing the natural background level of radiocarbon, its radiocarbon content would steadily drop by half every 5,730 years. His Nobel Prize-winning radiocarbon dating method has proven invaluable in archeology and anthropology to estimate ages of organic materials less than 100,000 years old. Organic material more than 100,000 years old, including coal, oil, and natural gas deposits, will have lost more than 99.999% of any original radiocarbon and for practical purposes can be considered radiocarbon free.

But radiocarbon decay does not only occur in dead organisms. Since carbon makes up about 18% of the human body by weight, an average 150 lb, adult contains about 6.2×10^{26} carbon atoms, of which 1 in 750 billion, or 8.2×10^{14}, are carbon-14. Such an individual therefore experiences about 3,150 radiocarbon decay events every second, 99 billion per year, and 6.8 trillion over an average 77.9 year lifetime (adjusting for fewer cells during the growing years), due solely to natural background radiocarbon.

More than 40 years ago, near the height of atmospheric nuclear weapons testing, Linus Pauling sounded an alarm concerning the health risks from radiocarbon in fallout, which nearly doubled the natural background level of radiocarbon in the mid-1960s (Pauling 1958; Pauling 1963). He estimated that general carbon-14 beta decay from this fallout, would ultimately be responsible for one and a half million viable children with gross physical or mental birth defects, and for nearly 15 million embryonic, neonatal, and childhood deaths. Within a few years after the signing of the "Treaty banning nuclear weapon tests in the atmosphere, in outer space and under water" in 1963, atmospheric radiocarbon began to fall back to its previous level, as did public concern over natural background radiocarbon.

But if doubling the natural background level of radiocarbon would cause such huge human health risks, why wasn't the natural background baseline itself considered hazardous? Perhaps the baseline natural background radiocarbon level was simply accepted as an unavoidable natural hazard. There may also have been belief in an implicit (and unproven) assumption that the detrimental effects of any form of radiation at or below certain levels were non-linear, and in particular sub-linear (i.e., that below a certain lower threshold it was completely harmless, or at least much less harmful), and that the natural background level of radiocarbon fell within that safe region. But a National Academy of Sciences panel recently completed a 5 year study which concluded that the evidence best supports a "linear no threshold" (LNT) model for extremely low doses of radiation (Board on radiation effects research 2006). The LNT model states there is no safe lower level or threshold for radiation, and that any amount, no matter how small, is potentially damaging to health. The panel also made particular note that radiation can alter the molecular structure of DNA, and that "some of these molecular changes are so complex that it may be difficult for the body's repair mechanisms to mend them correctly."

1.2 Fossil fuel CO_2 emissions

Radiocarbon in food is not unavoidable. One of the major environmental and political issues of our day is how to avert or minimize the potentially devastating effects of global warming. A leading option to reduce atmospheric carbon dioxide, the greenhouse gas believed most responsible for the anthropogenic heating of the earth's atmosphere, is to capture or sequester it from industrial fossil fuel combustion gas emissions. But carbon sequestration processes are quite expensive, and could produce billions of tons of useless and potentially harmful waste material each year--waste that would need to be permanently dumped deep into the oceans or abandoned mines and aquifers. Not surprisingly such efforts face considerable opposition. Finding a significant new use for carbon dioxide, and particularly low radiocarbon CO_2extracted from fossil fuel emissions or obtained as a byproduct from coal gasification or the refining of oil, could profoundly advance carbon sequestration efforts. One option deserving consideration is using this material for growing safer, low radiocarbon food.

2. THEORY AND CALCULATIONS

Pauling based his estimates for genetic damage from radiocarbon solely on general ionizing radiation arising from beta decay of all the carbon-14 scattered throughout the human body. However, in those particular cases where a radiocarbon atom incorporated into the genes undergoes decay, two additional mechanisms can produce genetic damage. The loss of a 156 kev beta particle from the nucleus during radiocarbon decay causes the transmutation of that carbon-14 atom to the chemically different nitrogen-14. The beta emission is also coupled with powerful recoil of the residual nitrogen-14, which will in most cases completely rip the atom out of its original molecule (the "Szilard-Chalmers Effect"). Asimov (1955) was the first to recognize that this transmutation and recoil effect guarantee that any carbon-14 decay occurring within the 39 different carbon atoms in the four DNA nucleotides (see Figure 1) will produce a chemical change, or mutation. Pauling was also aware of this "special mutagenic action of carbon-14." However, based on the limited knowledge of genetics available at the time (the genetic code had not even been worked out), he calculated that this special mechanism would amount to less than 10% of the genetic damage caused by general ionizing radiation from carbon-14, and was therefore not considered particularly significant.

Pauling's (1958) estimates assumed there were 50,000 genes in humans and 200,000 carbon atoms per gene, for a total of 1.0×10^{10} carbon atoms per cell. Although for its time this was an excellent estimate for the protein coding regions of the genome, it overlooked the rest of the genome, which contains regions now known to be critical to genetic regulation or the structural integrity of the chromosome.

We now know that human cells (with the exception of non-nucleated cells such as red blood cells) each have about 6.4×10^9 nucleotide base pairs (diploid number), or a total of 1.28×10^{10}nucleotide bases per cell. Of the four DNA nucleotides, three (adenosine, guanosine, and thymidine) have ten carbons each and the other (cytidine) has nine. Since the overall human AT and GC content is about 59 and 41%, respectively, this gives an average of 9.8 carbons per human nucleotide. Thus the chromosomal DNA in every nucleated human cell contains about 1.25 × 10^{11} carbon atoms, more than 12 times Pauling's estimate. Approximately one in six human cells contains a radiocarbon atom incorporated somewhere in its chromosomal DNA.

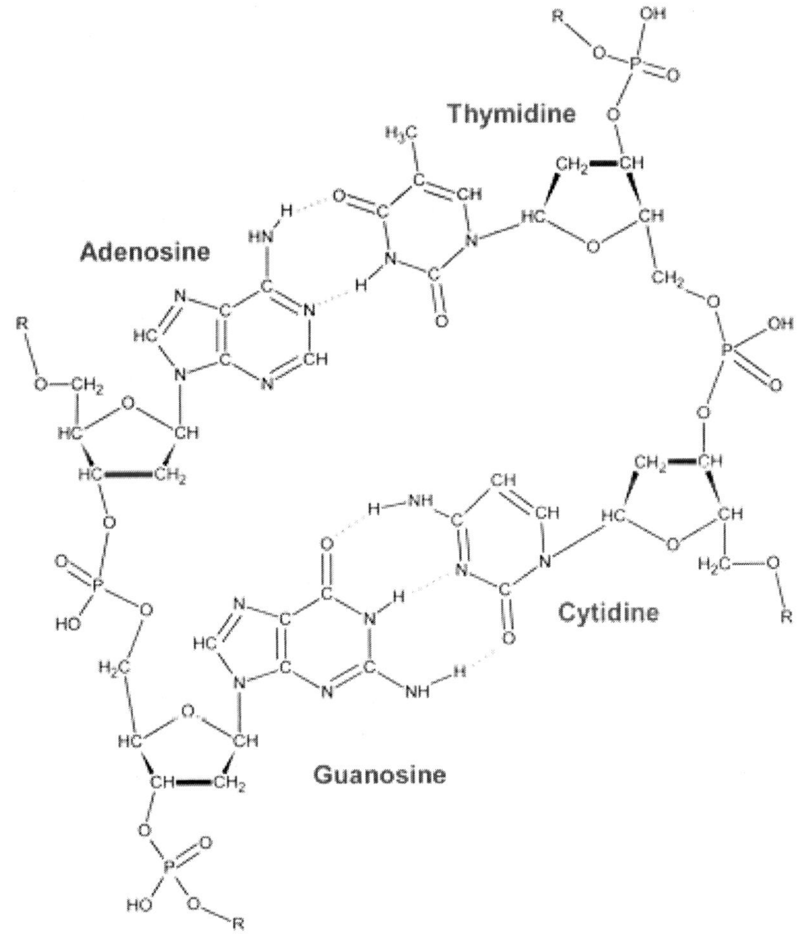

Figure 1. Possible DNA carbon positions. Radiocarbon atoms incorporated at any of these different nucleotide carbon positions may cause irreparable damage when they decay

In addition to DNA in the chromosomes, human cells contain extra-chromosomal DNA in the mitochondria, which are critical to cellular respiration. Each mitochondrial DNA (mtDNA) molecule is 16,569 bp long, with a variable copy number, typically between 1,000 and 10,000, per human cell. The total amount of mtDNA is only about 1.3% that of chromosomal DNA, but since mtDNA has less efficient repair mechanisms than chromosomal DNA, its damage by radiocarbon will have a greater tendency to go unrepaired. The high copy number of mtDNA allows more of such individual mutations to be tolerated. But as they accumulate in germ cells over many generations, such errors are more likely to afflict subsequent offspring. Indeed, numerous inherited mitochondrial disorders are known to exist (though their actual origins are unknown).

Furthermore, Pauling's calculations of chromosomal damage do not take into account the significance of carbon-14 in the histone proteins. In addition to DNA, the chromatin material of chromosomes also contains special proteins around which the DNA is tightly packed. These histone proteins are strongly basic (due to excess lysine and arginine residues) and thus neutralize the DNA's acidity (due to its linking phosphate groups). If the entire DNA in a single microscopic human cell were stretched out linearly, it would be almost 2 m (6 feet) long. Histones allow this large amount of DNA to be highly coiled and compacted in domains known as nucleosomes so the entire complement of human DNA will fit within a 10 μm cell nucleus less than 1/500,000th of its fully extended length. Incredibly, this highly coiled and compacted DNA retains full functionality and can be accurately replicated and transcribed as needed. Although the precise structure of chromatin varies according to tissue type, the position within the chromosome, and point of time within the cell cycle, on average every 200 bp of DNA is associated with a histone nucleosome core particle consisting of two molecules each of histones H2A, H2B, H3, and H4 along with a single histone H1 linker. Through extremely complex enzymatically controlled processes involving activities such as methylation and acetylation of specific basic histone residues, the cell is able to unwind and open up the correct sections of DNA precisely when needed for replication or for transcribing genes into messenger RNA. So not only are histones required for ultra compact DNA storage, they also are integrally involved in DNA processing. This critical role for histones most likely explains why their amino acid sequences are among the most highly conserved of all known proteins.

A carbon-14 decay that occurs within the histone protein portion of chromatin may not necessarily cause a direct DNA mutation. However, since the histone protein is in such close proximity to the DNA, there is a significant chance that the beta particle emitted during carbon-14 decay in a histone protein will directly strike the DNA. Additionally, the chemically modified amino acid residue (after the transmutation of carbon-14 to nitrogen-14) in the histone protein itself is likely to be highly reactive, or to generate reactive free radicals species nearby.

Because of the close proximity to the DNA, the chances of these reactive chemical species interacting with a DNA base and causing a mutation are significant. But even without actually damaging the DNA directly, a carbon-14 damaged histone protein itself may fold differently, or become cross-linked to itself or to an adjacent protein, or have a key residue modified. Any of these or other possible scenarios that cause structural alteration of a histone protein could result in abnormal cellular behavior, whether by preventing unfolding of the DNA when required, or by causing the unfolding to occur when it should not, or via some other as yet undefined mechanisms.

Because we know the amino acid sequences of the typical histone proteins and the general structure of nucleosomes, we can calculate that there are roughly 1.47 times as many carbon atoms in the histone proteins of the chromatin as there are in the DNA itself. Approximately one in four human cells contains a radiocarbon atom incorporated somewhere within its histone proteins.

Estimates for the number of nucleated cells in the average adult human body generally range from 10^{13} to 10^{14} cells. Using the most conservative estimate of 10^{13} cells, it can be calculated that the human body contains at least 1.25×10^{24} total carbon atoms within the chromosomal DNA, of which 1.67×10^{12} are carbon-14 atoms, and 1.84×10^{24} total carbon atoms within its histones, of which 2.46×10^{12} are carbon-14 atoms.

The number (N) of atoms that will decay during any time period (t) is given by the standard equation for radioactive decay:

$$N = N_0 e^{-\lambda t}$$

where

N_0 = initial number of carbon-14 atoms
$\lambda = 0.693/t_{1/2}$
$t_{1/2}$ = carbon-14 half-life (5,730 years, or 1.808×10^{11} s)
t = time period

Applying this equation for radioactive decay to the amount of radiocarbon incorporated in the various genetic materials gives us estimates of the cumulative number of human genetic damage events for various time intervals (see Table 1).

Table 1. Cumulative human genetic damage events from natural background radiocarbon

Time Interval	Chromosomal DNA	mtDNA	Histones	Total
1 s	6.4	0.083	9.4	16
1 min	380	5.0	570	950
1 h	23,000	300	34,000	57,000
1 day	550,000	7,200	810,000	1,400,000
1 year	200,000,000	2,600,000	300,000,000	500,000,000
Lifetime	14,000,000,000	180,000,000	20,000,000,000	34,000,000,000

The calculated number (within two significant digits) of radiocarbon decay events within each genetic component is based on the low end estimate of 10 trillion total human cells. mtDNA values assume 5,000 mtDNA copies per cell. Lifetime values are based on the 77.9 year average US life expectancy (equivalent to about 68.4 years at full adult size).

The calculated number (within two significant digits) of radiocarbon decay events within each genetic component is based on the low end estimate of 10 trillion total human cells. mtDNA values assume 5,000 mtDNA copies per cell. Lifetime values are based on the 77.9 year average US life expectancy (equivalent to about 68.4 years at full adult size).

3. RESULTS AND DISCUSSION

Although Pauling and other scientists were justly concerned about severe health risks from increasing the natural background level of radiocarbon, overlooked until now were the harmful health effects arising from the *baseline* natural background radiocarbon itself, and the subsequent benefits that might result if such levels could be reduced.

Our calculations show that an average person whose body contains 10^{13} cells will experience at least 3.4×10^{10} lifetime chromosomal damage events solely from natural background radiocarbon incorporated into DNA and histones. These are likely to be low-end estimates. The actual numbers could be as much as ten times higher if the average human body contains as many as 10^{14} cells. This significant amount of cumulative lifetime genetic damage should not be quickly dismissed as inconsequential, especially if the currently accepted LNT model for radiation damage is true and there is no safe lower level or threshold for radiation.

The total number of carbon atoms in the cell closely associated with genetic activity, including chromosomal DNA, histones and mitochondrial DNA, is more than 30 times greater than that originally estimated by Pauling. Thus, rather than damage from the special effect of carbon-14 in the genetic material being only a negligible 10% of the overall damage arising from carbon-14 beta decay, it could be as much as three times greater than general carbon-14 beta decay. This would nearly quadruple Pauling's already high (1963) estimates for overall human suffering resulting from atmospheric nuclear weapons testing. The number of grossly deformed children could go from 100,000 to 400,000, and the number of stillbirths and childhood deaths from 15 to 60 million.

Pauling also implicitly equated the severity of individual mutations arising from general ionizing radiation from beta decay of carbons outside the DNA to those caused by carbon-14 decay in the carbons within the DNA itself. This may not be true.

Human beings (as well as most other organisms) have complex systems for different types of DNA repair (Wood et al. 2001; Sancar et al. 2004), and not all mutations can be as successfully detected or repaired. For example, some mutations, such as thymine dimerization induced by overexposure to UV radiation, or the oxidation of a nucleic acid base by an oxidant, may involve no loss of DNA sequence information, and can be enzymatically repaired with near 100% efficiency. Likewise, other mutations that remove or modify a single DNA base, such as may occur from a free radical produced by beta decay, can also potentially be repaired with near 100% efficiency using the backup information from the complementary base on the opposite DNA strand.

However, there are three reasons why more severe mutations affecting multiple residues may be expected when a radiocarbon incorporated in a DNA nucleotide itself (see Figure 1) undergoes decay as compared to a mutation from stray ionizing radiation. First, the high-energy beta particle is emitted from within the genetic material itself, and thus has a much higher probability of striking multiple nearby DNA residues or forming free radicals in their vicinity than does a beta particle originating outside the chromosome. Second, there occurs a transmutation of carbon-14 to nitrogen-14, ensuring a significant chemical change in the affected DNA residue. Third, the beta emission produces a severe recoil in the new nitrogen atom, which is likely to result in a nitrogen free radical and to further increase the chance that the remaining nucleotide residue is converted into some highly reactive species. Any of these reactive species produced could subsequently attack adjacent nucleotide bases. Such complex damage involving multiple residues is more likely to be either unrepairable or else prone to erroneous repair. Overall long term damage and health risks are more closely related not to the total number of mutations which initially occur, but rather to the number of those that are ultimately unrepairable. In addition to birth defects and cancer, unrepaired genetic damage may also lead to genetic diseases and accelerated aging (Sancar et al. 2004; Park and Gerson 2005).

Significant reductions in human radiocarbon levels are theoretically possible using low radiocarbon foods produced from ancient subterranean sources of carbon, such as fossil fuels, which are known to be almost completely free of radiocarbon. The first such proposal involved the growing of low radiocarbon food in greenhouses or other closed systems using carefully mined coal and special handling processes (Matthews 1995). A subsequent variation was the use of ordinary greenhouses, alternatively using thermally degraded limestone (calcium carbonate) or directly burning fossil fuels on site to provide the requisite low radiocarbon CO_2(Miekka and Mackie 1999). To date there does not appear to have been any actual commercial application of these methods. This may be partially due to failure to fully recognize the potential damaging effects of radiocarbon as calculated here, and partially due to the apparent costs and difficulty of these specialized methods for producing low radiocarbon CO_2. However, a better alternative exists.

More than 27 billion (2.7×10^{10}) metric tons, or nearly 4 tons for every person in the world, of low radiocarbon CO_2 fossil fuel emissions from human activities are released into the atmosphere each year. These greenhouse gas emissions are generally considered the primary contributors to global warming and potential climate change (Oreskes 2004). Finding a safe and inexpensive way to minimize these CO_2 greenhouse gas emissions is one of today's most pressing political and environmental issues. Nearly all discussions involving CO_2 sequestration consider this material relatively useless and simply a waste that requires permanent disposal. Large scale recycling of fossil fuel CO_2 emissions as a means to grow safer, radiocarbon-reduced foods has not been considered until now and deserves further consideration. It is interesting to note that fossil fuel combustion gas emissions are already producing significantly radiocarbon-depleted vegetation, albeit unintentionally, with radiocarbon reductions of 12% or more found in grasses collected within a few meters of heavily trafficked French highways (Lichtfouse et al. 2005).

Further research involving mass spectrometry of small double stranded oligonucleotides or peptides enriched with radiocarbon could reveal the specific types of biochemical damage this substance can cause in DNA and histones. Studies using mice or other organisms with cohorts fed diets having normal (natural background), reduced, and elevated levels of radiocarbon could ascertain its overall impact on rates of cancer, birth defects, and aging. Although a low radiocarbon diet would be expected to primarily benefit organisms during their maximal periods of growth when new cells and DNA are being formed, studies are also needed to see if mature individuals could receive any benefit from a low radiocarbon diet.

Because it takes carbon materials nearly six half-lives, or more than 32,000 years, for their radiocarbon levels to drop by 98%, for all practical purposes we must consider fossil fuels and fossil fuel emissions a non-renewable low radiocarbon resource. Admittedly limestone and other ancient carbonate rocks are also potential sources of low radiocarbon CO_2, but they would likely be considerably more expensive and would only exacerbate the problem of rising atmospheric CO_2. If low radiocarbon food is found to be beneficial to human health, for the sake of future generations we should not waste the readily available low radiocarbon CO_2 in fossil fuel emissions by irretrievably releasing it to the atmosphere.

It also should be noted that the total amount of natural background radiocarbon in the atmosphere and in all land organisms is only about 3,200 kg, or 7,000 lbs. (Pauling 1958). That is why a relatively small amount, just 74 kg, of radiocarbon produced from the atmospheric testing of just one 30 megaton nuclear weapon could increase the worldwide atmospheric radiocarbon level by 2.3%. Although not pleasant to consider, an additional benefit of stockpiling low radiocarbon CO_2 in a recoverable manner is to provide for a safe emergency world food supply in the event of a catastrophe, such as a nuclear exchange or even a terrorist action, which could appreciably increase the background level of radiocarbon in the atmosphere for many years.

4. CONCLUSION

Low radiocarbon foods produced using CO_2 sequestered from fossil fuel emissions provide a means to reduce human chromosomal damage from natural background radiocarbon as well as from potentially elevated levels of radiocarbon resulting from future atmospheric nuclear explosions. Should a low radiocarbon diet ultimately prove to offer significant human health benefits, this would not only boost the current incentive for carbon sequestration of fossil fuel emissions beyond that of simply reducing the greenhouse gas effect, but would also shift carbon sequestration technologies toward methods, which permit the subsequent cost effective recovery and reuse of low radiocarbon CO_2 for food production.

REFERENCES

1. Anderson EC, et al (1947) Radiocarbon from cosmic radiation. Science 105:576–577
2. Asimov I (1955) The radioactivity of the human body. J Chem Educ 32:84–85
3. Board on radiation effects research (2006) Health risks from exposure to low levels of ionizing radiation: BEIR VII phase 2. National Academies Press, Washington
4. Libby, WF Radiocarbon dating. Nobel Lecture (1960) (http://www.nobelprize.org/nobel_prizes/chemistry/laureates/1960/libby-lecture.pdf)
5. Lichtfouse E, Lichtfouse M, Kashgarian M, Bol R (2005) ^{14}C of grasses as an indicator of fossil fuel CO_2 pollution. Environ Chem Lett 3:78–81
6. Matthews M (1995) Method for improving the chemical stability of living organisms. US Patent No. 5,471,785
7. Miekka F, Mackie P (1999) Methods and apparatus for reducing carbon 14 in living tissue. US Patent No. 5,956,896
8. Oreskes N (2004) The scientific consensus on climate change. Science 306:1686
9. Park Y, Gerson SL (2005) DNA repair defects in stem cell function and aging. Annu Rev Med 56:495–508
10. Pauling L (1958) Genetic and somatic effects of carbon-14. Science 128:1183–1186
11. Pauling L (1963) Science and peace. Nobel Lecture (http://nobelprize.org/ nobel_prizes/peace/laureates/1962/pauling-lecture.html)
12. Sancar A, Lindsey-Boltz LA, Keziban U, Linn S (2004) Molecular mechanisms of mammalian DNA repair and the DNA damage checkpoints. Annu Rev Biochem 73:39–85
13. Wood RD, Mitchell M, Sgouros J, Lindahl T (2001) Human DNA repair genes. Science 291:1284–1289

CHAPTER 21

Distribution of trace metals in surface water and sediments of Imo River Estuary (Nigeria): Health risk assessment, seasonal and physicochemical variability

Essien D. Udosen*, Nnanake-Abasi O. Offiong, Samuel Edem John B. Edet

Department of Chemistry, University of Uyo, P. M. B. 1017, 520001-Uyo, Akwa Ibom State, Nigeria.

ABSTRACT

The distribution of trace metals (Zn, Cu, Ni, Mn, Fe, Co, V, Pb and Cd) in surface water and sediments from Imo River estuary in relation with the physicochemical variables were studied for both dry and wet seasons. The results obtained revealed that trace metal levels displayed marked spatial and seasonal variations in surface water and sediments, which is attributed to both anthropogenic natural processes in the environment. Trace metal levels were generally higher in the sediments than overlying water column. Seasonal variability in trace metal distribution was mostly observed in sediments than in the water phase of the relatively anoxic estuarine ecosystem. The highest average concentration of trace metal in the sediments was recorded for Fe (1101.18 mg/kg) during the dry season while the lowest was recorded for Cd (0.08 mg/kg) during the wet season. For the surface water, highest average concentration was recorded for Zn (0.04 mg/L) for the dry season and the lowest level was recorded for Cd (0.002 mg/L) for both seasons. The levels of trace metals in the water were within the permissible limits of some regulatory bodies. Pearson correlation analysis revealed that some of the physicochemical variables were associated with patterns of trace metals distribution. Based on trace metal concentrations, the health risk assessment based on the chronic daily intake (CDI) and hazard quotients (HQ) were calculated. Zinc and cadmium contributed most to the total intake through drinking water. As HQ for all trace metals were less than unity, it is suggested that there is little or no significant health risk to the exposed population.

Keywords: Trace metals, surface water, sediments, seasons, Imo River estuary, Nigeria.

1. INTRODUCTION

In Nigeria, a large number of rural and suburban dwellers derive their drinking water from groundwater and surface water resources such as rivers, lakes and reservoirs. Therefore, the quality of drinking water remains a critical factor in the Nigerian public health (FMWR, 2000). More so, that Nigeria is among 80 countries of the world reported to be facing water deficit (Ahuja, 2012) calls for reliable information for effective management, preservation and utilization of available natural waters. As an index of contamination, the presence of trace metals in fresh water ecosystems has been a major concern to many (Iqbal et al., 2013; Muhammad et al., 2011; Tamasi and Cini, 2004). This concern is due to the fact that trace metals constitute one of the most insidious and dangerous contaminants known to man. The accumu-lation of these metals in the environment may constitute a threat to aquatic biota where they are often accumulated. This leads to an amplification of their levels along the food chain. This poses serious health implications as well as a threat to food security.

Sediments are considered as suitable medium to study the contamination of aquatic environment because they represent the sink of multiple contaminant sources (Sprovieri et al., 2007). Apart from being a final reservoir for these contaminants, sediments act as their secondary sources in relation to changes in environmental physicochemical characteristics and response to human activities such as dredging (Sprovieri et al., 2007;

Eggleton and Thomas, 2004). Sediments are important carriers of trace metals in the hydrological cycle and because metals are partitioned within the surrounding waters, they reflect the quality of aquatic system. As a result, pollutant species in sediment cores have been used as "pollutant records" in some studies (Chatterjee et al., 2007). In order to design abatement programmes for contaminated environments, it is often useful to under-stand the behaviour of the contaminants in relation to certain environmental conditions, both physical and chemical. This has necessitated the study of the influence of certain environmental factors on the distribution of legacy chemical toxicants in aquatic environments.

It is of interest to examine the distribution of trace metals in surface water and sediments of a turbulent estuary such as the Imo River in South Eastern Nigeria. The study area has witnessed massive changes in industry and shifts in population and social culture. A significant ecological change is pronounced in this estuarine environment due to huge discharges of domestic, agricultural, and industrial wastes coupled with increased agricultural activities and emergence of an aluminium smelting company. The estuary has become vulnerable to chemical pollutants. To the best of our knowledge, there has been no research on the distribution of trace metals in the surface water and sediments of the estuary in relationship with physicochemical variables as well as the consideration of risk of human exposures to the metals through drinking water.

2. MATERIALS AND METHODS

2.1 Study site

The study area is the southern segment of the Imo River estuary in Ikot Abasi, Akwa Ibom State (Southeast Nigeria). The area lies between latitudes 4°30' N and 4°45' N and longitudes 7°19' E and 7°45' E. The Imo River estuary opens into the Atlantic Ocean at Bight of Bonny. The estuary has a shallow depth ranging from 5 to 8 mm at flood and ebb tides though some spots may be deeper. The area is a multi-use resource for artisanal and commercial fishing and transportation. Other economic activities in and around the estuary include oil exploration and aluminium smelting. The climate of the area is characterised by distinct wet and dry seasons. The area is a mangrove estuarine ecosystem.

2.2 Sample collection

Fourteen sampling locations were strategically chosen along the length of the Imo River Estuary (Figure 1). At each sample stations, two subsamples were made into composites for dry and wet seasons, making a total of forty eight samples. The top 20 cm of the bottom sediments were collected using a Van Veen grab sampler and transferred into plastic containers (Loring and Rantala, 1992). Water samples were collected at the depth of 10 cm below the water surface using one litre polyethylene plastic containers. The samples were stored in ice-packed coolers and transported to the laboratory where they were kept in a freezer at -4°C before analyses. During sampling campaigns, some physicochemical parameters (pH, conductivity, salinity and redox potential) were measured *in situ* using a handheld multi-parameter probe (YSI 556, YSI Incorporated, USA).

2.3 Sample pretreatment and preparation

Water samples for trace metal analysis were filtered using a pre-weighed acid wash 0.45 μm pore size polycarbonate filters. Filtrates were acidified with concentrated trioxonitrate (HNO_3) (V) acid to a pH below 2 to minimise precipitation and adsorption of metals on the walls of the containers. Sediment samples were air-dried and ground using porcelain mortar and pestle. The texturally equivalent petite fractions (< 75 μm) of the sediments were separated by sieving through a 230 mesh nylon sieve. About 1 g of the powdered forms were digested with 20 dm^3 aqua regia ($3HCl:1HNO_3$) and concentrated to < 2 dm^3 volume on a silica beaker. The solutions were filtered into acid-washed 100 dm^3 volumetric flask after being rinsed with deionised water. Further, the solutions were made up to 50 dm^3 with deionised water prior to trace metal analysis.

2.4 Instrumental analysis by aas

Analysis of filtrates and digests for trace metals were done using atomic absorption spectrophotometer (UNICAM 939). Prior to analysis, 200 ml aliquot of the filtered water samples were extracted by chelation with ammonium pyrrolidine dithiocarbamate reagent (APDC) into 20 ml methyl isobutyl ketone (MIBK) as the extraction solvent as reported by Hatje et al., (2003). Standard solutions of each metal were prepared for calibration. Reagent blanks and the replicate samples were analyzed for quality control. The detection limits of the metals under study were: Cu 0.1 mg/L, Zn 0.01 mg/L, Ni 0.001 mg/L, Mn 0.01 mg/kg, Iron 0.1 mg/kg, Co 0.001 mg/kg, V 0.01 mg/kg, Pb 0.1 mg/kg and Cd 0.1 mg/kg. Other details of the analytical procedures are reported elsewhere (Hatje et al., 20013; Aktar et al., 2010).

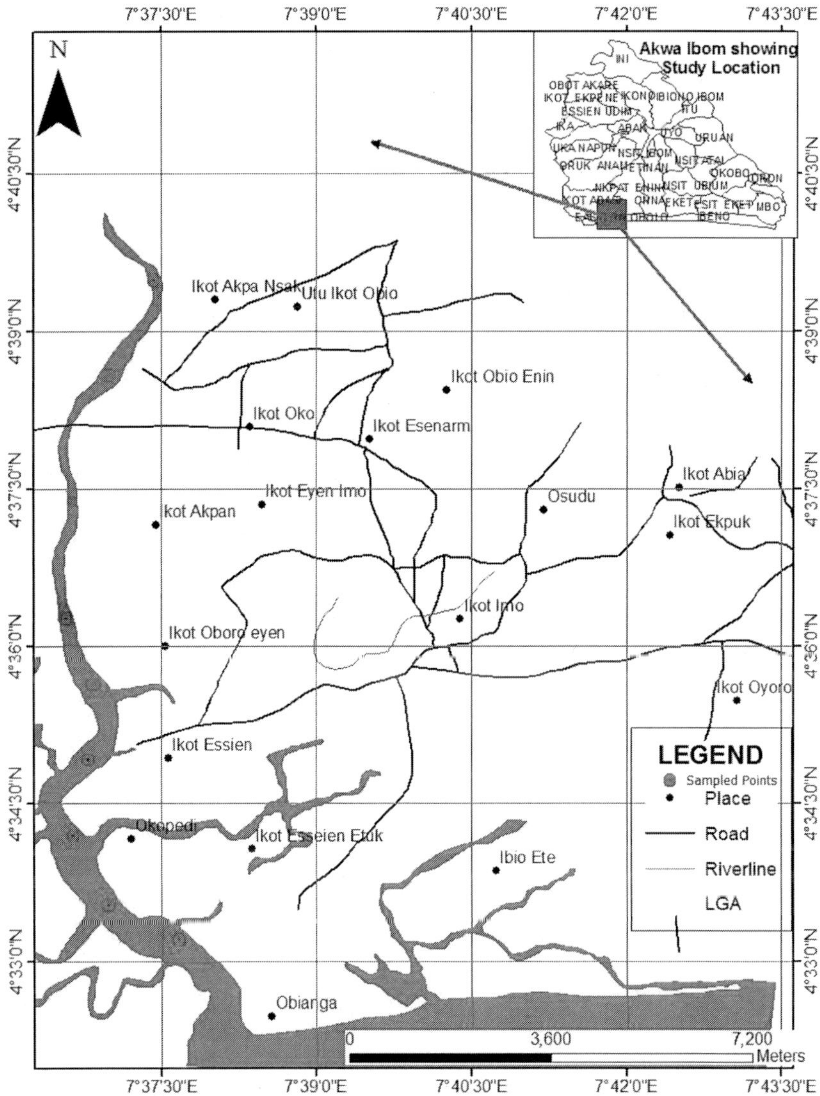

Figure 1. Imo River Estuary showing sample locations

2.5 Risk assessment procedure

Risk assessment of environmental contaminants is concerned first with identifying the hazards which substances or energy may pose threat to the health of people, animals and plants including the damages they may cause to structures and commodities and secondly, with quantifying as well as estimating the probability of these types of harm occurring (Alloway and Ayres, 1993).

Although the study area is an estuary characterised by mixing from both freshwater rivers and streams with inputs from the ocean, some local inhabitants use the source of water for drinking purposes. The major difference between risk assessment and mere comparison of monitoring data with permissible limits is that in risk assessment exposure routes, quantity of toxicant, frequency of exposure and body weight of the exposed individual is considered. It therefore becomes an imperative tool for assessing risk of target groups.

2.6 Daily intake of trace metals

This study focuses on the oral ingestion pathway. The chronic daily intake (CDI) through the drinking of water was calculated using a modified equation from Kavcar et al., (2009) as shown below:

$$CDI = \frac{C \times DI}{BW} \quad (1)$$

Where, C, DI and BW represent the concentration of trace metals in water (mg/L), average daily intake rate (2 l/day for an adult) and average body weight (70 kg for an adult) according to USEPA (1989).

2.7 Hazard quotient indices

The degree of harm is estimated in terms of the hazard quotient (HQ) as expressed in the equation from Shah et al., (2012):

$$HQ = \frac{CDI}{RfD} \quad (2)$$

Where, the oral toxicity reference dose (RfD) values are 5.0E-04, 3.7E-02, 1.4E-01, 2.0E-02, 3.6E-02, 3.0E-01, 3.0E-04, 7.0E-03, mg/kg-day for Cd, Cu, Mn, Ni, Pb, Zn, Co, and V respectively (USEPA, 2005; Iqbal et al., 2013; Muhammad et al., 2010; USEPA, 1995). The exposed population is assumed to be safe when HQ<1 (Khan et al., 2008; Muhammad et al., 2010).

2.8 Statistical analysis

Basic statistics were performed using Microsoft Excel while multivariate correlation matrix were performed using Statistical Package for Social Sciences (SPSS) version 15.

3. RESULTS AND DISCUSSION

3.1 Distribution and seasonal variation in trace metal levels

The levels of trace metals in sediment and surface water samples of Imo River estuary for both wet and dry seasons are presented in Table 1. Generally, the levels of trace metals were higher in sediment samples. This may be as a result of simple settlement from the water column or as a result of preferential partitioning of the trace metals to the sediments. Also, adsorption and complexation with organic matter in the sediment as well as precipitation may have been responsible for this observation (Eggleton and Thomas, 2004). As observed in Table 1, the trace metal levels in sediments appear to be generally higher during the dry season than in the wet season. The dry seasons are usually characterised by marked reduction in the water volume in rivers. This may have influenced the relatively higher concentration of the trace metals during the dry season.

The highest concentration was recorded for Fe; 1099.58 and 1101.18 mg/kg for wet and dry seasons respectively. The least concentration was recorded for Cd; 0.08 and 0.09 mg/kg for wet and dry seasons respectively. The trace metal levels in surface water samples were quite lower than the levels recorded for the sediments. The highest level in surface water was recorded for Zn (wet 0.04 mg/L; dry 0.03 mg/L). The levels of the trace metals in water were all within permissible limits (WHO, 2008).

There was no significant variation in the trace metals levels in the surface water for both seasons. This observation may be due to the fact that the water column is continually and rapidly affectedly by several environmental changes and flows. As a result, accumulation of trace metals at a point over a period of time is practically impossible, leading to essentially low levels of trace metals. Another factor that may have influenced the presence of trace metals in the water column was the pH, which was essentially at a neutral point (Table 2). This result indicates that in spite of the higher levels of trace metals in sediment cores, their dissolution into the water column would have been restricted by the almost neutral pH.

Table 1. Average trace metal levels in surface water and sediment in wet and dry season (mg/L).

Variables	Season	Zn	Cu	Ni	Mn	Fe	Co	V	Pb	Cd
Sediments	Wet season	75.00±0.04	7.75±0.05	8.22±0.05	243.76±0.04	1099.58±0.27	3.52±0.04	12.50±0.04	4.82±0.05	0.08±0.04
	Dry season	83.42±0.04	7.95±0.04	8.50±0.04	246.23±0.04	1101.18±0.04	3.90±0.04	12.99±0.04	5.08±0.04	0.09±0.04
Surface water	Wet season	0.04±0.04	0.02±0.04	0.01±0.03	0.01±0.04	NA	0.004±0.037	0.008±0.135	0.004±0.037	0.002±0.037
	Dry season	0.03±0.04	0.01±0.04	0.003±0.032	0.01±0.04	NA	0.004±0.037	0.008±0.135	0.004±0.037	0.002±0.037
*WHO		3.0	2.0	0.02	0.1		0.04	0.3^a	0.01	0.003

NA = not analysed, *WHO permissible values for drinking water, [a]USEPA (1995).

Table 2. Average physicochemical variables of surface water in wet and dry season.

Season	pH	Conductivity ($mscm^{-1}$)	Salinity (%)	Redox potential (mV)
Wet season	7.16 ± 0.03	41.63 ± 0.04	26.23 ± 0.04	-9.57 ± 0.04
Dry season	7.20 ± 0.03	43.21 ± 0.04	28.03 ± 0.04	-13.93 ± 0.04
WHO (2008)	6.0-9.0	4000		

Table 3. Correlation matrix for trace metal levels in sediment during dry season.

Metal	Zn	Cu	Ni	Mn	Fe	Co	V	Pb	Cd
Zn	1								
Cu	0.707**	1							
Ni	0.725**	0.789**	1						
Mn	0.706**	0.492	0.628*	1					
Fe	0.393	0.648*	0.595*	0.000	1				
Co	-0.333	-0.260	-0.056	-0.341	0.173	1			
V	0.641*	0.646*	0.841**	0.722**	0.287	0.023	1		
Pb	-0.064	-0.018	0.061	-0.376	0.174	0.479	0.065	1	
Cd	-0.220	0.068	-0.224	-0.065	-0.007	-0.390	-0.123	-0.138	1

**Correlation is significant at the 0.01 level (2-tailed). *Correlation is significant at the 0.05 level (2-tailed).

3.2 Physicochemical variability in relation with trace metal levels and distribution

The results of the physicochemical variables are presented in Table 2. Physicochemical parameters recorded for the study location did not show any significant variation between seasons. Conductivity and pH values were within the permissible limits of WHO (2008). The low conductivity values of the water samples are remarkably reflected in the low levels of soluble trace metals in the water column (Table 2). The high percentage of salinity observed in the water body is indicative of possible mixing of freshwater tributaries and saline Atlantic Ocean. This further suggests high inputs of metals into the water body. Also, since salinity is mostly related to chloride content (IEPA, 2001), the result indicates possible risk for agricultural and domestic use of the water.

On the other hand, the high salinity of the river may help in limiting the bioavailability of trace metals in the estuarine ecosystem thereby protecting estuarine organisms (Lores and Pennock, 1998).

The redox potential characterizes the oxidation-reduction state of natural waters. Redox reactions in aquatic systems are central in element cycling, sorption processes, trace metals mobility and toxicity (Grundl et al., 2011). In this study, the negative redox potential values indicate that the estuarine ecosystem was reducing in nature during the period of investigation. This represents anoxic conditions with low oxygen content. The implication of this is that mobility of trace metals would be limited. This is in good agreement with the observed high but somewhat immobile trace metal levels in the sediment cores (Table 1). As observed by De Jonge et al. (2012), increasing oxygen concentrations due to general water quality improvements can enhance trace metal mobility through oxidation which may release sediment-bound metals to the overlying surface water.

In order to understand the observed distribution of trace metals in relation to their possible sources or inter-relationship, experimental data were subjected to Pearson correlation analysis. In both dry and wet seasons, significant and positive correlations were found for majority of the metals in the sediments, except for Co, Pb and Cd (Tables 3 and 4). Alloway and Ayres (1993) had suggested that positive correlation may result from their co-precipitation in marine and aquatic environments. Vanadium showed significant positive correlation with most of the metals investigated. This result indicates similarity in source of contamination in the study area.

Table 4. Correlation matrix for trace metal levels in sediment during wet season.

Metal	Zn	Cu	Ni	Mn	Fe	Co	V	Pb	Cd
Zn	1								
Cu	0.683**	1							
Ni	0.702**	0.758**	1						
Mn	0.699**	0.454	0.549*	1					
Fe	0.361	0.678**	0.599*	0.031	1				
Co	-0.287	-0.167	-0.074	-0.192	0.226	1			
V	0.654*	0.619*	0.803**	0.688**	0.278	0.037	1		
Pb	-0.043	-0.014	0.012	-0.355	0.179	0.389	0.042	1	
Cd	-0.247	0.061	-0.218	-0.041	-0.024	-0.432	-0.137	-0.166	1

**Correlation is significant at the 0.01 level (2-tailed). *Correlation is significant at the 0.05 level (2-tailed).

Table 5. Correlation matrix between trace metal levels and physicochemical parameters in surface water during dry season.

Metal	Zn	Cu	Ni	Mn	Co	V	Pb	Cd	pH	Conductivity	Salinity	R.P
Zn	1											
Cu	0.403	1										
Ni	0.564*	0.471	1									
Mn	-0.051	0.059	-0.258	1								
Co	0.502	0.629*	0.418	0.166	1							
V	0.041	0.456	-0.047	0.848**	0.296	1						
Pb	0.149	0.427	0.126	-0.130	0.221	0.164	1					
Cd	0.079	0.358	-0.190	0.173	0.094	0.476	0.412	1				
pH	0.315	0.685**	0.552*	0.087	0.351	0.534*	0.317	0.476	1			
Conductivity	0.445	0.746**	0.399	0.235	0.383	0.620*	0.528	0.345	0.781**	1		
Salinity	0.514	0.742**	0.407	0.207	0.505	0.578*	0.573*	0.325	0.773**	0.973**	1	
R.P	-0.201	-0.618*	-0.530	0.193	-0.245	-0.264	-0.214	-0.477	-0.920**	-0.588*	-0.564*	1

*Correlation is significant at the 0.05 level (2-tailed). **Correlation is significant at the 0.01 level (2-tailed). RP= Redox potential.

Table 6. Correlation matrix between trace metal levels and physicochemical parameters in surface water during wet season.

Metal	Zn	Cu	Ni	Mn	Co	V	Pb	Cd	pH	Conductivity	Salinity	R.P
Zn	1											
Cu	0.015	1										
Ni	0.162	0.544*	1									
Mn	0.142	0.029	0.306	1								
Co	0.380	0.187	0.273	0.124	1							
V	0.355	0.224	0.617*	0.782**	0.333	1						
Pb	0.288	0.050	0.148	-0.068	0.483	0.252	1					
Cd	0.192	-0.053	0.060	0.133	0.129	0.415	0.500	1				
pH	0.641*	0.230	0.594*	-0.016	0.486	0.449	0.426	0.278	1			
Conductivity	0.713**	0.495	0.606*	0.185	0.378	0.607*	0.513	0.348	0.775**	1		
Salinity	0.750**	0.315	0.526	0.234	0.294	0.652*	0.518	0.460	0.805**	0.951**	1	
R.P	-0.193	-0.566*	-0.676**	0.241	-0.207	-0.302	-0.278	-0.422	-0.678**	-0.646*	-0.577*	1

*Correlation is significant at the 0.05 level (2-tailed). **Correlation is significant at the 0.01 level (2-tailed).

Unlike the similarities and correlation between trace metals in both wet and dry seasons observed for sediments, significant difference could be observed in the case of surface water (Tables 5 and 6). Negative correlation was observed for redox potential. Strong and positive correlation was observed between Zn, pH, conductivity and salinity in the wet season while V and Cu correlated positively with pH, conductivity and salinity in the dry season.

Observed variability in trace metal distribution in relationship with the physicochemical parameters may be due to the possible differences in the pattern of human activities in the two seasons. Also, characteristics in water chemistry in estuaries are said to be more complicated than in rivers and lakes since they are influenced by the inflow of freshwater from rivers and streams, by tides of the oceans, large salinity, stratification, while the back-and-forth movement of the water during the tides enhance dispersion and mixing (Hemond and Fechner-Levy, 2010). These factors may have greatly affected the divergent patterns in trace metals distribution and relationship with physicochemical parameters in the two seasons studied.

4. RISK ASSESSMENT

4.1 Chronic daily intake (CDI) and hazard quotient (HQ) indices

The estimated CDI indices for ingestion of drinking water are presented in Table 7; showing values predicted for an exposed adult. The result reveals that Zn contributed most to the CDI of trace metals through drinking water from the Imo River for wet season while Cd contributed most for dry season calculations. Zinc and cadmium usually occur together (Baird and Cann, 2008). This may be responsible to the high CDI values of the two metals in addition to anthropogenic inputs. There is call for concern over the high intake of cadmium in the study area since the metals belong to the most dangerous class of metal toxicants. The HQ indices are also summarised in Table 7. The highest HQ value was recorded for Cd while the least was recorded for Mn. The HQ indices for all trace metals, which were less than unity, suggest ignorable or no significant risk to the local population drinking water from the studied area.

5. CONCLUSION

The study revealed that the Imo River estuary is impacted with very low contamination of trace metals largely influenced by both natural and anthropogenic inputs. The study also revealed that the water body is an anoxic estuarine ecosystem influenced by dynamics both freshwater characteristics and oceanic inputs. The high levels of trace metals recorded in sediment are of major concern since there is possibility of dissolution into the water column and possible uptake and bioaccumulation by aquatic organisms over time. The physicochemical variables also exerted certain influence on each other as well as the trace metal distribution and levels. Trace metal levels in the drinking water were below permissible limits set by WHO and USEPA. Risk assessment based on consumption of water by adults revealed no significant health risk.

Table 7. Summary chronic daily intake (CDI) and hazard quotient (HQ) indices for drinking water from Imo River Estuary.

Metal	Zn	Cu	Ni	Mn	Co	V	Pb	Cd
Wet season								
CDI	1.19E-03	4.84E-04	1.35E-04	3.69E-04	1.12E-04	2.39E-04	1.22E-04	6.94E-05
HQ	3.97E-03	1.34E-02	6.75E-03	2.64E-03	3.74E-01	2.44E-02	3.40E-03	1.39E-01
Dry season								
CDI	7.46E-04	4.18E-04	6.74E-05	2.24E-04	7.95E-05	1.78E-04	8.16E-05	4.11E-03
HQ	2.49E-03	1.13E-02	3.37E-03	6.59E-04	2.67E-01	2.19E-02	2.55E-03	8.15E-02

CONFLICT OF INTERESTS

The authors have not declared any conflict of interest.

REFERENCES

1. Ahuja S (2012). Water sustainability and reclamation. Paper presentation at the Spring 2012 American Chemical Society National Meeting, San Diego, California.
2. Aktar MW, Paramasivam M, Ganguly M, Purkait S, Sengupta D (2010). Assessment and occurrence of heavy metals in surface water of Ganga river around Kolkota: A study for toxicity and ecological impact. Environ. Monit. Assess. 160:207-213.
3. Alloway BJ, Ayres DC (1993). Chemical Principles of Environmental Pollution. London, UK: Champman & Hall, pp. 44-153.
4. Baird C, Cann M (2008). Environmental Chemistry. 4th ed., New York, England: W.H. Freeman, P. 692.
5. Chatterjee M, Silva EV, Sarka SK, Sella SM, Bhattacharya A, Satpathy KK, Prasad MVR, Chakraborty S, Bhattacharya BD (2007). Distribution and possible source of trace

elements in the sediment cores of a tropical macrotidal estuary and their ecotoxicological significance. Environ. Int. 33(3):346-356.

6. De Jonge M, Teuchies J, Meire P, Blust R, Bervoets L (2012). The impact of increased oxygen conditions on metal-contaminated sediments part I: Effects on redox status, sediment geochemistry and metal bioavailability. Water Res. 46:2205-2214.
7. Eggleton J, Thomas KV (2004). A review of factors affecting the release and bioavailability of contaminants during sediment disturbance events. Environ. Int. 30:973-980.
8. FMWR (Federal Ministry of Water Resources), (2000). National Water Supply and Sanitation Policy, 1st ed., Abuja: Federal Republic of Nigeria. Grundl T, Haderlein S, Nurmi J, Tratnyek PG (20011). Introduction to Aquatic Redox Chemistry. In: Tratnyek P, Grundl TJ, Haderlein SB. Aquatic Redox Chemistry. ACS Symposium Series; American Chemical Society: Washington, DC, pp. 1-14.
9. Hatje V, Apte SC, Hales LT, Birch GF (2003). Dissolve Trace Metal Distribution in Port Jackson Estuary (Sydney Harbour), Australia. Mar. Pollut. Bull. 46(2003):719-730.
10. Hemond HF, Fechner-Levy EJ (2010). Chemical Fate and Transport in the Environment. 2nd Ed., San Diego: Academic Press, pp. 88-89.
11. IEPA (Irish Environmental Protection Agency) (2001). Parameters of Water Quality: interpretation and standards. Environmental Protection Agency, Wexford, Ireland, P. 93.
12. Iqbal J, Shah M, Akhter G (2013). Characterization, source apportionment and health risk assessment of trace metals in freshwater Rawal Lake, Pakistan. J. Geochem. Explor. 125:94-101.
13. Kavcar P, Sofuoglu A, Sofuoglu SC (2009). A health risk assessment for exposure to trace metals via drinking ingestion pathway. Int. J. Hyg. Environ. Health 212:216-227.
14. Khan S, Cao Q, Zheng YM, Huang YZ, Zhu YG (2008). Health risk of heavy metals in contaminated soils and food crops irrigated with wastewater in Beijing, China. Environ. Pollut. 152:686-692.
15. Lores EM, Pennock JR (1998).The effect of salinity on binding of Cd, Cr, Cu and Zn to Dissolved organic matter. Chemosphere 37(5):861-874.
16. Loring DH, Rantala RT (1992). Manual for geochemical analysis of marine sediments and suspended particulate matter. Earth-Science Rev. 32(4):235-283.
17. Muhammad S, Shah MT, Khan S (2011). Health risk assessment of heavy metals and their source apportionment in drinking water of Kohistan region, northern Pakistan. Microchem. J. 98:334-343.
18. Muhammad S, Shah MT, Khan S (2010). Arsenic health risk assessment of drinking water and source apportionment using multivariate statistical techniques in Kohistan region, northern Pakistan. Food Chem. Toxicol. 48:2855-2864.
19. Shah MT, Ara J, Muhammad S, Khan S, Tariq S (2012). Health risk assessment via surface water and sub-surface water consumption in the mafic and ultramafic terrain, Mohmand agency, northern Pakistan. J. Geochem. Explor. 118:60-67.
20. Sprovieri M, Feo ML, Prevedello L, Manta DS, Sammertino S, Tamburrino S, Marsella E. (2007). Heavy metals, polycyclic aromatic hydrocarbons and polychlorinated biphenyls in surface sediments of Naples Harbour (South Italy). Chemosphere 67:998-1009.
21. Tamasi G, Cini R (2004). Heavy metals in drinking waters from Mount Amiata (Tuscany, Italy). Possible risks from arsenic for public health in the province of Siena. Sci. Total Environ. 327:41-51.
22. USEPA (U.S. Environmental Protection Agency), (2005), Guidelines for carcinogen risk assessment. Risk Assessment Forum, Washington, DC. EPA/630/P-03/001F.
23. USEPA (United States Environmental Protection Agency) (1989). Risk Assessment Guidance for Superfund. Volume 1: Human Health Evaluation Manual (Part A). Interim Final. Office of Emergency and Remedial Response, U.S.EPA, Washington, D.C. 20450. EPA/540/1-89/00.
24. USEPA (United States Environmental Protection Agency) (1995). Regulatory and Human Health Benchmarks for SSL Development.
25. WHO (World Health Organisation) (2008). Guidelines for drinking water quality, 3rd ed. Recommendations, vol. 1. WHO Press, World Health Organisation, Geneva, Switzerland.

CHAPTER 22

Appraisal of Fluoride Concentration, Distribution and Geogenic Origin in Ground and Surface Water from Semi-Arid Region, Part of Yobe State North-eastern Nigeria

Mohammad Yerima Kwaya[1], Hamidu Hassan[1,*], Jibrin Umar[2], Kachalla Musa[3]

[1]*Department of Geology, Faculty of Sciences, Usmanu Danfodiyo University, Sokoto, Nigeria*

[2]*Department of Civil Engineering Technology, Faculty of Engineering, Federal Polytechnic, Damaturu, Nigeria*

[3]*African Development Bank-Assisted Rural Water Supply and Sanitation, Yobe State Sub-Programme, Damaturu, Nigeria*

ABSTRACT

The Fluoride concentration of water sources from part of Yobe State has been investigated, Fluoride levels in Pond water is between 0.71 and 1.36 mg/l, Dugwells concentration is from 0.00 to 1.5mg/l. The sampled water from tube wells and boreholes have concentrations ranges of 0.65-1.5 mg/l and 0.00- 2.2 mg/l respectively with the level of concentration being a function of depth. Health risk assessment have revealed three classes of risk in terms of Fluoride concentration levels in the water with 16 samples representing 14.3% falling within the Risk class of $1.5 \leq R \leq 2.5$(risk). Individual consuming such water have the tendency of getting dental fluorosis. Origins of Fluoride in the water are from Fluoride bearing minerals like Micas, Amphiboles, Illites, Apatite, Topaz, and Cryolite which occur in the sandy and clayey layers, anthropogenic sources from agricultural and irrigated land and leachate from dumps are the other possible sources of Fluoride in the tube wells and ponds water respectively. The dissolution, distribution and concentration of Fluoride in groundwater in the area are favoured by the high temperature high evapotranspiration rate and low rainfall situation of the semi-arid climate. Geochemical situation that allowed the leaching of Fluoride into the water are; high Alkaline or soda water environment, anion exchange of OH for F of the aquifer material, water residence time in the aquifer and type of climate. Drab and Sierozem Soil types control and facilitate the mobility and leaching of Fluoride in arid and semi-arid area. Generally the result has shown the area to have medium to average concentration of Fluoride in its waters with the highest value slightly above the 1.5 mg/l of WHO (2011) safe limit for drinking water. Periodic monitoring of Fluoride in the water sources is recommended so as to minimize the chances of people taking in water with high undesirable harmful Fluoride concentration.

Keywords: Geogenic Origin, Semi-Arid Area, Groundwater, Fluoride Bearing Minerals, Fluvial-Lacustrine Sediment, Yobe State

1. INTRODUCTION

Fluorine is a member of the Halogen family of group (VII) of the periodic table of elements. Being the most chemically active element among its pairs and a strong oxidant, it does not occur in the free elemental state in nature. It account for about 0.3 g/kg of the earth's crust were it exists as Fluoride in a number of minerals like Fluorspar, Cryolite, Fluoroapatite, Fluorite, Villiaumite, Silicates Micas, Apatite, Amphiboles and Topaz,[1], [2], [3] and [4]. These minerals can occur either as evaporates or detrital grains in sedimentary rocks or as Composition of crystalline rocks in bedrock aquifers, still it is considered a trace or less abundant element on earth because it constitute only 0.06-0.09% of the earth crust, [5]. Fluoride concentration in groundwater or in an aquifer will solely depend on factors which includes among others, fluoride concentration of the minerals in the surrounding, mineral decomposition, dissociation and dissolution rates, resident time of groundwater in the rock or aquifer and the reaction kinetics, these factors will determine the solubility of Fluorite in the water in a given type of aquifer, [6]. [7] Pointed out that more than 260 million people in the world take in water with Fluorite concentration of more than 1.0 mg/l with majority of them residing in the tropical countries with arid and humid climate conditions.

The study of Fluoride in water sources is vital because of it positive and negative side effects on both human and animal health. The positive benefits of fluorides intake in drinking water is that of prevention of tooth decay and Caries in children, Concentration level of less than 0.5 mg/l in water can lead to the risk of tooth decay. [4] Showed that Fluoride concentration bellow 0.5 mg/l in water is completely absorbed by the body hence leading to dental fluorosis which causes the staining, weakening and eventual loss of teeth. Concentration of between 0.5 and 1.00 mg/l of Fluoride can prevent or reduces tooth decay and Caries. For tooth paste preparation concentration of 0.25-1.00 mg/l (1000-1500 per kg of tooth paste is needed for prevention of Caries, A Fluoride intake of 3 mg f/day has being recommended by [1]. Negative harmful effects of high Fluoride consumption in water studied, elevated Fluoride intake of 3-6 mg/l can lead to skeletal fluorosis (change in bone structure) and crippling skeletal fluorosis will occur when concentration of 10 mg/l is taken in water [8].

[9] evaluated the contamination of groundwater from Ibadan town; they recorded a Flouride concentration level range of 1.82 to 6.38 mg/l with an average of value of 3.37mg/l for the groundwater from the crystalline aquifer. [10] investigated the occurrence of Flouride in the drinking water of Langtang, a concentration range of 0.12 to 10.30 mg/l with a mean of 2.42 mg/l was obtained, they identified sources of the Flouride in the water to be mica mineral present in the host rocks and pegmatite because of the positive correlation between the Flouride and lithium minerals. Cases of dental fluorosis and leg bowing common in the area indicated the high Flouride from drinking water. [11] also studied the distribution of Flouride in the major aquifers of Northern Nigerian; they recorded a concentration range of 0.03 to 10.30 mg/l for the basement aquifers, 0.00 to 5.00 mg/l was recorded for the sedimentary aquifers while the younger granites Fluoride concentration in the granite aquifers range between 0.00 and 0.89mg/l. They discovered a positive correlation between dental fluorosis and high fluoride water intake, with the Basement and Sedimentary aquifers of Northeastern Nigeria having the higher Flouride concentration level above the 1.5 maximum permissible levels, generally the northwestern sedimentary basin have low Flouride concentration level in their water.

[12] investigated the distribution of Flouride in the drinking water of Kaltungo area, they obtained a concentration that ranges between 0.52 to 4.4 mg/l and 0.1 to 3.95 mg/l for both surface and groundwater sources with an average of 2.8 mg/l, their findings revealed that people in the area are at a high risk of consuming water with high fluoride and has resulted in cases of dental fluorosis in individuals in the area. [13] studied the suitability of both surface and groundwater for different uses in Boh community of Shongom local government area in Gombe state, they obtained a Fluoride concentration range of between 0.42 to 1.72 mg/l which is above the background value of 1.00 mg/l and the 1.5 mg/l maximum permissible limit, they indicated the origin /sources of these ions in the water to be the different rock units in the area. [14] in their study on lithological influence on the hydrogeochemical characteristics in Zango Katsina State; presented a fluoride concentration range of 0.10 and 3.16 mg/l with a mean of 0.94 mg/l for the dry season period while a range of 0.10 to 1.47 mg/l was obtained for the wet period with an average of 0.523 mg/l. Concentrations above the 1.5 mg/l was obtained from groundwater in the younger granites aquifers, they also identified the fluoride source to be from the dissolution of the granite rock, however cases of dental fluorosis were observed in the area.[15] and[16] recorded a Fluoride concentration of 5.6 mg/l in the groundwater from the upper aquifer of Chad basin in Maiduguri area of Borno State.

[17], determined the trace elements in the municipal water supply in Damaturu town, fluoride concentration measured for a dug well at Powari community was 0.48 mg/l and the value obtained for a borehole at Hausari area was 0.35 mg/l these values indicates a low and safe fluoride in the analysed samples. [18] Assessed the fluoride levels in the surface and groundwater from Borno and Yobe states in the Northeastern Nigeria, their findings showed that the groundwater fluoride concentration is 0.02-2.42 mg/l while the surface water has a concentration range of between 0.03 and 2.07 mg/l with 43% of the samples having their concentrations above the 1.5 mg/l limit. It is based on this background information on the implication of Flouride in north eastern Nigeria that prompt this present research work to be carried out in the proposed area because of its location within the zone of high Flouride concentration in drinking water and with prevailing cases of dental fluorosis and other related abnormalities cause by high Flouride intake from drinking water.

2. MATERIALS AND METHODS

2.1 The Study Area

All the nine local governments' areas which comprise the study area are located in the northeastern state of Yobe. Yobe State is located within the semi-arid region of northeastern Nigeria, with the following coordinate: Latitudes 10° 55′ and 13° 40′ and Longitudes 9° 45′ and 12° 29′ Figure 2. There are two distinct seasons, a short wet season of four months which is between June and October, [19], with temporal and spatial distribution of low rainfall of about 1000 mm at the southern part of the state to less than 500 mm at extreme northern part of the state, while the long dry season which normally lasts for seven months is characterized by dry and hot climatic conditions with an annual average temperature of about 32°C. Mean annual evaporation is 1600 mm, [20]. Vegetation is typical of Sudan type with sparsely distributed short trees and shrubs, soil type is the vertisols which is heavy and dark with the land elevation ranging between 100 to 300 m above sea level.

2.2 Geology and Hydrogeology of the Area

The geology of the area is similar with that of Chad Basin, the geology of Yobe area consist of Precambrian Basement complex rocks and the cretaceous Sedimentary Chad basin, [21 distinguished three different phases for the Pan African Older granites in the Northeastern Nigeria these are Basic and intermediate plutonic, the fine grained granites and the syntectonic granites. These are rich in potash which occurs as microcline phenocryst. The plutonic variety occur as irregular bodies of gabbro, quartz-diorite and granodiorites, while the fined grained granites occur as minor discordant intrusions, the syntectonic type of granites occur as anatectic migmatites or equigranuler and porphyritic semi concordant granites. The Volcanics in the area are represented by basalts, trachyte and rhyolites. The Cretaceous sedimentary rocks in the study area from the top to bottom are represented by the following sequence. Chad formation which consist of sands and clays units of upper pliocene- pleiscene in age it is a fluviatile and lacustrine deposit of thickness of 1,500 m deep, this is follow by the Kerri-Kerri formation unconformably overlies the Gombe sandstone this consists of grits, sands, clays and oolitic sands deposited in a lacustrine/deltaic environment this formation is Paleocene in age with estimated thickness of about 130 m, [21]. Gombe sandstone underlies the Kerri-Kerri and this is mainly composed of sandstone, siltstone, shale and ironstone it is Maastrichtian in age. A transitional sequence of Gongila formation underlies the Gombe sandstone it is mainly a bed of sandstone, calcareous shale, limestone and Silty sandstone deposited under shallow marine environment during the Turonian, well log interpretation by various researchers has revealed a thickness range of 420 -1,410m for the Gongila. The Fika shales are Senonian in age and are made up of blue shales, poorly exposed with a cotton soil feature and gypsiferous, [22], deposited under deep marine environment, a thickness range of between 430 to 1,453m have being recorded for this formation by different authors, Obaje [23]. The Aptian-Albian Bima Sandstone which was deposited in fluvial environment is the oldest sedimentary formation in the area, it unconformably lies on the Pre Cambrian Basement in the area. In most parts of the study area the Gombe Sandstone and the Chad formation are either very thin or completely absent thereby exposing the Kerri-Kerri formation to the surface, figure 1.

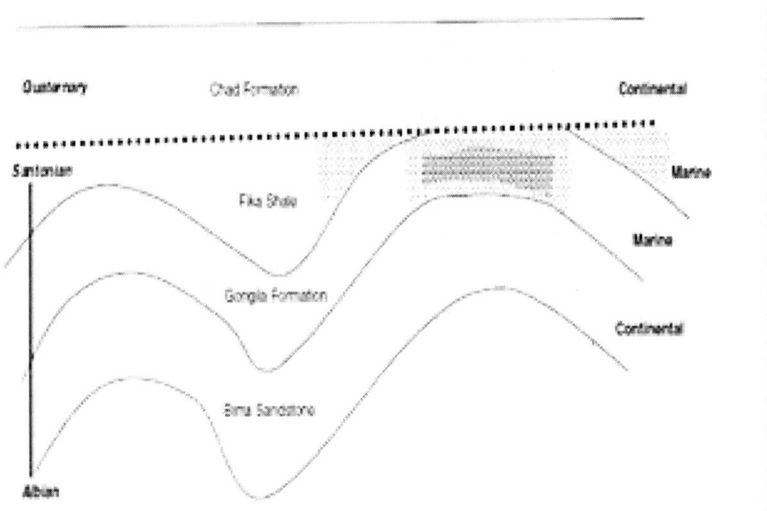

Figure 1. Stratigraphic successions in the Nigerian sector of the Chad Basin adopted from Obaje(2009).

For groundwater system in the area four Aquiferous layers have been identified, out of these only two contain appreciable quantity of water these include a second layer of sandy soil within the Chad formation with thickness range of between 80- 150 m, boreholes tapping water from this formation have a yield that ranges between 8 and 10l/s with conductivity of less than 10 m/day to 65 m/day, [19]. The second water bearing layer in the area is within the Kerri- Kerri formation and is made of aquifer materials that consists of sandstones. These aquifers are recharge through precipitation with a value range of 16- 30 mm/year, [19]. For areas like Machina and Gulani with younger granite and volcanic intrusion of basement rocks, groundwater occurs in the fractured bedrock and weathered mantle, Yield of wells is control by depth of weathering intensity and connectivity of fractures, geology and mineral composition of the parent rock. As far as groundwater potentials is concern the Chad and Kerri-Kerri formations are the highly water bearing formations.

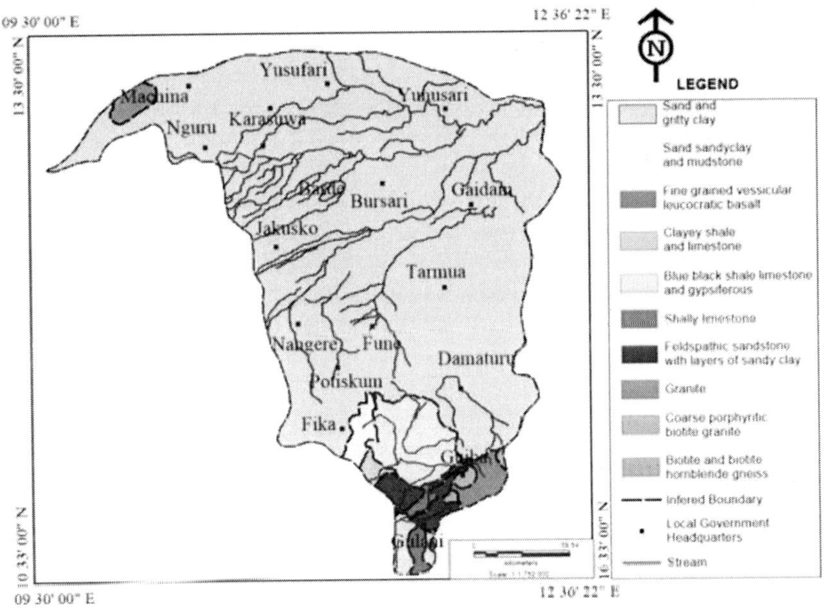

Figure 2. Geologic Map of Yobe State showing the local government areas of study modified from the geology and mineral map of Yobe State of the Nigerian Geological Survey Agency (2006).

2.3 Water Sampling and Analysis

Water samples were collected from four different sources which includes Boreholes, Tube wells, Dug wells and Ponds and analysed for the presence of Fluoride ion. In all 112 water samples were collected, each in a little plastic bottle container from nine local government areas of Yobe State. Each sampling container was rinsed at the sampling point with the sources water before collecting and transported to water control laboratory of Yobe State ministry of water resources in Damaturu for analysis. A Dr/200(Hach) Spectrophotometer was used for the detection of Fluoride in the water samples after preparation. The analysed samples were then interpreted.

3. RESULTS AND DISCUSSIONS

The summary of the result in given in Table 1

Out of the 112 water samples analysed Flouride was not detected in only ten samples with four from boreholes and six from dug wells. The four Pond water analysed has Fluoride concentration between 0.71 and 1.36 mg/l while Dug well values ranges between 0.00 to 1.5mg/l for the 18 samples collected. Concentration for the three Tube wells water samples collected from Potiskum local government area ranged between 0.65-1.5 mg/l. Water samples analysed from the 86 Boreholes water samples analysed contain Fluoride that ranges in concentration from 0.00 to 2.2mg/l. Both the least and highest detectable Fluoride concentration of 0.01 and 2.2 mg/l were detected from boreholes in the area of study. The result shows the background concentration of 1.00 mg/l was recorded from the four types of water sources sampled in this study.

3.1 Health Implication

A total of 34 samples out of the samples analysed have a fluoride concentration of bellow 0.6 mg/l. By implication and going by the [1], [8] drinking water standard, there is the likelihood for people taking this water for a longer period of time to develop Dental Caries, especially young Children. Concentrations of 0.6 to 1.5 mg/l were obtained from 64 samples this represent 57% of the analysed water; waters with this range of Fluorite concentration are beneficial because if taken into the body it help in the prevention of Dental Decay and bone development. [24] gave different risk classes based on Fluoride concentration as $0.5 \leq R \leq 1$ (risk free): $1 \leq R \leq 1.5$(slight risk): $1.5 \leq R \leq 2.5$(risk): $2.5 \leq R \leq 3.5$(high risk) and $R > 3.5$ (very high risk), based on this classification 16 samples representing 14.3 % of the water from this study have Fluoride values that are within the risk class of $1.5 \leq R \leq 2.5$(risk) with the probability of individual consuming the water developing dental Fluorosis, while a total 19 of samples representing 16.9% are within the slightly risk class of $1 \leq R \leq 1.5$ and the highest number of 85 samples representing 75% of the water analysed in the area falls in the $0.5 \leq R \leq 1$ of risk free class. It was observed that the fluorite concentration in the area increases with depth of water sources.

Table 1. Summary of Fluoride Concentration Obtained from the Study Area.

S/NO	Local Government Area	Concentration and Ranges of Fluoride in Water Sources From the Study Area in (mg/l)				
		0.00-0.52	0.60-0.98	1.00-1.50	2.00-2.20	Total no. of Samples Analysed
1	Barde	9	-	-	-	9
2	Bursari	3	5	-	-	8
3	Damaturu	8	3	5	-	16
4	Fika	-	2	2	-	4
5	Gulani	-	1	14	-	15
6	Karasuwa	9	8	6	-	23
7	Nangere	4	-	-	3	7
8	Nguru	10	10	2	1	23
9	Potiskum	2	2	3	-	7
	Total					112

3.2 Origin and Sources of Fluoride in the Area

The two main formations forming the aquifers in the study area are the Chad formation (Upper zone) and Kerri-Kerri formation aquifers both deposited under fluvial-lacustrine environmental conditions. The semi-arid climate in the area with an extreme high temperature of 42°C that result to high evapotranspiration of more than 1600, an unpredictable and variable low rainfall that falls for 120 to 140 days with an annual range of < 1000 to <500 from the southern to the extreme northern part of the state, is favourable for dissolving fluoride bearing minerals. The most probable origin of Fluoride detected in the water from the area of study area is Geogenic, originating from the dissolution and ejection of Fluorite bearing minerals like micas, amphiboles Illites, Apatite, Topaz or Cryolite in the sandy and clayey layers in these aquifers. Apatite contains florite and occurs as an accessary mineral in sedimentary horizon, [25], weathering process and the circulation of water in rocks and soils leach out Fluoride into the groundwater. Factors that favour the release of Fluoride into groundwater in sediments are the mineralogical composition of the sediment, the extent of sediment alteration and the Geochemistry of the surrounding water, [26]. They authors went further to point out that high Fluoride of clastic sediment is concentrated in micas and clay mineral like Illites, in fluvial sediments the release of Fluoride under acidic condition of low PH condition is controlled by the dissolution/ alteration of Fluorapatite and biotite minerals originating from fluoride rich igneous and metamorphic rocks. For silicates rich sediments Fluoride dissolution and enrichment in groundwater is principally control by the alkaline PH, anion exchange of OH^- for F^- capacity of the aquifer materials because of their almost similar ionic radii of F (1.23-1.36 Å) and OH (1.37-1.40 Å), cation exchange capacity, and residence time in the aquifer of the groundwater and the climate type, [6], [27].

Fluoride in groundwater can originate from Clays within the aquifer, factors like elevated alkalinity and soil grain size, initial Fluoride concentration in minerals and the salinity favours the desorption of Fluoride from the clay minerals into groundwater, the adsorption capacity of different clay minerals containing Fluoride is given in the order *Bentonite > halloysite > gibbsite. Kaolinite >> soap alkaline soil, vermiculite. Goethite; layered silicate mineral > various oxides* [26]. Base on the findings of [28] the dominant water type in the Chad basin in Nigeria is the sodium bicarbonate type, this will dissolve and concentrate Fluoride in the groundwater, the mixing up of bicarbonate groundwater with aquifer sediments triggers a reaction that lead to the ejection of Fluoride from the fluorite bearing minerals resulting in the Fluoride becoming enriched in the groundwater, this is represented in equation 1.

$$CaF_2 + 2NaHCO_3 \rightarrow CaCO_3 + 2Na^+ + 2F^- + H_2O + CO_2 \quad (1)$$

The Sodium carbonate water specie facilitate the precipitation of Ca_2^+ and CO_3^{2+} which quickly dissolve the Fluorite (CaF_2) from the sediment and rock matrix to facilitate the release of Fluoride in the water.

Enrichment of Fluoride into the groundwater is dependent on Soil type, according to [29] the rate of leaching of Fluoride is related to its movement in different soil type, Drab and Sierozem types of soil favours the leaching of Flouride in arid and semi-arid areas. Beside natural sources anthropogenic sources may be responsible for adding Flouride into the surface Pond water and tube wells water in the area, leached refused dump into these ponds and fertilizer added to farm lands and irrigated areas to increase the soil fertility for the purpose of busting crop yield can add appreciable quantity of Fluoride to these water sources and increase its concentration in water.

4. CONCLUSION

Out of the four sources of water analysed in the area only ten samples are without Fluorite, and this constitute only 8.9% of the water samples analysed, the remaining 92 samples have concentration of Fluoride that ranges between 0.01 and 2.2 mg/l with the highest value of 2.2 mg/l obtained from three boreholes at Nangere local government area. The analysed tube well samples have a Fluoride concentration range of 0.65-1.5mg/l while surface Pond water has its highest Fluoride concentration to be 1.36 mg/l. The result shows that there is a good correlation between fluoride concentration and depth. Base on the concentration of the Fluoride detected in the water from this study and the higher concentration of 5.6 mg/l obtained elsewhere in the same Chad basin by [15] and [16], it can be concluded that the area has medium to average Fluoride level in the water sources. The Fluoride detected in the water originated from the Florite bearing minerals from within the aquifer materials and from other anthropogenic sources.

REFERENCES

1. WHO (2004) Fluoride in Drinking- water. Background document for development of WHO guidelines for drinking water quality: WHO/SDE/WSH/03.04/96.
2. Veressinina, Y., Trapido,V.A., Munter, R. (2001) Fluorine in drinking water, the problem and its possible solution. Proce: Estonia Acad. Sci. Chem. vol 50:2 pp 81-88.

3. Alabdulaay,I.A., Al-Zarah, I.A., Khan, and A.M. (2013) Occurrence of Fluoride in ground waters of Saudi Arabia: Appl water Science journal, 3: 589-595.
4. Lar, U.A., Dibal, H., Schoneich, K.(2014) Fluoride in groundwater in Nigeria: origin and human health impact: American journal of Environmental protection 3 (6-2) pp 66-69.
5. Fawell, J. Bailey, K.,Chilton, J., Fewtrell,L., Magar, Y. (2006) Fluoride in drinking water WHO, IWA Publishing, 1-144.
6. Sexena, V.K., Ahmaed,S.(2001) Dissolution of Fluoride in groundwater: a water- rock interaction study. Environ geol. 40:1084-1087.
7. Shreya, D., Nag, S.K. (2016) Geochemistry appraisal of Fluoride-laden groundwater in Suri I and II Block, Birbhum district, west Bengal. Journal of earth science and climate change, vol: 7 issue 6 pp 1-68.
8. WHO (2001) Revision of WHO guidelines for drinking water quality world health organization Geneva.
9. Egbinola, N.C, Amanambu, C.A (20014) Groundwater contamination in Ibadan, Southwest Nigeria: Springer open journal 3:448 pp1-6.
10. Dibal, H.U., Schoenich, k., Garba, I. Lar, U.A.,Bala E.A. (2012) Occurrence of Fluoride in the drinking waters of Lang tang area North central Ngeria: Health vol4 No 11 pp 1116-1126.
11. Dibal, U.H., Schoenich., k., Garba, I. Lar, U.A., Bala E.A. (2012) Overview of Fluoride distribution in major aquifer units of northern Nigeria: Health vol 4 No 12 pp 1287-1294.
12. Haruna, I.A., Muhammed,.U.,Muhammed, A.A (2014) Environmental distribution of fluoride in drinking waters of Kaltungo area northeastern Nigeria: American journal of environmental protection. 3(6-2) 19-24.
13. Lovelyn, S.K., Hamidu, H., Mbiimbe, E.Y., Sidi, M.W., Farida,G.I. (2016) Suitability of ground and surface water resources for different uses in Boh community Gombe state Northeastern Nigeria: nature and science journal 14 (2) pp 22-31.
14. Amadi, A.N., Aminu,T., Okunlola, I.A., Olasehinde,P.I., Jimoh, M.O. (2015) Lithological influence on the hydrogeochemical characteristics of groundwater in Zango, northwestern Nigeria, natural resources and conservation 3(1): pp 11-18.
15. Edmunds, W.M., Fellman, E., Goni, I.B., McNeil, G., Harkness, D.D. (1998) Groundwater palaeoclimate and recharge in the southwest Chad Basin, Borno state, Nigeria. In. Isotopes technique in the study of environmental change. Proceedings of an international symposium, IAEA, Vienna, 1997.
16. Edmunds, W.M., Fellman, E., Goni, I.B. (1999). Lakes, groundwater and palaeohydrology in the Sahel of NE Nigeria: Evidence from hydrogeochemistry. Journal of geological society of London, 156. pp 345-355.
17. Emeka,D.O., Weltime, O.M. (2008) Trace elements determination in municipal water supply in Damaturu metropolis Yobe state: Bayero journal of pure and applied sciences, 1(1): pp58-61.
18. Waziri, M., Musa,U., Hatis, S.S. (2012) Assessment of Fluoride concentration in surface waters and groundwater sources in northeastern Nigeria: resources and environment 2 (2) pp 67-72.
19. Dawoud, A.M.,Raouf, A.R.A (2008) Groundwater exploration and assessment in rural communities of Yobe state northern Nigeria; water resources management vol. 23 issue 3 pp 581-601.
20. Mante, M.A. (1986) Groundwater exploitation and drilling technology in Arid areas of Nigeria, Borno as a case study, proceedings of the first annual symposium and training workshop on groundwater resources in Nigeria pp141-173.
21. Carter, J.D., Barber, W., Tait, E.A., Jones,G.P. (1963) The geology of parts of Adamawa, Bauchi and Borno provinces in northeastern Nigeria: Geol. Survey Niger Bull 30: 108 p.
22. Offodile, M. E. (2002): An Approach to Groundwater Study and Development in Nigeria Second edition: Published Mecon Geology and Engineering Services Ltd. Jos, Nigeria. 453p.
23. Obaje, N.G. (2009) Geology and mineral resources of Nigeria: springer 219 p.
24. Linthurst, R.A., Bourdeau, P., Tardill, R.G. (1995) Methods to assess the effects of chemicals on ecosystems, chichester. UK; John Wiley and Sons.

25. Kundu, N., Panigrahi, M.K.,Tripathy, S., Munshi, S., Powell, M.A.,Hart, B.R.,(2001) Geochemical appraisal of Fluoride contamination of groundwater in the Nayagath district of Orissa. India environ geol vol. 41 (3) pp 451-460.
26. Sivasankar, V., Darchen, A.,Omine, K.,Sakthivel, R., (2016) Fluoride: a world Ubiquitous compound, its chemistry, and ways of contamination. Springer international publishing Switzerland. pp 6-32.
27. Reddy, D.V., Nagabhushanam, P., Sukhija, B.S., Reddy,A.G.S., Smedley,P.L. (2010) Fluoride dynamics in the granitic aquifers of the Wailapally watershed. Nalgonda district: India chem geol: 269 (3-4): pp 278-289.
28. Yusuf, A.K., Goni, I.B., Hassan, M. (2014) Hydrochemical studies of the groundwater resources of the middle zone aquifer in southeastern Chad basin: IOSR journal of applied geology and geophysics vol. 2issue 2 ver.1 pp61-66.
29. Wang, W., Li, R., Tan, J., Luo, K., Yang, L., Li, H., Li,Y. (2002) Adsorption and leaching of Fluoride in soils of China: Fluoride, 35 (2): pp 122-129.

CHAPTER 23

Utilization of Recycled Polypropylene for Production of Eco-composite Reinforced with Leaf Sheath Fiber of Betel-Nut Plant

Pathik Shah*, Rataram Prajapati, Bornali Dey, Dipsankar Maiti

Central Institute of Plastics Engineering and Technology (CIPET), Vatva GIDC, Ahmedabad, India

ABSTRACT

The interface modification of leaf sheath fiber of betel nut was accomplished by NaOH & Silane coupling agent. The chemically treated (alkaline and silane) fibers are analyzed by FT-IR. The incorporation of these bio-fibers with Recycled Polypropylene (5-20%wt fiber) was carried out by twin screw extruder. The standard test specimens were prepared by injection molding. Mechanical and thermal characteristics of the blends systems were studied to evaluate the effect of chemically treated fibers content on recycled polypropylene. It has been found that as there is a decrease in tensile strength and elongation, the tensile modulus was found to be increased with the chemically treated fiber content. There is a good dispersion of both type treated fibers materials in recycled PP matrix result in an expansion in their use into more diverse, and technically demanding, application areas. Biodegradability (Soil Burial Test), properties of green composite is also tested. It suggest that these Eco-composite moderately biodegradable. This research paper seeks to provide use of recycled pp with non-food renewable material and consider the environmental and economic benefits of the materials.

Keywords: Biocomposites, Mechanical Strength, Recycled Polypropylene, Mercerization and Silanation

1. INTRODUCTION

Biocomposites offer a significant non-food market for crop-derived fibers and resins. Considerable growth has been seen in the use of Biocomposites in the automotive and decking markets over the past decade. Over the last decades, research has increasingly been conducted on renewable materials from Sustainable resources for a variety of applications. This has been influenced by the ever increasing demand for newer, stronger, stiffer, recyclable, fire repellent, less expensive and yet lighter-weight materials. Nowadays, polyethylene (PE) and polypropylene (PP) based plastic materials are being extensively used for commercial and household purposes. However, these polymer based plastics are substantially resistant to biodegradation. Thus their increasing accumulation in the environment is proving to be an ecological threat to the world. Consequently, over the last couple of decades exhaustive studies on the biodegradation of plastics have been carried out in order to overcome the environmental problems associated with synthetic plastic waste [1]. The improvement in the biodegradability of PP was observed by blending it with small additions (5%) of wood sawdust and wheat flour. Betel nut fiber (Bn)/polypropylene (PP) composites were prepared in the different ratio of 10:90, 20:80, 30:70, 40:60 (Bn wt%: PP wt%) using extruding and hot press molding technique. From the results, it can be inferred that Bn30: PP70 mixture composite (BnPP) showed better performance among the composites prepared [2].

The natural Activated Carbon (AC) composites prepared from carbon coconut shell reinforced with polypropylene (PP) [3]. The results showed that the tensile stress was increased when AC is increased specifically for sample 4 wt% and 8 wt%. Maximum tensile stresses lead by sample 4 wt% with 30 MPa. Two types of short natural fibers, namely jute and betel nut fibers, are employed as fibers for the formation of short fiber reinforced polypropylene composites in difference ratios (fiber content: 3, 5, 10 and 20 wt%) by hot-press molding technique [4]. For better compatibility, jute and betel nut fibers were subjected to alkalization and the hybrid composites of alkalized jute/betel nut and PP showed much improved performance with respect to other composites. However, this relies on having reasonable fiber/matrix interfacial strength, and strength can reduce with strongly hydrophobic matrices such as polypropylene with increasing fiber content unless coupling agents or some other interfacial engineering method is used [5].

The mechanical performance of polypropylene (PP) composites filled with betel nut (Areca catechu) short fiber (Bn) at different compositions using extruding and hot press moulding technique was examine by M. Masudul Hassan et al.[6] Results showed that Bn10:PP90 mixture composite (BnPP) had better performance among the composites prepared. Seaweed is subjected to hybridize with betel nut short fiber in PP composite to achieve superior mechanical performance. Bagasse filled recycled polyethylene bio-composites were produced by the compounding and compressive molding method. Two sets of composites were produced using uncarbonized (UBp) and carbonized (CBp) bagasse particles by varying the bagasse particles from 10 to 50 wt%. The bagasse particles added to the RLDPE polymer improved its rigidity and the hardness values of the composites. [7]

There are lots of waste materials available during fabrication process like Polyethylene, Polypropylene, polystyrene etc. These wastes constitute nuisance to the environment not only in India but the world at large. The ability to convert these wastes into useful engineering materials e.g. composites sharpens the focus of this present research work. From the available literature no investigation has been conducted on the application of the leaf sheath of betel-nut plant (Areca Catechu) in polymer composite materials. Fiber modification includes mercerization and Silanation for enhancing better surface interaction towards polymer matrix. Based on the above-mentioned situation, the study described in this work intends to study the mechanical properties of recycled PP composites reinforced with Leaf Sheath Areca Fiber.

2. MATERIALS AND METHOD

2.1 Materials and Equipment

The betel Nut tree which grow abundantly in the eastern & Northern region of India scientifically belongs to 'Areca' species named as "Areca Catechu". The leaf sheath is a basically waste material or used as fuel (fire resource) in rural areas. The Recycled PP used was collected literally from the industrial waste. The chemicals used are Sodium

Hydroxide (NaOH), Methyl Alcohol (MeOH), and Acetic Acid (AcOH), Silane coupling agent (Vinyl Tri-methoxysilane) was from MERCK India Ltd. The equipment used in the research were: metal mold, sieves, digital weighing balance, hack saw, grinding machine, Vernier caliper, compounding machine, molding machine and bending and tensile testing machines, Rockwell hardness tester and Charpy impact tester.

2.2 Alkali Treatment (UT)

Generally, the first step is the mercerization process (pretreated process) for all of the fiber surface treatments. Mercerization causes the changes in the crystal structure of cellulose and then the different chemicals can be used on the fibers surface in order to improve the interfacial properties. Chopped fiber (leaf sheath of betel-nut plant) was pre-treated with 4% NaOH for about half an hour in order to activate the -OH groups of the cellulose and lignin in the fiber. Fabric was then washed many times in water and finally dried (Alkaline treated-UT). This treatment removes a certain amount of lignin, wax and oil covering the external surface of the fiber cell wall, depolymerizes cellulose and exposes the short length crystallites [8]. It is reported that alkaline treatment has two effects on the fiber: (1) it increases surface roughness resulting in better mechanical interlocking. (2) It increases the amount of cellulose exposed on the fiber surface, thus increasing the number of possible reaction sites.

2.3 Silane Treatment (TM)

1% of the vinyl- triethoxysilane was mixed with a Methanol/water mixture in the ratio 6:4 mixed well and was allowed to stand for an hour. The pH of the solution was carefully controlled to bring about the complete hydrolysis of the silane by the addition of acetic acid. Fiber was dipped in the above solution and was allowed to remain there for 1 and 1/2 hours. The Methanol/water mixture was drained out and fiber was washed in water. Silane treated fiber (TM) was then dried at room temperature for 2hour, followed by drying in the oven at 70°C for 24 hours [8]. Silane used in this work has two functional groups, a hydrolysable group which can condense with the hydroxyls of the Areca fiber and an organo functional group capable of interacting with the matrix. The hydrolyzed silane can undergo the condensation and bond formation stage when influenced by acid or base-catalyzed mechanisms. Besides these reactions, the silanols can condense to give polysiloxanes. The silanol groups are chemically attached to the fiber through an ether linkage [9].

2.4 Preparation of Bio-composites

At first the fiber from leaf sheath of betel-nut plant (UT& TM) was preconditioned in the hot air oven at the temperature 70°C for 24 hours and then the recycled PP and the fiber was compounded using Twin Screw Extruder. There are five temperature zones in extruder from hopper to die: melting begins from the second zone. The following temperatures were set at respective zones: 180°C, 190°C, 205°C, 210°C, 225°C with the die temperature 210°C The molten strand is then cooled by chilled water and chopped into pellets using pelletizing machine and then left for cooling. The compounded is then again dried in the hot air oven at temperature 70°C and then for preparing the test specimens for analysis of the mechanical properties and the thermal properties of the composites were injection molded at 190°C, 200°C, 210°C and 220°C temperatures in the barrel zones of injection molding machine.

Table 1. Different batches of Biocomposites.

Batch	Recycled PP (%wt)	Fiber (%wt)	Note
0% Fiber	100	0	Non treated
5% UT	95	5	
10% UT	90	10	Only NaOH
15% UT	85	15	treated
20% UT	80	20	
5% TM	95	5	
10% TM	90	10	NaOH and
15% TM	85	15	Silane treated
20% TM	80	20	

2.5 FTIR Analysis

The FTIR imaging was performed in Agilent Resolution PRO, Cary 600 series. The generated bio-composite fibers were characterized using Fourier transform infrared spectroscopy (FTIR) to understand the nature of chemical interaction between fiber reinforcement and PP matrix. Changes in absorption peaks were observed in FTIR spectra of Biocomposites fibers as compared to the pure fiber which indicated possible chemical linkages between the fiber and polymer matrix.

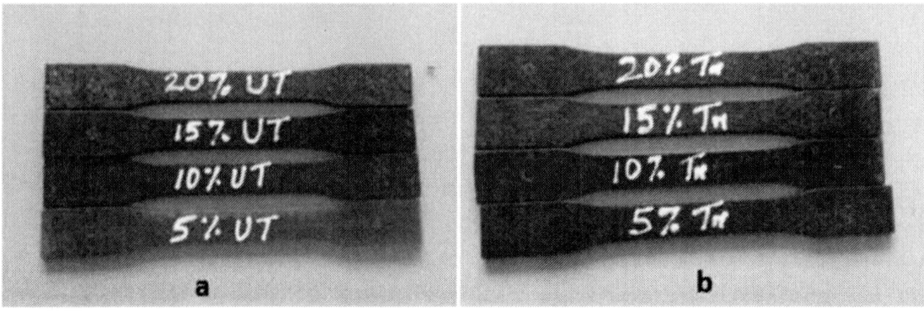

Figure 1. Photograph of (a) Alkaline treated (UT) & (b) Silane Treated (TM) Biocomposites.

2.6 Differential Scanning Calorimetry (DSC)

Differential Scanning Calorimetry (DSC) is a thermo-analytical technique in which the difference in the amount of heat required to increase the temperature of a sample and reference are measured as a function of temperature. Typical applications include determination of melting point temperature, heat of melting, measurement of the glass transition temperature, curing and crystallization studies and identification of phase transformations.

2.7 Mechanical Properties

The tensile test and flexural test were conducted according to ASTM standard testing methods D638 and D790, respectively. The tests were conducted on an INSTRON - Universal Testing Machine (Model-3382).

2.8 Biodegradable Test (Soil Burial Test)

Soil burial test methods have been established and standardized for testing resistance of plastics to micro-organisms. The aim was to access their resistance in soil, rather than the degradability. The test material is buried under laboratory or field conditions. Visual assessment of exhumed materials is carried out and mass loss measurements are also performed. Soil burial tests are used to give an indication of the duration of the test material in a given soil under given conditions. The choice of location can affect the test results.

3. RESULTS AND DISCUSSION

3.1 FTIR Analysis

Untreated, alkali (NaOH) and silane treated leaf sheath fiber were investigated by FTIR analysis. Presence of alcohol in untreated fibre results peak at 1030 cm^{-1} which will not be shown due to removal of alcohols –OH group in both alkali & silane treated fiber. The C=O stretching vibration peak which shows presence of pectin is observed in both treated & untreated fibre, but their peaks slightly shifts from 1735 cm^{-1} to 2348 cm^{-1} due to intermolecular H-bonding, which dominates. Peak near to the 1654 cm^{-1} corresponds to–C=O stretching due to the presence of aliphatic carboxylic acid in cellulose chain, & the peak is eliminited due to removal of lignin & pectin which contain ester –C=O group. Surface modification of leaf sheath fiber exhibits the absence of characteristic peaks at 1643 and 1156 cm^{-1}. This is because of the decomposition of hemicellulose and partial leaching out of lignin by sodium hydroxide [2]. The disappearance of the peak at 1643 cm^{-1} is due to the dissolution of a portion of uranic acid, a constituent of hemicelluloses xylan. The shift of the peak from 2900 cm^{-1} to 2698 cm^{-1} indicates participation of some free hydroxyl groups in the chemical reaction. Band at 758 cm^{-1} correspond to silane bride.

The DSC analysis of all samples is carried out using differential scanning calorimetric by heating the sample at 10°C /min from 50°C to 250°C in argon in a thermal analyzer (Perkin Elmer). The plots are shown in the figure 2. The DSC curves in Figure 2 show endothermic melting peaks at 163.78°C for PP. Decreasing amounts of PP gave rise to decreasing melting enthalpies. This was to be expected, since the melting enthalpies are related to the amounts of polymer in the sample, and since the amounts of polymer decrease with increasing fiber content in the sample, the melting peak sizes and related enthalpies should correspondingly decrease. It is, however, possible that the presence of solid fiber may influence the crystallization behaviour of either or both polymers.

3.2 Differential Scanning Calorimetry (DSC)

Table 2. DSC melting results of Biocomposites.

Bio-composite	Peak (Tm) °C	Area under melting curve (mJ)	Melting Euthalpy (ΔHm) J/g
5% UT	163.45	301.761	73.600
10% UT	163.43	233.72	64.924
20% UT	162.81	128.578	47.621
5% TM	164.78	273.350	71.934
10% TM	164.44	257.560	59.898
20% TM	164.47	146.705	48.902
0 % PP	163.78	346.738	78.804

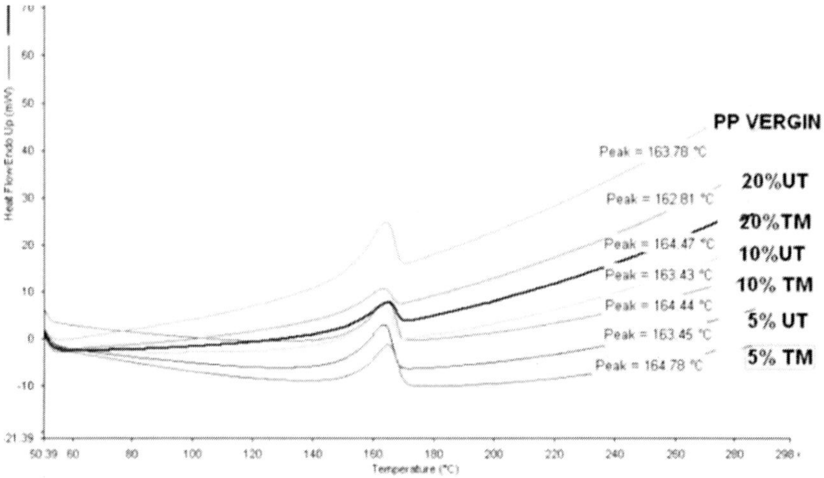

Figure 2. DSC heating curves of Recycled Polymer (RPP) with fiber Biocomposites.

3.3 Mechanical Properties

3.3.1 Tensile Test

Tensile properties such as tensile strength of the Leaf sheath fiber / Recycled PP composites containing 5% wt, 10% wt, 15% wt and 20% wt were measured and the results are presented in the Figure 3. It is observed that with an increase in fiber content from 5% wt to 20%, wt the tensile strength gradually decreased for alkaline treated biocomposite. But for 10% wt loading in silane treated biocomposite shows highest tensile strength and then decreases. This may be due to lack of stress transfer from the matrix PP to fiber.

3.3.2 Elongation Test

The elongation at break of the composite shows a similar trend as shown for the 10% wt Leaf Sheath fiber Composite (silane Treated) sample for tensile strength performance and maximum elongation at break. An increase of the elongation at break of the composites increases the toughness and ductility of the composite.[10]

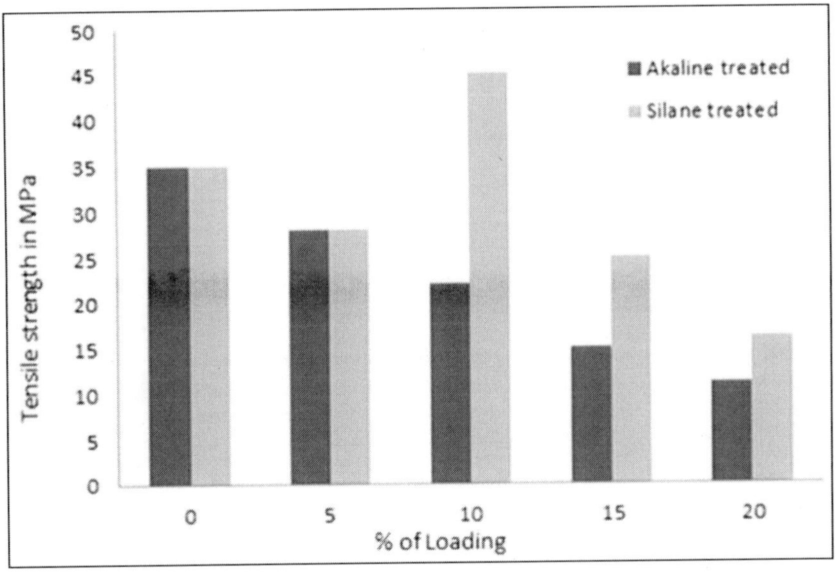

Figure 3. Tensile strength for alkaline treated and silane treated biocomposites.

The composite might be present as moderate physical-mechanics adhesion (better known as inter-diffusion that allows a kind of bonding between two polymers surfaces via diffusion of the macromolecules both polymers.

3.3.3 Flexural Strength & Flexural Modulas

When compared between alkaline Treated & silane treated, flexural strength is highest in 10% wt alkaline treated composite. The increase in flexural strength is highest in 10% wt treated composite than other loading. So 15% wt loading for composite shows most efficient result on treatment with silane. Incase of Flexural modulas similar pattern is also observed. The highest flexural modulas increase is observed in 10% wt loading on treatment with silane.

But the Flexural modulas shows a anomolous result which shows highest value in 20% wt loading of alkaline treated biocomposite. It shows a moderate value in both 10% wt & 15% wt treated composite. Recycled polymer shows lower value than composite with loading percentage. The flexural strength & modulas value is least in virgin polymer.

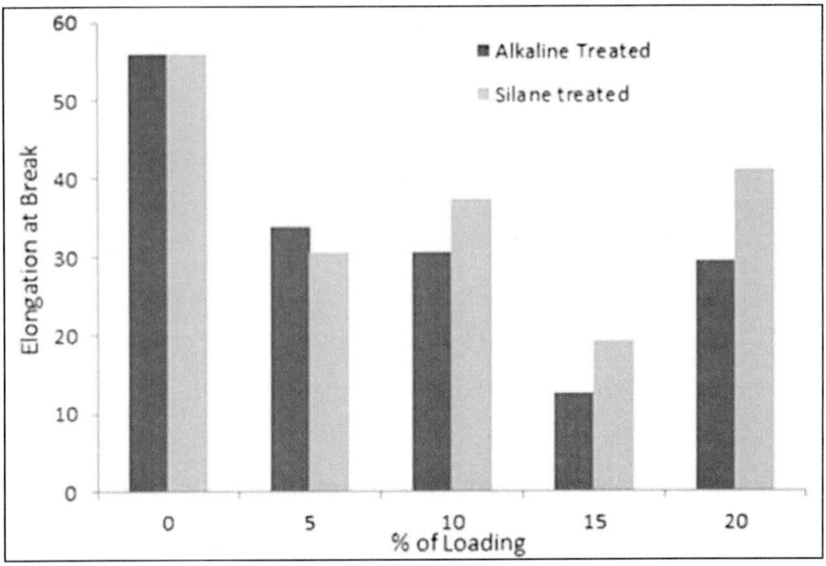

Figure 4. Elongation at break of alkaline treated and silane treated biocomposite.

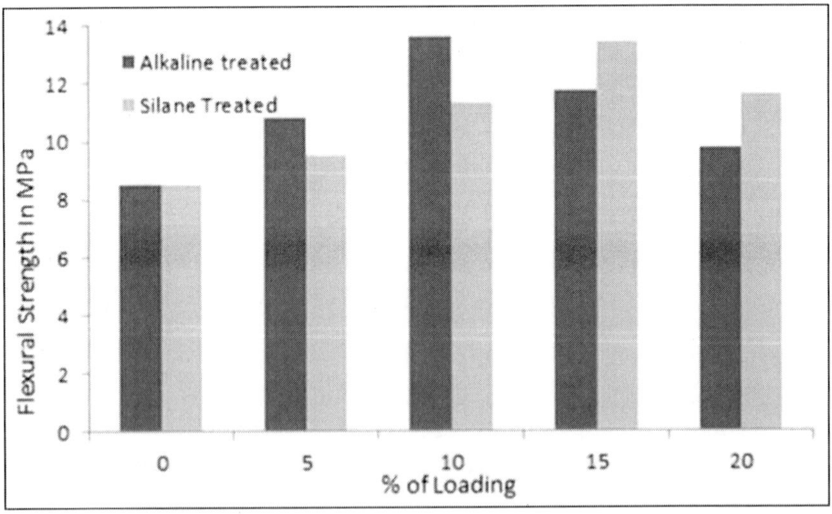

Figure 5. Flexural Strength of Alkaline treated and Silane treated biocomposites.

Flexural strength of fibre reinforced composites depend on the nature of matrix material and the distribution and the orientation of the reinforcing fibre, as well as the nature of the fibre- matrix interfaces and the interphase region.

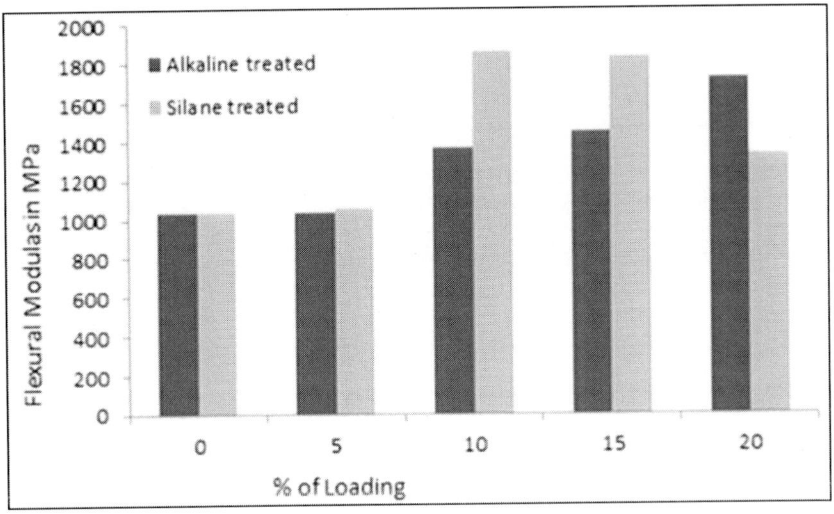

Figure 6. Flexural Modulas of Alkaline treated and Silane treated biocomposites.

Even a small change in the physical nature of the fibre for a given matrix result in prominent changes in the overall mechanical properties of composites. It is well known that different degrees of reinforcement effects are achieved by the addition of hydrophilic fibres to different polymers. This may be due to different adhesion strength between matrix and fibres.

3.3.4 Compression Strength

Compression Strength is an important factor to consider where structural stability is required. As one of the objective of this study is to utilize this composite in structural application, so that require measurement of Compression Strength for evaluation.

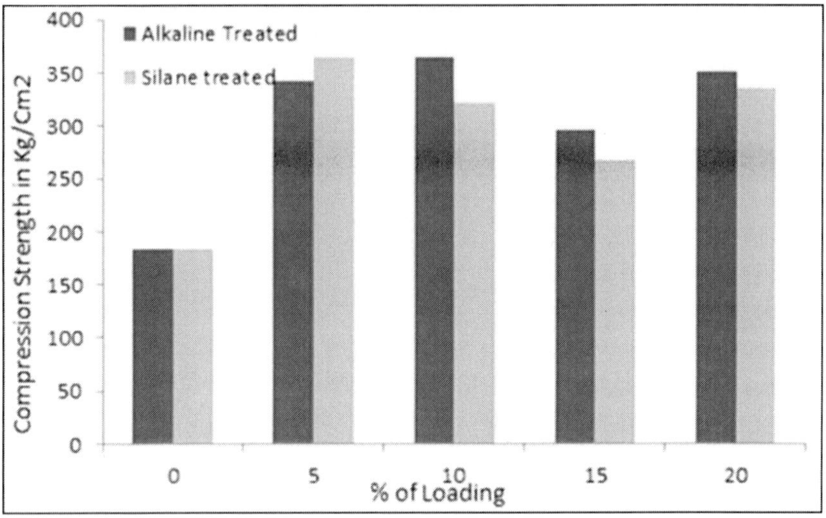

Figure 7. Compression Strength of Alkaline treated and Silane treated biocomposites.

As from graph plotted, in alkaline treatedbiocomposite it shows a little decreasing trend in compression strength with increase in loading, until 15% wt, but it suddenly increase in 20% wt loading. This may be due to more better packing of fibre results in better Compression stregth.Increase in fibre content increase the amorphous part & increse free space between polymeric chains, which imparts better chain movement & more dynamic dimension change which increase Compression Strength. The virgin polymer shows least Compression strength than composites.

In silane treated composite 10% wt & 20% wt loading shows better compression strength with moderate test value. This may be due to better interfacial bonding betweem fibres & matrix which imparts better stress transfer through composite.

3.3.5 Rockwell Hardness

Rockwell Hardness is a mesurement of resistance to penetration, indentation, scratching & surface deformation. It is basically measured for surface durability of composite & plastics. As shown in graph it initially marginally increase with fibre loading percent upto 10% wt in both silane treated & alkaline treated biocomposite the gradually decreases with increase in loading percent.

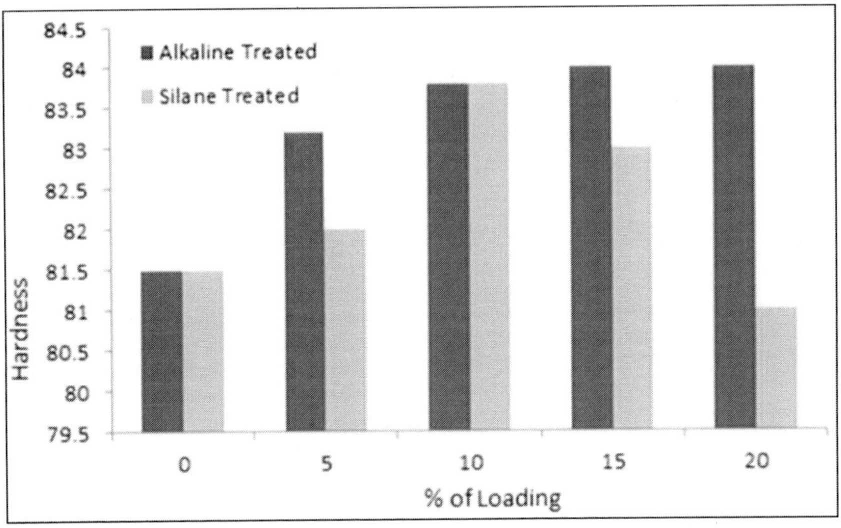

Figure 8. Hardness of Alkaline treated and Silane treated biocompostes.

As fiber percentage increase the surface stiffness increases due to decresing percentage of visco elastic nature plastic, & as Hardness measures the non-recoverable visco-elastic property, so the general trend should be with incresing fiber percent Hardness also increses. Here also this trend is followed in both treated & untreated composite with an exception in 15% in treated composite. This may be due to lack of proper dispersion of fibre in composite matrix, which can impart better visco elastic property decrease in hardness.

3.3.6 Izod Impact Strength

Impact strength is defined as the ability of a material to resist fracture under stress applied at high speed. The impact properties of composite materials are directly related to its overall toughness.

Figure 9 shows the influence of the alkali treatment, different matrix materials and fiber loading on the Izod impact strength of the areca fiber composites. Figure 9 shows the variation of impact strength of areca composites under Izod impact test. It is seen that with the increase in fiber content in the composite, the energy absorption improves initially but then gradually decreases with increase in loading in case of treated composites.

It is imagined that as the size of the fiber becomes smaller, greater interaction between the fiber and matrix could result in better and more efficient stress transfer which intern could increase the impact strength of the composite [11]. The optimum fiber content varies with the nature of the fiber and matrix, fiber aspect ratio, fiber/matrix interfacial adhesion etc. the low value at high fiber content may be due to the presence of so many fiber ends in the composites, which could cause crack initiation and hence potential composite failure [12]. In view of the above results, it is noted that the composite of 10% wt leaf sheath content exhibits better mechanical behavior.

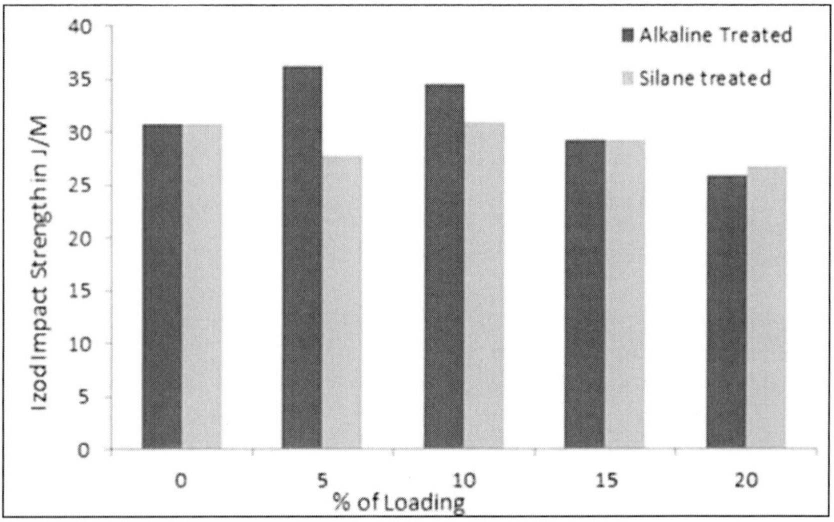

Figure 9. Izod Impact Strength for alkaline Treated & silane treated biocomposite.

3.3.7 Soil Burial Test for Biodegradation

Cellulose possesses the tendency to be degraded when buried in soil. For this purpose, the composite samples were weighed individually and buried in soil for 1-4 weeks. Thereafter, samples were carefully withdrawn, washed with distilled water and dried at 105°C for 20 min and kept for 24h and then weight is recorded. Finally weight loss of various degraded samples is calculated. The weight loss indicating that these Biocomposites are partially biodegradable. Soil environment contains different kind of microorganism which leads to weight loss of Biocomposites.

4. CONCLUSIONS

Chemical treatment can reduce the hydrophilicity of the fiber to decrease the hydroxyl groups from the fibers and improved coupling between treated leaf sheath fiber and recycled PP which improved interfacial adhesion and so increases the mechanical properties. The tensile strength has reduced for the Biocomposites when compared to recycled polypropylene. In this study the benefits are regarded from an Economic, an Ecological and a Technical point of view for various parts of society. It will be an alternative way to develop the Biocomposites which can be particularly used for daily needs of common people whether it is house hold furniture, house, fencing, decking, flooring, and light weight car components or sports equipment. Waste disposal is becoming increasingly important with the recognition that landfill is not sustainable and as such costs are increasing, with more responsibility being placed on producers. These Biocomposites have low cost, easy availability and aesthetic designs will be the main driving force to transform the depended present to sustainable future. These kinds of Biocomposites are likely to see a period of sustained growth.

REFERENCES

1. Takian Fakhrul, M. A. Islam. "Degradation Behavior of Natural Fiber Reinforced Polymer Matrix Composites". Procedia Engineering Vol. 56, Pages 795-800, 2013.
2. Chakra arty, J.; Hassan, M.; Khan, Mubarak. "Effect of Surface Treatment on Betel Nut (Areca catechu) Fiber in Polypropylene Composite". Journal of Polymers & the Environment. Vo l. 20 Issue 2, pp 501-506, Jun2012.
3. Z. Salleh. M. M. Islam, M. Y. M. Yusop, M. A. Mun'aim M. Idrus. "Mechanical Properties of Activated Carbon (AC) Coconut Shell Reinforced Polypropylene Composites Encapsulated with Epoxy Resin". APCBEE Procedia. Vol. 9, Pages 92-96, 2014.
4. M. MasudulHassan, Manfred H. Wagner, Hayder U. Zaman & Mubarak A. Khan. "Study on the Performance of Hybrid Jute/Betel Nut Fiber Reinforced Polypropylene Composites". Journal of Adhesion Science and Technology. Vol. 25, Issue 6-7 pp 615-626, 2011.
5. K. L. Pickering, M. G. Aruan Efendya, T. M. Lea, "A review of recent developments in natural fiber composites and their mechanical performance". Composites Part A: Applied Science and Manufacturing. Vol. 83, Pp. 98–112, April 2016.
6. M. Masudul Hassan, Manfred H. Wagner, H. U. Zaman& Mubarak A. Khan. "Physico-Mechanical Performance of Hybrid Betel Nut (Areca catechu) Short Fiber/Seaweed Polypropylene Composite". Journal of Natural Fibers. Vol. 7, Issue 3, pp.165-177, 2010.
7. J. O. Agunsoyea, V. S. Aigbodion. "Bagasse filled recycled polyethylene bio-composites: Morphological and mechanical properties study". Results in Physics. Vol. 3, Pages 187–194, 2013.
8. Nor Azowa Ibrahim, Kamarul Arifin Hadithon, Khalina Abdan. "Effect of Fiber Treatment on Mechanical Properties of Kenaf Fiber-Ecoflex Composites." Journal of Reinforced Plastics and Composites. Vol. 29 (14), pp. 2192-2198·July 2010.
9. K. Mohanty, M. Misra, L. T. Drza."Surface modification of natural fibres and performance of the resulting bio-composite". Composite Interfaces. Vol. 8, pp. 313-343. 2001.

10. Thabang Hendrica Mokhothu. "Preparation and Characterization of Natural Fibre/Co-Polyester Bio-Composite". Department of Chemistry, Faculty of Natural and Agricultural Sciences at the Univerity of the Free State (Qwaqwa Campus) dated-7 December 2010.
11. Dhanalakshmi Sampath kumar, Ramadevi Punyamurthy, Basavaraju Bennehalli, Raghu Patel Ranganagowda, Srinivasa Chikkol Venkateshappa "Natural Areca Fiber: Surface Modification and Spectral Studies" Journal of Advances in Chemistry. Vol. 10, No. 10, pp-3263-3272, December 2014.
12. G. Ramachandra Reddy, M. Ashok Kumar, K. V. P.Chakradhar."Fabrication and performance of hybrid Betel nut (Areca catechu) short fiber/ Sansevieriacylindrica (Agavaceae) epoxy composites". International Journal of Materials and Biomaterials Applications. Vol. 1 (1), pp. 6-13, 2011.

INDEX

N

O

P

R

S

T